Engineered Approaches for In Situ Bioremediation of Chlorinated Solvent Contamination

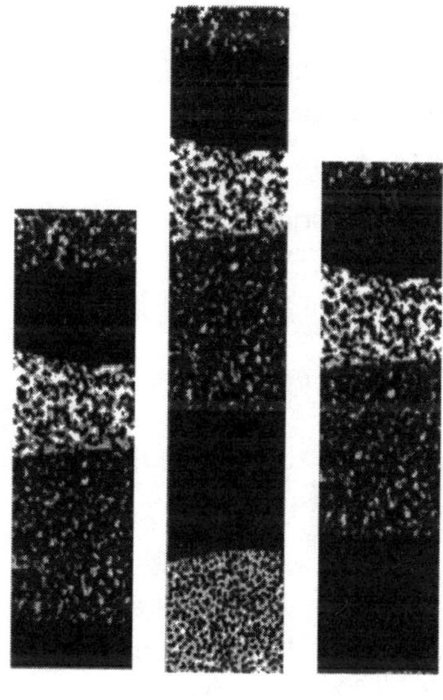

Editors

Andrea Leeson
and Bruce C. Alleman
Battelle

The Fifth International In Situ and On-Site Bioremediation Symposium

San Diego, California, April 19–22, 1999

BATTELLE PRESS
Columbus • Richland

Library of Congress Cataloging-in-Publication Data

Engineered approaches for in situ bioremediation of chlorinated solvent contamination / editors, Andrea Leeson and Bruce C. Alleman
p. cm.
Proceedings from the Fifth International In Situ and On-Site Bioremediation Symposium, held April 19–22, 1999, in San Diego, California.
Includes bibliographical references and index.
ISBN 1-57477-075-6 (hardcover : alk. paper)
1. Solvents--Biodegradation Congresses. 2. Organochlorine compounds--Biodegradation Congresses. 3. In situ bioremediation Congresses.
I. Leeson, Andrea, 1962– . II. Alleman, Bruce C., 1957– .
III. International Symposium on In Situ and On-Site Bioremediation (5th : 1999 : San Diego, Calif.)
TD196.S54E53 1999
628.5'2--dc21 99-23400
CIP

Printed in the United States of America

Battelle Press
505 King Avenue
Columbus, Ohio 43201, USA
614-424-6393 or 1-800-451-3543
Fax: 1-614-424-3819
Internet: press@battelle.org
Website: www.battelle.org/bookstore

For information on future environmental conferences, write to:
Battelle
Environmental Restoration Department, Room 10-123B
505 King Avenue
Columbus, Ohio 43201-2693
Phone: 614-424-7604
Fax: 614-424-3667
Website: www.battelle.org/conferences

CONTENTS

FOREWORD

The Fifth International In Situ and On-Site Bioremediation Symposium was held in San Diego, California, April 19–22, 1999. The program included approximately 600 platform and poster presentations, encompassing laboratory, bench-scale, and full-scale field studies being conducted worldwide on a variety of bioremediation and supporting technologies used for a wide range of contaminants.

The author of each presentation accepted for the program was invited to prepare a six-page paper, formatted according to specifications provided by the Symposium Organizing Committee. Approximately 400 such technical notes were received. The editors conducted a review of all papers. Ultimately, 389 papers were accepted for publication and assembled into the following eight volumes:

Natural Attenuation of Chlorinated Solvents, Petroleum Hydrocarbons, and Other Organic Compounds – Volume 5(1)
Engineered Approaches for In Situ Bioremediation of Chlorinated Solvent Contamination – Volume 5(2)
In Situ Bioremediation of Petroleum Hydrocarbon and Other Organic Compounds – Volume 5(3)
Bioremediation of Metals and Inorganic Compounds – Volume 5(4)
Bioreactor and Ex Situ Biological Treatment Technologies – Volume 5(5)
Phytoremediation and Innovative Strategies for Specialized Remedial Applications – Volume 5(6)
Bioremediation of Nitroaromatic and Haloaromatic Compounds – Volume 5(7)
Bioremediation Technologies for Polycyclic Aromatic Hydrocarbon Compounds – Volume 5(8)

Each volume contains comprehensive keyword and author indices to the entire set.

This volume deals with efforts to remediate sites contaminated with chlorinated compounds, such as tetrachloroethylene, trichloroethylene, carbon tetrachloride, pentachlorophenol, chlorinated benzenes, and various pesticide/herbicide compounds. Not only do these compounds carry health risks, but they also are challenging and often expensive to treat in the field. However, progress is being made, and this volume brings together the most up-to-date laboratory findings and the latest full-scale results from bioremediation field efforts. Engineering approaches discussed include biobarriers, cometabolism, bioaugmentation, in situ oxidation, Fenton's Reagent, in situ bioremediation, and more.

We would like to thank the Battelle staff who assembled the eight volumes and prepared them for printing. Carol Young, Lori Helsel, Loretta Bahn, Gina Melaragno, Timothy Lundgren, Tom Wilk, and Lynn Copley-Graves spent many hours on production tasks—developing the detailed format specifications sent to each author; examining each technical note to ensure that it met basic page layout requirements and making adjustments when necessary; assembling the volumes; applying headers and page numbers; compiling the tables of contents and author and keyword indices, and performing a final check of the pages before submitting them to the publisher. Joseph Sheldrick, manager of Battelle Press, provided valuable production-planning advice and coordinated with the printer; he and Gar Dingess designed the covers.

The Bioremediation Symposium is sponsored and organized by Battelle Memorial Institute, with the assistance of a number of environmental remediation organizations. In 1999, the following co-sponsors made financial contributions toward the Symposium:

Celtic Technologies
Gas Research Institute (GRI)
IT Group, Inc.
Parsons Engineering Science, Inc.
U.S. Microbics, Inc.
U.S. Naval Facilities Engineering Command
Waste Management, Inc.

Additional participating organizations assisted with distribution of information about the Symposium:

Ajou University, College of Engineering
American Petroleum Institute
Asian Institute of Technology
Conor Pacific Environmental Technologies, Inc.
Mitsubishi Corporation
National Center for Integrated Bioremediation Research & Development (University of Michigan)
U.S. Air Force Center for Environmental Excellence
U.S. Air Force Research Laboratory Air Base and Environmental Technology Division
U.S. Environmental Protection Agency
Western Region Hazardous Substance Research Center (Stanford University and Oregon State University)

The materials in these volumes represent the authors' results and interpretations. The support of the Symposium provided by Battelle, the co-sponsors, and the participating organizations should not be construed as their endorsement of the content of these volumes.

Andrea Leeson and Bruce Alleman, Battelle
1999 Bioremediation Symposium Co-Chairs

INDUCED COMETABOLISM OF CHLORINATED VOCs USING PROPANE

Peter I. Dacyk (Parsons Engineering Science, Inc., Cincinnati, Ohio)
William D. Hughes (Parsons Engineering Science, Inc., Cincinnati, Ohio)

ABSTRACT: Propane is being injected into an aquifer contaminated with chlorinated volatile organic compounds (VOCs) to induce cometabolic degradation of the chlorinated compounds. Cometabolism of the chlorinated VOCs was observed within the aquifer when fuel hydrocarbons migrated into the aquifer from an upgradient bulk fuel terminal. The cometabolic degradation of the VOCs appears to have ceased concurrent to the disappearance of the fuel hydrocarbons. Propane was selected to replace the fuel hydrocarbons as a source of anthropogenic carbon in an effort to stimulate the cometabolic process. A field-scale pilot test was implemented at the laundry facility utilizing propane as an alternative carbon source. Propane was injected in pure form at a rate of 4 cubic feet per minute (CFM) for two minutes every six hours. The propane injection is followed by three hours of air injection for an air to propane injection ratio of approximately 100:1. Preliminary and follow-up groundwater sampling results showed propane to be dissolving into the groundwater and a decrease in dissolved chlorinated VOC concentrations was observed in wells surrounding the propane injection well. A temporary disruption in propane injection showed a rapid disappearance of the dissolved propane and a minor rebound in dissolved chlorinated VOC concentrations.

INTRODUCTION

The solvent release site is a commercial laundry located in southwestern Ohio. The laundry facility sits atop a buried valley aquifer composed primarily of glacial sand and gravel deposits with interbedded clay, silt, and sand stringers. The buried valley aquifer typically contains low dissolved oxygen (D.O.) concentrations (< 1 mg/L) and is reducing in nature. The aquifer is the primary potable water source for the surrounding communities. A groundwater production well field is located approximately 1070 meters north (downgradient) of the site.

The laundry facility is located immediately downgradient of a bulk fuel storage complex consisting of four bulk storage terminals. Each of the four fuel terminals has reported hydrocarbon releases. The terminals have implemented combinations of active and passive remediation technologies. The fuel terminal that is immediately upgradient of the laundry facility uses pump and treat technology coupled with soil vapor extraction as its primary remediation technologies. Fuel hydrocarbons from the terminal were detected in groundwater sampled at the laundry facility during the early stages of remediation monitoring. The fuel hydrocarbons (benzene, toluene, ethylbenzene, and xylenes -BTEX) diminished to non-detect levels in the laundry facility's monitoring wells in late 1992; however BTEX concentrations at the fuel terminal's downgradient perimeter wells persisted into mid-1996.

Dry cleaning operations at the facility used tetrachloroethene (PCE) as the primary cleaning solvent through 1986. A release of PCE occurred in 1985 when a 1000-gallon above ground storage tank (AST) ruptured during refilling. The released solvent flowed into a nearby storm sewer which empties into an on-site infiltration pond. An additional chlorinated compound release of undetermined quantity may have occurred from underground storage tanks (USTs) located adjacent to the AST. Active remediation of the chlorinated VOCs began in 1989 with the installation of groundwater extraction wells and an air stripper to treat the extracted groundwater. Soil vapor extraction (SVE) wells were added to strip VOCs from contaminated vadose zone soils. Groundwater treatment and SVE were subsequently terminated in the mid-1990's due to poor VOC removal rates. The groundwater recovery wells were retained to provide hydraulic control of the solvent plume. Two additional groundwater recovery wells were added in May 1998 to control off-site plume migration. An air-sparge system with vapor extraction was installed and began operation in April 1997. VOC removal rates improved but initial results indicate that mechanical removal of the chlorinated VOCs by air sparging and SVE will not decrease VOC concentrations to drinking water standards. Figure 1 is a site plan showing the solvent source area, propane injection area, and select monitoring wells.

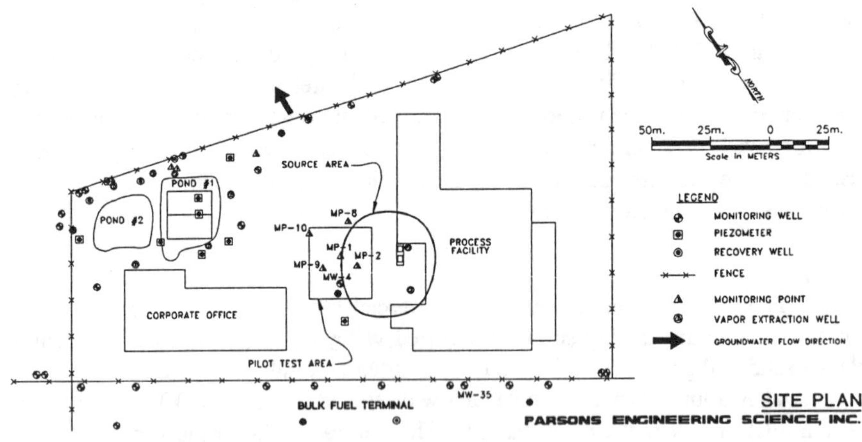

FIGURE 1. Site plan of industrial laundry facility showing location of solvent source area, groundwater recovery wells, and select monitoring wells.

Evidence of Previous Cometabolic Activity. PCE as well as TCE and cis1,2-dichloroethene (1,2-DCE) were detected in monitoring wells located down and cross-gradient from the solvent release point. No chlorinated hydrocarbon, other than PCE, is known to have been released at the site; however, the commercial-grade PCE may have contained impurities that include TCE and 1,2 (DCE). The lesser-chlorinated TCE and 1,2-DCE hydrocarbons are typical byproducts of PCE degradation (Guest et al., 1995). The presence of both TCE and 1,2-DCE in concentrations greater than PCE was the first indication that the PCE was undergoing dechlorination to it's lesser-chlorinated degradation byproducts. Additionally, several groundwater samples revealed the presence of minor (< 5 µg/L) concentrations of vinyl chloride (VC). Groundwater sampling at the laundry facility during the early and mid-1990's revealed the presence of BTEX compounds which, suggested the migration of fuel hydrocarbons from the adjacent and upgradient fuel terminal onto the laundry facility property.

Dissolved oxygen (D.O.) data collected during groundwater sampling events show the aquifer beneath the laundry facility to have D.O. concentrations generally greater than 1 mg/L and concentrations of 3 mg/L are not uncommon. The high D.O. concentrations show the aquifer beneath the laundry facility to be aerobic rather than anaerobic. Lower D.O. (< 1 mg/L) conditions prevailed at the facility concurrent with BTEX detection in the upgradient fuel terminal's perimeter groundwater monitoring wells (pre-1996). Dissolved oxygen data collected at the site prior to 1994 show unusually high (> 3 mg/L) D.O. concentrations which were determined to be questionable based on unreliable instrumentation used when these data were collected.

Aerobic conditions within the aquifer beneath the laundry facility preclude the occurrence of reductive dechlorination of PCE. The presence of the less chlorinated TCE and 1,2-DCE degradation byproducts suggest degradation of PCE was occurring under aerobic conditions, which is uncommon for highly chlorinated compounds such as PCE (Weidemeier et al., 1996). Figure 2

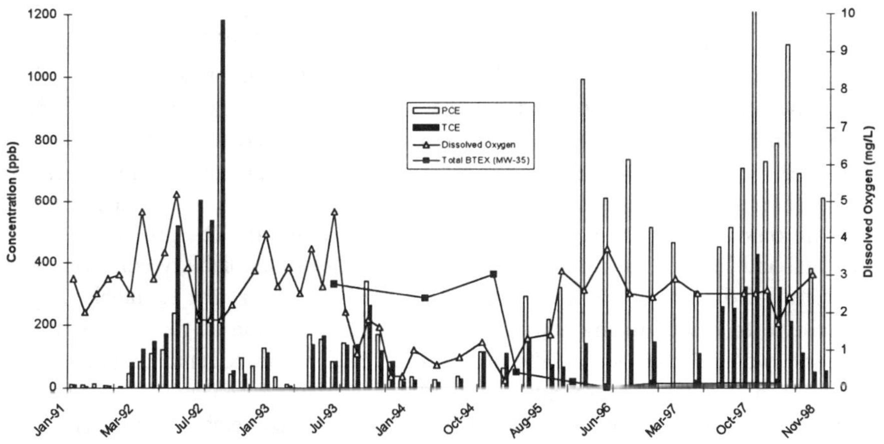

FIGURE 2. PCE, TCE, BTEX, and dissolved oxygen concentrations in groundwater sampled from monitoring wells MW-4 and MW-35 and located near the solvent source area and immediately up-gradient of the laundry facility.

represents groundwater analytical results taken from a monitoring well (MW-4) located near the solvent release area. Groundwater BTEX (dissolved) data collected from a monitoring well (MW-35) located on the downgradient edge of the fuel terminal is superimposed on the Figure 2 data. The degradation of PCE to less chlorinated degradation byproducts may have occurred as a fortuitous byproduct of the biodegradation of the BTEX compounds present in the aquifer beneath the laundry facility. The degradation of PCE may have been catalyzed by an enzyme or cofactor released by the microbes during degradation of the BTEX (McCarty and Semprini, 1994). Groundwater analytical results from the laundry facility taken primarily in 1992 show a TCE/PCE ratio greater than 1. This ratio corresponds to the appearance of BTEX compounds in the laundry facility's wells and in the high BTEX concentration in monitoring wells adjacent to the laundry facility (MW-35 on Figure 2). Post 1992 groundwater results show a decreasing TCE/PCE ratio with PCE becoming the dominant chlorinated compound after 1994. The reduction in TCE and 1,2-DCE concentrations relative to PCE concentrations corresponds to the absence of BTEX compounds in groundwater sampled from the fuel terminal's downgradient perimeter wells and a corresponding increase in D.O. concentrations within the laundry facility's groundwater. Minor favorable fluctuations in the PCE/TCE ratio (increasing TCE concentrations) were observed in several wells at the site after disappearance of the fuel hydrocarbons. The increase in TCE (and 1,2-DCE) and decrease in PCE correlates to low D.O. concentrations measured in these wells. The low D.O. may have produced transient anaerobic conditions that stimulated reductive dechlorination; however, the aquifer has remained primarily aerobic and frequent and wide-spread monitoring of the laundry facility's wells show no evidence of periodic anaerobic episodes.

Source of Anthropogenic Carbon. The favorable TCE/PCE ratio observed in groundwater samples when BTEX compounds were also present in the site's groundwater, indicates conditions at the site may favor cometabolism of PCE when sufficient carbon is available for microbial biodegradation. An attempt to convince local regulatory agencies and the operators of the adjacent fuel terminal to cease active remediation at the fuel terminal and allow fuel hydrocarbons to migrate onto the laundry facility property was unsuccessful. The proximity of the terminal and laundry facility to the production well field and liability concerns with the terminal owners precluded a shutdown of remediation at the terminal. As an alternative to discontinuing active groundwater remediation at the fuel terminal, propane was selected as an alternative source of anthropogenic carbon. The solubility of propane in groundwater is less than that of benzene but should be sufficient to provide a source of carbon to stimulate the cometabolic process.

MATERIALS AND METHODS

A pilot test evaluating the feasibility of injecting propane through an existing air sparge well located near the solvent release area was performed in June 1998. The propane injection feasibility test was performed over a period of two days to determine if the existing air-sparge system was a suitable and safe propane delivery system. Results of the pilot test showed a favorable distribution of propane into the aquifer without creating a potentially hazardous condition. Groundwater sampling of the injection well and several surrounding monitoring points revealed the presence of propane within the groundwater. The air sparge system with the designated propane injection well and propane monitoring points is illustrated on Figure 3.

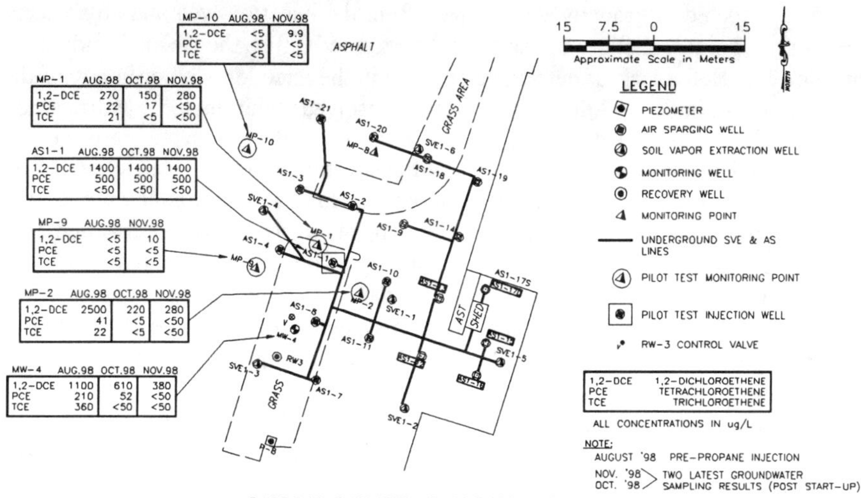

FIGURE 3. Air sparge injection well layout showing designated propane injection well and propane monitoring wells. Groundwater analytical results from pre and post-propane injection monitoring are shown.

The favorable short-term results allowed the existing sparge system to be modified to allow automated injection of propane into the test well. The injection was planned to take place over a three-month period beginning in September 1998. The system was designed to perform propane injection at a ratio of 100:1 (air to propane). The propane source was a 100-pound propane cylinder (850 cubic feet) with a delivery pressure of approximately 15 psi. The propane cylinder was plumbed into the sparge manifold leading to the designated injection well. The other four sparge wells on the manifold were shut-off at the well head to isolate the injection well. A solenoid controlled by a programmable controller was placed in-line with the propane piping to control the timing of the propane injection. The cylinder was replaced at approximately three week intervals. The propane was injected in pure form for two minutes followed by several hours of air injection. For safety purposes, extracted vapors were monitored continuously by a lower explosive limit (LEL) detector located within the remediation enclosure. The injection wellhead and monitoring points were monitored periodically for propane accumulation within the subsurface.

RESULTS AND DISCUSSION

The groundwater sampling results from groundwater samples collected after approximately one month of propane injection were encouraging. All wells that were sampled in the propane injection area showed a decrease in dissolved chlorinated hydrocarbon concentrations. One well in particular (MP-2) showed an order of magnitude decrease in the PCE concentration (2500 ug/L to 220 ug/L) relative to its pre-propane injection concentration (Figure 3). Dissolved propane was also detected in groundwater sampled from the injection well and an adjacent monitoring point. Follow-up sampling performed after the second month of injection revealed a lack of detectable propane in the injection well or in any of the monitoring points. Additionally, the dissolved chlorinated hydrocarbon concentrations rebounded slightly in the monitoring wells located between the injection well and source area. A check of the injection system revealed the system to have stopped injecting propane between the first and second month of the planned three-month injection period. The propane injection was interrupted by a series of programming failures caused by multiple site-wide power outages. The programming failures prevented propane from injecting at designated time intervals. Difficulties in bringing the injection system on-line delayed the restart of propane injection until mid-January 1999. Follow-up sampling was scheduled for March 1999.

The lack of consistent propane injection after the initial propane injection sampling prevented the collection of meaningful analytical data. The March 1999 results were pending during the preparation of this paper. Due to the relatively low cost of propane and maintaining the propane injection system, propane injection will continue through May 1999. If groundwater sampling reveals a continuing decline in chlorinated hydrocarbon concentrations in the propane injection area, a decision will be made whether to expand propane injection to all sparge wells located near the chlorinated hydrocarbon source area.

REFERENCES

Guest, P. R., Benson, L. A., and Rainsberger, T. J., 1995. *Inferring Biodegradation Processes for Trichloroethene from Geochemical Data.* Presented at Battelle International Symposium on In Situ and On-Site Bioreclamation. April 24-27. San Diego California.

McCarty, P.L. and L. Semprini. 1994 "Ground-Water Treatment for Chlorinated Solvents" In Norris, R.D. (Ed.) Handbook of Bioremediation. Lewis Publishers, Boca Raton, Fl. Pp. 87-116.

Wiedemeier, T. H., Swanson, M. A., Moutoux, D. E., Wilson, J. T., Kampbell, D. H., Hansen, J. E., and P. Haas. 1996. *Overview of the Technical Protocol for Natural Attenuation of Chlorinated Aliphatic Hydrocarbons in Groundwater Under Development for the Air Force Center for Environmental Excellence.* Presented at Symposium on Natural Attenuation of Chlorinated Organics in Groundwater. September 11-13. Dallas, Texas.

BIOVENTING NONPETROLEUM HYDROCARBONS

James T. Gibbs, Bruce C. Alleman, Rick D. Gillespie, Eric A. Foote, Sarah E. McCall, Frances A. Snyder. James E. Hicks (Battelle, Columbus, Ohio), R. Kenneth Crowe (TRW Inc., Tyndall Air Force Base, Florida), and Jon Ginn (United States Air Force, Hill Air Force Base, Utah)

INTRODUCTION

Bioventing is an in situ remedial technology that enhances aerobic biodegradation of contaminants by supplying oxygen to anoxic vadose zone soils via air injection and/or extraction. The technology was developed to treat fuel hydrocarbons, including gasoline, diesel, and various jet fuel mixtures. It has been proven successful for achieving in situ treatment and meeting site closure criteria for these types of hydrocarbon contamination under varying geologic and climatic conditions.

Although bioventing is a mature technology currently being employed at sites worldwide to treat fuel contamination, research continues to expand the application to other contaminants and in geologic settings that provide challenges for effective air movement. Efforts in cometabolic bioventing (Sayles, 1997a) and anaerobic bioventing (Sayles, 1997b) have expanded the list of contaminants to include chlorinated solvents. Bioventing proved successful at treating PAH contamination at the Reilly Tar Site in St. Louis Park, Minnesota (McCauley et al., 1999).

The focus of the effort described in this paper was to demonstrate the effectiveness of bioventing for remediating soils contaminated with a variety of nonpetroleum contaminants, including 1,2-dichlorobenzene (DCB). Table 1 lists the suite of compounds that were monitored and against which bioventing effectiveness was evaluated. The effort included designing, installing, operating, and monitoring a pilot-scale bioventing system. The effectiveness of the technology was determined based on mass removal of the selected compounds of interest (COIs) from within the radius of influence of the system over approximately 1 year.

SITE DESCRIPTION

The study was conducted in Chemical Disposal Pit (CDP) 1 at Operable Unit (OU) 1 at Hill Air Force Base (AFB), Utah. The Base is located approximately 25 miles (40 km) north of Salt Lake City and five miles (8 km) south of Ogden.

TABLE 1. List of Compounds of Interest for the Hill AFB Nonpetroleum Bioventing Study at CDP 1.

Compound	Formula	Compound	Formula
cis-1,2-dichloroethylene	$C_2H_2Cl_2$	1,3,5-trimethylbenzene	C_9H_{12}
1,1,1-trichloroethane	$C_2H_3Cl_3$	1,2,4-trimethylbenzene	C_9H_{12}
Trichloroethylene	C_2HCl_3	1,2-dichlorobenzene	$C_6H_4Cl_2$
Toluene	C_7H_8	1,3-dichlorobenzene	$C_6H_4Cl_2$
Tetrachloroethylene	C_2Cl_4	1,4-dichlorobenzene	$C_6H_4Cl_2$
Chlorobenzene	C_6H_5Cl	1,2,4-trichlorobenzene	$C_6H_3Cl_3$
Ethylbenzene	C_8H_{10}	1,2,3-trichlorobenzene	$C_6H_3Cl_3$
m,p-Xylene	C_8H_{10}	Naphthalene	$C_{10}H_8$
o-Xylene	C_8H_{10}		

Site Selection. The objective of the site selection process was to identify a site impacted with nonpetroleum hydrocarbons that was amenable to bioventing (i.e., was sufficiently permeable to allow air delivery). The main COI was 1,2-DCB. During the site selection process, site data were solicited from Air Force Bases previously identified as being contaminated with nonpetroleum hydrocarbon compounds, where some level of site characterization already had been performed. The data were reviewed and the list of candidate sites was reduced based on a set of screening criteria.

CDP 1 at Hill AFB was selected for this project because the site met most of the criteria and provided the most beneficial environment for this study. Although the organic compound mixture impacting this site was very complex and petroleum hydrocarbons also were present, it was determined that this scenario was more common than finding sites contaminated with either pure compound or solely with a mixture of nonpetroleum compounds. Based on this understanding, the list of COIs was developed based on the organic compounds found at this site (Table 1).

Site History. CDP 1 is part of OU 1, which is located close to the northeastern boundary of Hill AFB, approximately 300 ft (90 m) above the Weber River on the edge of a steep hillside that forms the southern boundary of the Weber River Valley (Montgomery-Watson, 1995). CDP 1 was used as a disposal area for liquid industrial wastes from the early 1950s to 1973. Results from analyses of soil samples taken from throughout the area showed 1,2-DCB present at concentrations as high as 170 mg/kg.

Site Geology. Hill AFB is situated within the Lake Bonneville Basin, which is characterized by alternating, isolated, north-trending, block-faulted mountains and intermontane basins flanked by alluvial slopes. The near-surface geology of OU 1 is characterized by approximately 40 ft (12 m) of interfingered sand, gravelly sand, and gravel underlain by approximately 200 ft (60 m) of silty clay containing interbeds of silt and very fine to fine sand. The northern boundary of OU 1 is a steep escarpment formed through the erosion of poorly consolidated sediments by the Weber River.

SYSTEM DESIGN, INSTALLATION, OPERATION, AND MONITORING

System Design. Separate bioventing systems were designed and installed in each of two plots, an active plot that was aerated and a control plot that was not aerated to provide an experimental control soil volume (Figure 1). Each system included a single vent well constructed from 2-inch (5-cm), Schedule 40 polyvinyl chloride (PVC) and screened between 10 and 20 ft (3 and 6 m) below ground surface (bgs). The system in the active plot was designed with eight tri-level in situ soil-gas monitoring points located at distances of 5, 10, and 20 ft (1.5, 3, and 6 m) from the vent well, with four on each side of two perpendicular axes intersecting at the vent well. The system in the control plot included four tri-level monitoring points located 20 ft (6 m) from the system vent well. Each of the soil-gas monitoring points was designed with three soil-gas sampling probes placed at 7, 12, and 17 ft (2.1, 3.7, and 5.2 m) bgs. The sample lines from each probe in the active plot

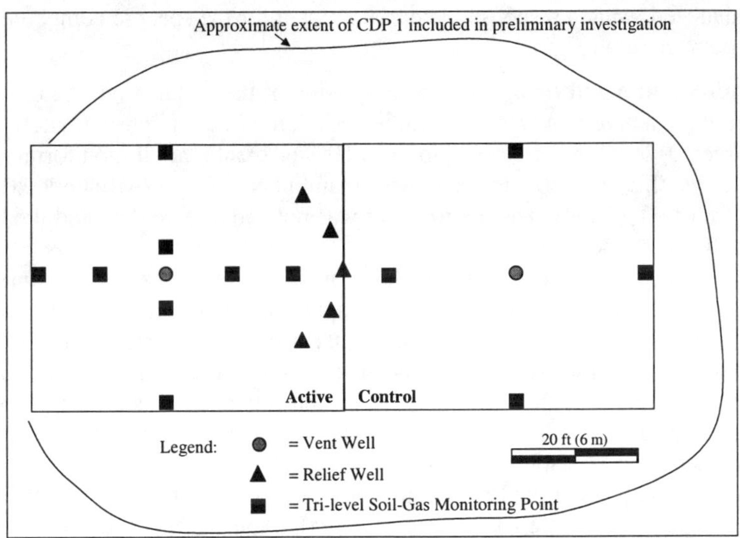

FIGURE 1. Layout of Bioventing System at CDP 1 Hill AFB, UT

were wrapped in heating cord and fed to a sampling panel located in a field trailer placed adjacent to the vent well in the active plot. The sample lines from the probes in the control plot were fed to the surface, fitted with pneumatic quick-connect couplers, and housed in flush-mount well vaults. Because of the close proximity of both plots, five relief wells were placed between the two plots to hinder oxygenation in the control plot during air injection into the active plot.

Installation. The system was installed using a hollow-stem auger drill rig. The holes for each in-ground component were advanced to the desired depth, and the flights were removed to facilitate completion of the installation. Sand pack was placed around the vent well screens and soil-gas sampling probes, and bentonite borehole seals were used to isolate probes and well screens. A concrete pad was poured around the vent well and well vaults in the control plot. All the tubing from the probes in the active plot was fed to a central manhole, then through a conduit into the field trailer and connected as previously described.

During installation of the in-ground components of the bioventing systems, soil samples were collected with a 2-ft (0.6-m) split spoon loaded with stainless steel sleeves. Soil samples were weighed and preserved in a brine solution in the field, then sent to the laboratory for analysis. The samples were analyzed for the COIs listed in Table 1 by heated headspace/gas chromatography, following a procedure based on EPA Method 5021. A Hewlett Packard 5890 gas chromatograph (GC) fitted with a 60 m SPB-1 wide-bore capillary column (Supelco, Bellefonte, PA) connected to a flame ionization detector (FID) was used. A Tekmar Model 7000 equilibrium headspace autosampler was programmed to equilibrate the sample vials at 95°C for 55 minutes, then pressurized to 10 psi prior to injection into the GC. The initial GC oven temperature was held at 35°C for 2 minutes, then ramped at 8°C/min to 200°C and held for 2 minutes. The concentration of each compound was calculated using a response factor determined for the specific compound using a five-point calibration procedure.

Operation and Monitoring. Upon completion of the installation, the bioventing system in the active plot was operated in injection mode for approximately 1 year. The target O_2 level in the active plot was 10% or greater at all monitoring points. To achieve this, the air flowrate was maintained at approximately 200 scfh (3.3 scfm or 93.4 Lpm). The control plot was not actively vented, and the control vent well was capped.

System parameters, including blower temperature, feed air temperature, manifold pressure, and air flowrate, were monitored to ensure proper operating conditions of the equipment. Soil-gas O_2, CO_2, and total petroleum hydrocarbon (TPH) concentrations were measured to ensure adequate aeration, and to monitor for biological activity and vapor transport. One initial and four quarterly in situ respiration tests were conducted to monitor microbial activity and to predict the total contaminant mass biodegraded (Leeson and Hinchee, 1997). A surface emission test was conducted prior to operation and following startup to measure the mass of contaminant volatilized from the vadose zone due to venting (Leeson and Hinchee, 1997).

After approximately 1 year of air injection in the active plot, a final set of soil samples was collected from both plots for contaminant analysis. The samples were collected and processed following the same procedures used for the initial soil samples. The data were compared to the initial data to assess the effectiveness of the technology.

Laboratory Study. The demonstration included both field and laboratory components. Laboratory experiments were conducted to support the findings from the field effort. The laboratory experiments allowed for more controlled tracking of the fate of the COIs, including volatilization and biodegradation removal mechanisms.

A laboratory-scale column study was performed to track the fate of the COIs under more controlled conditions in support of the results obtained in the field. Soil cores collected from CDP 1 during installation of the bioventing system were aerated, and influent and effluent COI concentrations were monitored to establish mass balances.

DATA ANALYSIS

The data generated during this demonstration were evaluated to determine the total mass of contaminant removed over 1 year, and the individual portions of the total mass removed due to biodegradation and to volatilization. The difference between the total mass removed and the mass removed by volatilization was attributed to biodegradation.

Approach for Evaluating Biodegradation. A site-specific mass ratio comparing the mass of compounds being degraded versus the mass of oxygen utilized was calculated. Each of the compounds present at the site has a distinct stoichiometric ratio to oxygen in the oxidation reaction. That site-specific mass ratio was determined from the balanced oxidation reaction and by using the molecular weights of the compounds involved.

Oxygen utilization rates measured at soil-gas monitoring probe locations during in situ respiration tests were used to calculate the biodegradation rate at each location for each test using the site-specific mass ratio. The resulting rates were used to evaluate biodegradation at the three levels within the test plot and to identify changes in the rates over the year of operation.

The total mass of aerobically biodegradable organic compounds destroyed in situ was estimated by calculating in situ biodegradation rates for each of three 5-ft (1.5-m)-thick soil layers at CDP 1: 4.5 to 9.5 ft (1.4 to 2.9 m) bgs, 9.5 to 14.5 ft (2.9 to 4.4 m) bgs, and 14.5 to 19.5 ft (4.4 to 5.9 m) bgs. Soil-gas monitoring probes placed in those layers collected representative data from the midpoint depths of each layer, at levels of 7, 12, and 17 ft (2.1, 3.7, and 5.2 m) bgs.

Approach for Evaluating Total Mass Removal. EarthVision® geospatial graphics software (Dynamic Graphics, Inc., Alameda, CA) was used to contour soil volumes of similar concentration in three dimensions, and then calculate the volume of soil within each isopleth shell. Each volume of soil was converted to a mass of soil using the average dry bulk density calculated from selected soil samples obtained during system installation. The COI concentration (on a mass/mass basis) for each volume of soil was multiplied by the mass of soil within each concentration shell, and then the individual masses were summed to obtain the total mass of that compound. This same procedure was used for both the initial and final soil sampling events, and the mass of COIs lost during the test period was taken to be the difference between those computed values.

RESULTS AND DISCUSSION

The results from the soil-gas monitoring showed that the majority of the monitored volume of vadose zone soils was effectively oxygenated; however, the deeper soils at the outer fringe of the volume were not (Figure 2). It should be noted that the oxygenated volume included the control plot, indicating that the relief wells did not prevent that soil volume from being oxygenated. All mass removal calculations were performed on the combined volume of the active and control plots.

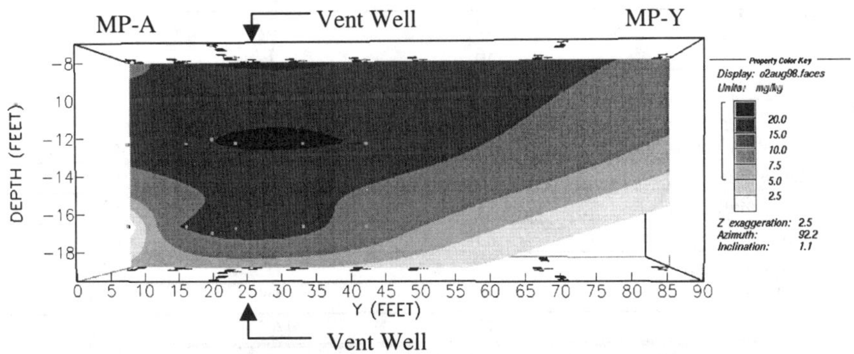

FIGURE 2. Oxygen Profile Showing 5% Oxygen Contour During Active Venting at CDP 1.

Respiration rate data indicated that a total mass of 3,400 lb (1,490 kg) of organic contaminants were biodegraded over the 1 year of operation within the volume of soil that was monitored (10 to 20 ft [3 to 6 m] bgs). This total mass of organic contaminants includes all of the contaminants present at CDP 1, both petroleum hydrocarbons and nonpetroleum COIs.

The soil contaminant data and the surface emission data were used to calculate the total mass of COIs removed and the mass that could be attributed to volatilization. The mass that was biodegraded was determined to be the difference between the two calculated values. The results are presented in Table 2.

The soil data indicated that significant reduction in mass of all COIs occurred in 1 year. The difference between initial and final soil sampling results indicated that 1,2-DCB, 1,3-DCB, and 1,4-DCB were reduced by 74%, 42%, and 81%, respectively. The resulting reduction in total DCB distribution is illustrated in Figure 3. The results for the chlorinated solvents showed that, based on the difference in soil concentrations, more than 100 percent of the mass removed was attributable to volatilization.

The mass losses for the COIs that could be attributed to biodegradation were calculated as the difference between the total mass removed, as measured through initial and final soil sample analyses, and the mass volatilized from the system, as determined through surface emission testing. Note that only the 12- and 17-ft (3.7 and 5.2-m)-bgs layers were included in the soil mass loss calculation because the soil-sampling interval was between 10 and 20 ft (3 and 6 m) bgs;

TABLE 2. COI Mass Balances During Bioventing at CDP 1, Hill AFB

Compound	Mass Removal, %	Mass Volatilized, g	Mass Biodegraded, g
cis-1,2-dichloroethylene	>100	410	—
1,1,1-trichloroethane	28	130	—
Trichloroethylene	42	910	—
Toluene	27	2.5	190
Tetrachloroethylene	9	66	—
Chlorobenzene	30	1.7	110
Ethylbenzene	36	0[(a)]	60
m,p-Xylene	27	0.9	99
o-Xylene	56	1.3	280
1,3,5-trimethylbenzene	64	1.4	1,600
1,3-dichlorobenzene	42	0[(b)]	1,100
1,2,4-trimethylbenzene	58	3.1	830
1,2-dichlorobenzene	74	77	9,100
1,4-dichlorobenzene	81	0[(a)]	1,400
1,2,4-trichlorobenzene	73	7.5	3,100
1,2,3-trichlorobenzene	79	0[(b)]	1,400
Naphthalene	93	0[(b)]	5,500
TPH	37	NA	NA

(a) = Concentrations were below detection limits during venting.
(b) = Compounds were not detected during surface emission testing.
— = Volatilization accounted for ≥100% of the mass loss for this COI.
NA = Not available because TPH is not included in EPA Method TO-14.

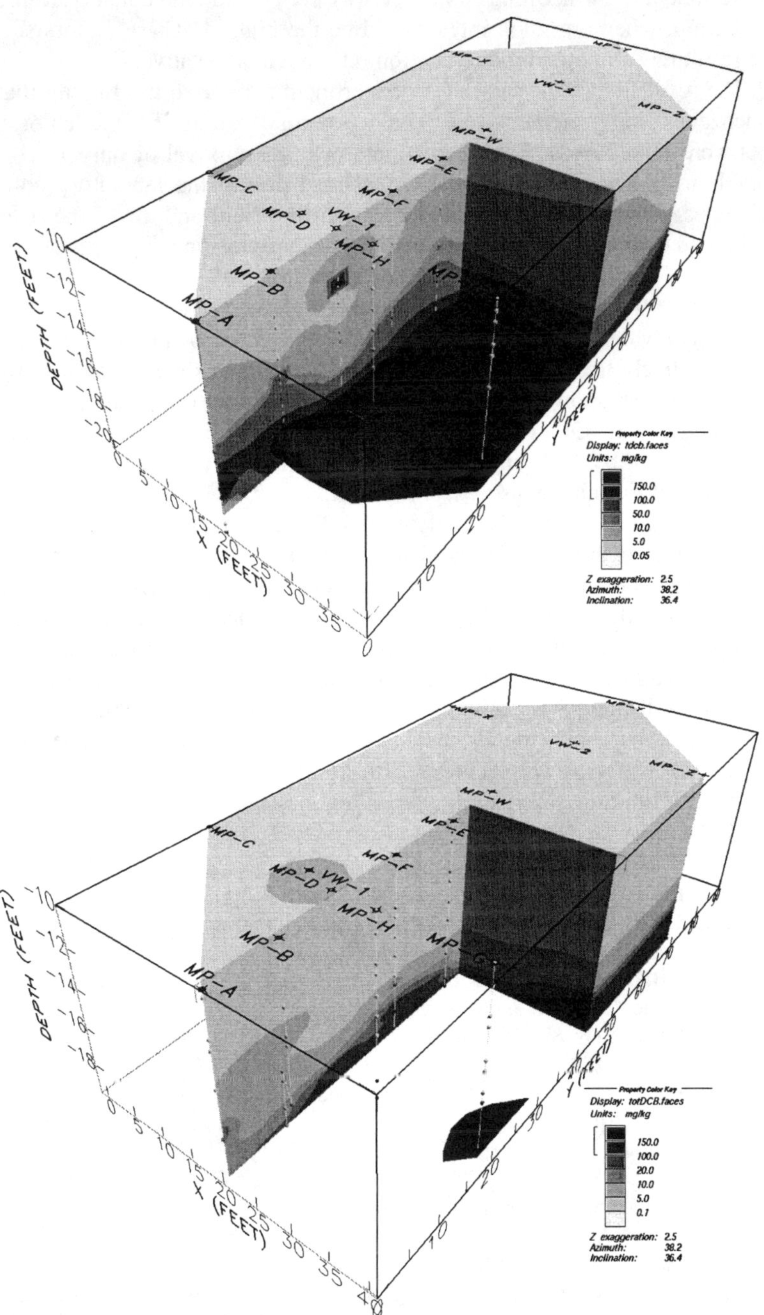

FIGURE 3. Total DCB Concentration Profile (a) Before Bioventing, and (b) After One Year of Bioventing.

also, note that the surface emission test was conducted immediately following system startup, when emission rates would be the highest. These factors suggest that the resulting estimated biodegradation rates are conservative.

The results from the laboratory experiment supported the finding that the chlorobenzenes were biodegraded. The laboratory results for the chlorinated solvents were mixed, with some percentage of mass removal of only 1,1,1-TCA attributable to biodegradation. The air exchange rate in the laboratory columns was 15 times greater than the exchange rate in the field and all of the detected removal could be accounted for in the off-gas from the columns.

CONCLUSIONS

This study demonstrated that nonpetroleum hydrocarbon organic compounds could be effectively treated by bioventing. The focus in this demonstration was on 1,2-DCB, which was shown to be removed 74% over 1 year of operation of a standard bioventing system. Other DCB isomers also were effectively removed, with 1,3-DCB being reduced by 42%, and 1,4-DCB being reduced by 82%. Removal rates of the same order of magnitude also were demonstrated for many other compounds that were tracked. Based on the results described above, it was concluded that DCB should be added to the list of candidate compounds for bioventing applications. Because DCB-degrading capabilities are not inherent in soil bacteria (see Nishino et al., 1994; van der Meer et al., 1998), it was concluded that the age of the DCB spill must be taken into account when considering the application of bioventing for DCB remediation. Significant periods of time may be required for the indigenous microbial population to exchange the necessary genetic material to obtain the ability to degrade DCB. At CDP 1, disposal activities ceased in the mid-1970s, providing a period of at least 25 years for the soil bacteria to acclimate. This acclimation period may have allowed enough time for soil microbes to obtain the ability to metabolize DCB and the other COIs that were found to biodegrade at CDP 1.

REFERENCES

Leeson, A. and R.E. Hinchee. 1997. *Soil Bioventing: Principles and Practice.* Lewis Publishers, Inc., Boca Raton, FL.

McCauley, P.C., B.C. Alleman, J. Abbott, and R.C. Brenner. 1999. "Bioventing PAH Contamination at the Reilly Tar and Chemical Corporation Site." Presented at the 5th International In Situ and On-Site Bioremediation Symposium, San Diego, CA., April 19-22.

Montgomery-Watson, 1995. *Final Comprehensive Remedial Investigation Report for Operable Unit 1, Hill Air Force Base, Utah.* December.

Nishino, S.F., J.C. Spain, and C.A. Pettigrew. 1994. "Biodegradation of Chlorobenzene by Indigenous Bacteria." *Environ. Toxicol. Chem. 13*(6):871-877.

Sayles , G.D., Moser, L.E., Gannon, D.J., Kampbell, D.H., and C.M. Vogel. 1997a. "Development of Cometabolic Bioventing for the In Situ Bioremediation of Chlorinated Solvents." In B.C. Alleman and A. Leeson (Eds.), *In Situ and On Site Bioremediation: Volume 1.* Battelle Press Columbus, OH.

Sayles, G.D. 1997b. "Anaerobic Bioventing of PCE." In B.C. Alleman and A. Leeson (Eds.), *In Situ and On Site Bioremediation: Volume 1.* Battelle Press Columbus, OH.

van der Meer, J.R., C. Werlen, S.F. Nishino, and J.C. Spain. 1998. "Evolution of a Pathway for Chlorobenzene Metabolism Leads to Natural Attenuation in Contaminated Groundwater." *Appl. Envion. Microbiol. 64*(11):4185-4193.

VINYL CHLORIDE ATTENUATION BY DIRECT INJECTION OF PEROXYGEN

Peter D. Tacy, Jr. (STS Consultants, Ltd., Detroit, Michigan, United States)

ABSTRACT: This case study shows that there is potential to remediate dissolved vinyl chloride by oxygenation in the field. Residual vinyl chloride concentrations have been remediated in the field at a mid-western site with a phosphate-intercalated peroxygen (ORC® is a proprietary formulation of magnesium peroxide that releases oxygen slowly). The site was formerly operated as an aluminum coating facility that stored aromatic and chlorinated solvents in a underground storage tank (UST). The UST created a narrow plume in a prolific sandy aquifer less than twenty feet thick. The plume was approximately 350 ft. long by 40-60 ft. wide at the completion of the site investigation and prior to the start of remediation. Remediation of the solvent plume was completed in on-site areas. An off-site plume of vinyl chloride remained with dissolved concentrations just above generic residential clean-up criteria.

The rate of natural attenuation of vinyl chloride had stalled during the previous year and vinyl chloride reached the first sentinel well RL-5A in October 1997. At the same time, prior to treatment, the off-site vinyl chloride concentrations were between 2.4 to 37 parts per billion (ppb) over five locations.

The objectives were to complete the remediation of the off-site vinyl chloride with ORC through oxidation and enhancement of the indigenous bacteria to expedite vinyl chloride attenuation. Groundwater sampling results prior to and following injection of the ORC indicated a significant reduction of vinyl chloride concentrations within the first month in the treated areas. Rebound was observed downgradient of areas that did not receive full treatment. The ORC project was tailored to the site's unique conditions that involved multiple contaminants, and other sources adjacent to the subject plume. In this instance, this technology provided a surgical tool to address contaminants within the plume without risking the hydraulic disturbance of contaminants from the unrelated areas. With the use of ORC, the regulators were able to approve the shut down of the advanced oxidation and SVE systems to reduce operational costs.

SITE BACKGROUND

When ORC injection was initiated, only the off-site solvents plume remained for remediation. The wells impacted included RL-1A, RL-5A, and RL-7A. Unimpacted off-site wells included NW-5, MW-7, MW-9, and MW-10. This remaining impacted area, measured approximately 250 ft. long by 40 ft. at the time of ORC injection in March of 1998 (see Figure 1). Since the on-site area has been remediated, only the off-site portion of the plume is shown on Figure 1. Off-site sources in and adjacent to the solvents plume included a former gasoline retail station, petroleum bulk storage area, and a former coal gasification facility. Non-chlorinated petroleum compounds associated with the off-site sources remain co-mingled with the off-site vinyl chloride plume. These sources are located along the vinyl chloride plume or just to the east of the vinyl chloride plume shown on Figure 1. This co-mingling of non-chlorinated compounds in these areas was justification alone for the use of a non-invasive remedial technology.

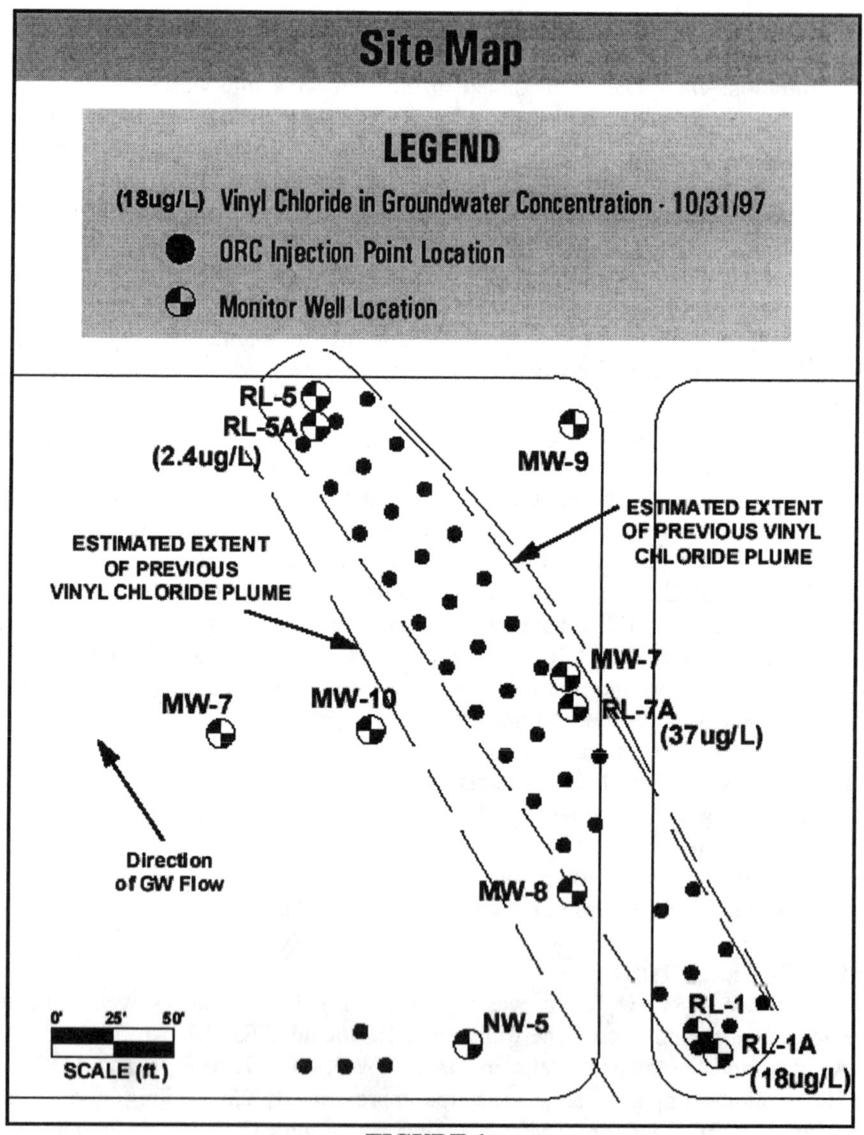

FIGURE 1.

In 1994 all parties historically involved with the on-site activities opted for generic residential clean-up criteria to eliminate future liabilities. These established remedial objectives included active remediation for only the on-site portion (source) of the solvent plume. The off-site dissolved plume was not initially targeted due to the co-mingled nature of the downgradient portion (off-site) of the solvents plume and the lack of involvement of the other parties in the clean-up.

During the site investigation groundwater was encountered in medium sands at 13 to 19 feet (3.96 to 5.79 meters) and bounded on the bottom with a silty clay at approximately 30 feet (9.14 meters). The groundwater flow in this area is to the north-northwest at a velocity of 0.012 feet/day (4.2 x 10^{-7} cm/sec) with a nearly flat gradient at the site.

Active remediation of the site was initiated in the Fall of 1995. Closed loop soil vapor extraction was employed at the site in conjunction with advanced oxidation of the dissolved solvent plume (subject plume) located on-site. During the operation of the remediation system groundwater was extracted downgradient of the source and highly oxygenated groundwater was reinjected upgradient of the dissolved plume. The groundwater system reduced dissolved concentrations on-site to less than generic residential criteria by April of 1997.

RESULTS

Magnesium Peroxide Slurry Injection: Direct injection of the specially formulated magnesium peroxide manufactured by Regenesis of San Juan Capistrano, California was utilized within an area of residual vinyl chloride contamination found off-site. The specially formulated, phosphate intercalated magnesium peroxide is sold as Oxygen Release Compound (ORC®). The material is supplied in powder form and is combined with water to create a slurry prior to injection.

Based on the vertical, aerial extent of the plume, the ORC was injected with a grout pump and direct push drilling technology within the vinyl chloride plume in grid with a roughly 12.5 foot spacing. Each boring was injected with the ORC slurry from the aquitard at the base of the aquifer to approximately one foot above the water table. STS utilized approximately twenty-five percent more ORC that recommended by Regenesis. This approach was designed to attenuate vinyl chloride through direct oxidation and through enhancing the activity of the indigenous bacteria to expedite the clean-up in those areas.

The ORC injection was performed in March of 1998, the results as presented on Table 1 per EPA Method 8260. The results suggest that it is feasible to remediate vinyl chloride in the field by delivering oxygen to the saturated zone. In areas with adequate spacing for ORC addition (ie. RL-5A), complete remediation of vinyl chloride was observed within three to four months. Table 1 shows the vinyl chloride concentrations in the wells located off-site prior to and following treatment.

TABLE 1. Vinyl Chloride Concentration (ppb)

Date	RL-1A	RL-5A	RL-7A
4/17/97	$15^{3.7}$	BQL	34
10/31/97	18	2.4	37
12/17/97	NS	5.0	NS
March 1998	**Inject ORC**	**Inject ORC**	**Inject ORC**
4/15-16/98	5.0	BQL	24
5/18/98	$BQL^{8.2}$	$BQL^{7.8}$	$13^{8.3}$
6/24/98	$BQL^{7.8}$	$BQL^{8.9}$	$BQL^{8.2}$
7/23/98	6.0	BQL	20
8/21/98	$BQL^{1.2}$	$BQL^{2.5}$	$BQL^{2.5}$
9/25/98	$BQL^{2.5}$	$BQL^{1.2}$	$6.0^{1.7}$

NOTE: Dissolved oxygen is shown as a superscript in parts per million
NS=Not Sampled BQL=Below Quantitation Limit

The monthly analytical results indicate that these areas have attenuated at an accelerated rate since addition of ORC. Rebound was observed after four months in areas monitored immediately downgradient of untreated areas. The right of way associated with RL-1A and the roadway associated with RL-7A are the primary areas not treated in the first phase of ORC injection. At this time the sentinel well is consistently below generic residential criteria for vinyl chloride. Measurement of dissolved oxygen in the monitor wells located along the spine of the plume indicates that the effectiveness of the magnesium peroxide at this site was limited to approximately three to four months, possibly due to impact from other known contaminant sources in the system.

CONCLUSION

The method of oxygenation was tailored to the existing site conditions, multiple contaminants, and orientation of sources adjacent to the subject plume. This technology provided a controlled means of remediating contaminants within the site's plume without the risk of hydraulically influencing extrinsic contaminants from unrelated source areas. The use of this technology also facilitated regulatory approval to shut down the advanced oxidation system and the SVE system to reduce overall operational costs. The migration of the plume beyond the sentinel wells was effectively stopped.

Results suggest that spacing and the appropriate quantity of the ORC is important for success. Rebound occurred where planned spacing was not adhered to. The presence of roadways, buildings, and utility corridors, which precluded a uniform application, may have increased the amount of time required to remediate contaminants. Sparse placement of injection points and reduction of stoichiometrically sound quantities is not recommended.

It is important to remember that the primary focus of these activities was to remediate the plume and do this at a reasonable cost. This study did not therefore include bench studies, microbial population assays, or other ancillary field testing.

STS has recently reapplied ORC in December 1998 to the area in the same locations. This application has also included additional injection points in the roadway that bisects the body of the plume. The data will be evaluated in the months following the December 1998 treatment to further evaluate the effectiveness of this technology.

REFERENCES

Bell, P. K. Casper and P. McIntire. 1997. "Treatability Evaluation of In-Situ Biodegradation of Chlorinated Solvents in Groundwater." Proceedings from Air & Waste Management Association 90th Annual Meeting and Exhibition. Toronto, Canada.

Davis, J. and C. L. Carpenter. 1990. "Aerobic Biodegradation of Vinyl Chloride in Groundwater Samples." Applied Environmental. Microbiology 56: 3878-3880.

Hartmans, S. and J. A. M. DeBont. 1992. "Aerobic Vinyl Chloride Metabolism in Mycobacterium Aurum L1." Applied Environmental Microbiology 58: 1220-1226.

Martinovich, B. and A. Putscher. 1997. "Aerobic Biodegradation of Vinyl Chloride in Groundwater." In-situ and On-site Bioremediation (5): 469. Alleman and Leeson eds. Batelle Press, Columbus, Ohio.

McCarty, P. and L.Semprini. 1994. "Groundwater Treatment for Chlorinated Solvents." Handbook of Bioremediation, Chapter 5. Matthews, T.E. ed. Lewis Publishers, Boca Raton, Florida. 257.

Suthersan, S. 1997. Remediation Engineering. Design Concepts. Chapter 5. Lewis Publishers, Boca Raton, Florida. 132.

CHLORINATED SOLVENT BIOREMEDIATION: A NOVEL APPROACH TO MEASURING BACTERIAL KINETICS

Brennan, R.A.[1]; Swanson, A.R.[1]; Loffler, F.E.[2]; and Sanford, R.A.[1]
[1]University of Illinois at Urbana-Champaign, Urbana, IL, USA
[2]Michigan State University, East Lansing, MI, USA

Abstract: The industrial solvent, perchloroethylene (PCE), can be degraded by a variety of anaerobic microorganisms *in situ*. Kinetic analyses of these cultures are required to determine the optimal growth conditions for remediation. Because of its volatility, it is difficult to supply PCE at the constant concentration required for kinetic analyses. To overcome this problem, a bioreactor with a bed of the solid-phase sorbent Tenax was used to maintain steady-state concentrations of PCE in the aqueous phase. Our data confirm that steady-state PCE concentrations can be maintained for several months in halorespiring cultures using Tenax as a source. We have also shown that Tenax adsorbs the byproducts of dechlorination before they can accumulate to toxic levels. The anaerobic bacterial strain BB1, which gains energy by dechlorinating PCE to 1,2-cis-dichloroethene (DCE), was used in this study. 1.0 mM radiolabeled acetate with a specific activity of 0.03 μCi/ml was used as the sole carbon and energy source to quantify increases in biomass. Finding optimum PCE concentration ranges using the methods presented in this study may help to design more effective bioremediation strategies for the remediation of chlorinated solvents.

INTRODUCTION

The industrial solvent tetrachloroethene (perchloroethylene, PCE) is one of the most frequently detected groundwater contaminants in the United States. This dense nonaqueous phase liquid (DNAPL) is difficult to remediate using standard "pump and treat" technologies; however, a variety of anaerobic microorganisms are capable of degrading this compound in situ. By studying pure cultures, a better understanding of the requirements of individual organisms can be achieved.

The microbial culture used in this study is the anaerobic bacterial strain BB1, which gains energy by reductively dechlorinating PCE to 1,2-cis-dichloroethene (DCE). Strain BB1 was isolated from sediments of the Père Marquette River near Baldwin, Michigan (Löffler, 1998), and found to be a gram-negative bacterium belonging to the *Desulfuromonas* group. Microbial growth kinetics of a pure culture, such as strain BB1, can be determined by measuring either the change in substrate concentration or change in biomass concentration with time. As a new approach to quantify biomass increase, ^{14}C-labelled acetate was used as the sole carbon source for strain BB1 in this study.

Supplying the electron acceptor, PCE, at the constant concentration required for kinetic analyses is made difficult by its volatile and toxic nature. To overcome this problem, a new design incorporating the solid-phase sorbent Tenax was used to maintain steady-state concentrations of PCE in the aqueous phase. Tenax is a hydrophobic polymer with a strong affinity for volatile organic compounds (VOCs), such as PCE. In these experiments, a Tenax bed is one component of a self-contained, closed-loop bioreactor developed specifically for this study. By supplying PCE via Tenax desorption and by using radiolabled acetate to quantify growth rates, kinetic parameters can be determined for a range of PCE concentrations. This information can be used to help understand the subsurface concentrations that are conducive to the growth of similar halorespiring microoganisms.

METHODS

Medium preparation and source cultures. A reduced anaerobic basal salts medium with reduced chloride content was used for all experiments (Löffler, 1998). After autoclaving, Wolfe's vitamin solution (Atlas, 1997) and Na_2S (0.2 mM) were added from sterilized anaerobic stock solutions. 100 ml source cultures of BB1 were maintained in 160 ml serum bottles with 1 mM acetate. PCE was added to these cultures to give an aqueous concentration of 20 mg/l and was replenished periodically after it was consumed.

Analytical methods. The concentrations of PCE, TCE, and DCE were determined from both liquid and headspace samples. Liquid samples were analyzed using a Tekmar purge and trap unit connected to a HP 5890A gas chromatograph (GC) equipped with a photoionization detector (PID) and an electrolytic conductivity detector (ELCD). Headspace samples were analyzed using a Perkin Elmer AutoSystem GC equipped with a flame ionization detector (FID). Carbon dioxide, H_2, and O_2 were analyzed using a GOW-MAC series 580 GC equipped with a thermal conductivity detector (TCD).

Chloride ion concentration was determined using the colorimetric assay described by Bergmann and Sanik, 1957. The measurement of chloride production was used to quantify the extent of dechlorination without measuring the PCE concentration directly.

Liquid samples were analyzed for acetate concentration using a Waters high pressure liquid chromatograph (HPLC) using a mobile phase of 0.005 M H_2SO_4.

To quantify increases in biomass, 0.22 μm filters were used to trap cells from 0.5 ml of radiolabeled culture, and rinsed with 30 ml distilled deionized water. Filters were then placed in glass scintillation vials with 5 ml of scintillation coctail and analyzed using a Packard Tri-Carb 1600A Scintillation Counter. Cells were calculated by converting the total acetate consumed into cells produced using the stoichiometry given in the results section. To determine CO_2 production, 0.50 ml of unfiltered sample was divided equally into two scintillation vials, one containing 100 μl 2N HCl, and the other containing 100 μl 2N NaOH. The acidic sample was purged with air to drive off any residual CO_2. 5 ml of scintillation solution was added to both vials, and analyzed as described above.

Batch Experiment. 30 ml balsch tubes (nominal capacity) containing 20 ml of medium and a nitrogen headspace were inoculated with 2% source culture. Each vial was amended with 1 μl neat PCE, and ^{14}C-acetate with a concentration of 1 mM and a specific activity of 0.1 μCi/ml. Triplicate cultures and duplicate controls (no cells) were incubated at 25, 30, and 35°C for 11 weeks.

Bioreactor Experiment. A schematic of the bioreactors designed and constructed for these experiments is shown in Figure 1. All reactor components were made of glass, stainless steel, teflon, or viton to minimize organic solvent adsorption.

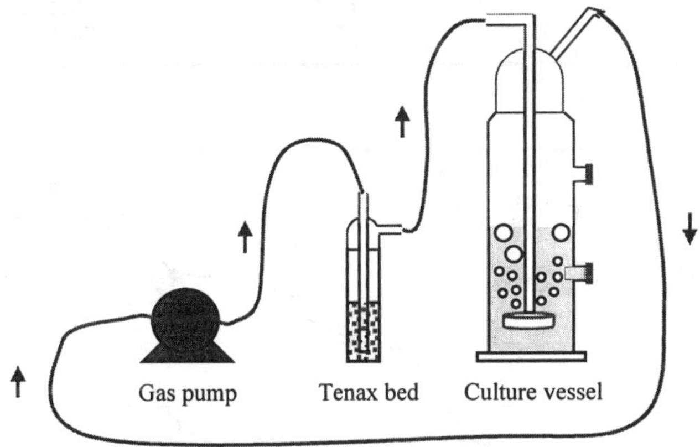

FIGURE 1. Bioreactor schematic

Both bioreactors were sterilized and pressure tested prior to use, and then placed in an anaerobic chamber to safeguard against future leaks. The neck of the culture vessel was opened to allow for the addition of 400 ml of medium, and then re-secured with anaerobic tape. Neat phase PCE was added through the liquid sampling port to give an aqueous PCE concentration of 4 mg/l according to the Tenax loading ratios presented by Pignatello, 1991. After 2 weeks of equilibration with a pumping rate of 20 ml/min, 4 ml of vitamins and ^{14}C-acetate (1.6 mM, 0.03 μCi/ml) were added through the liquid sampling port to each reactor. 4% source culture was added to Reactor 1 and not to the Control. The reactors were incubated at 24°C (± 1°C) for 40 days. Every 5 days, triplicate liquid samples were taken and analyzed for acetate, chloride, and biomass concentrations. Headspace was also monitored for chlorinated ethenes by GC.

RESULTS

Batch Experiment. The results of the batch experiment are summarized in Table 1 and Figure 2. Cultures were allowed to grow until all the PCE was utilized and the cells achieved a steady concentration. A second spike of 1 μl neat PCE was added to show that growth and ^{14}C-acetate uptake are dependent upon the dechlorination reaction. Temperature appeared to only affect the length of the lag phase and not the rate of growth. The higher growth rate observed at 35°C is most likely due to the higher PCE concentration present at the time of the exponential phase. The decrease in biomass after 65 days was probably due to the exhaustion of available acetate and endogenous respiration.

TABLE 1. Batch experiment: maximum growth rates observed at varying temperatures.

Temperature (°C)	PCE Concentration (mg/l)	Maximum growth rate (mg cells/l/day)
25	37	0.210
30	37	0.191
35	74	0.360

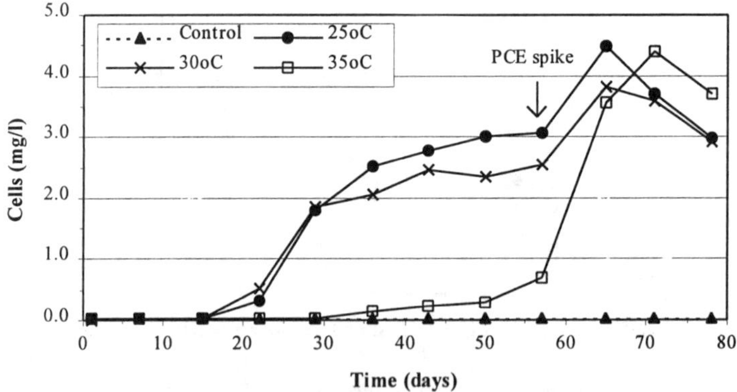

FIGURE 2. BB1 cell yield derived from ^{14}C-acetate and PCE utilization in a batch experiment at varying temperatures over time.

A similar curve was observed for CO_2 production over time (results not shown). From this experimental data, the fraction of electrons derived from acetate that are used for the reduction of PCE, fe, and the fraction that are used for cell synthesis, fs, can be calculated (McCarty, 1971). At all temperatures and time points, fe and fs were very similar, with an average fe = 0.915 and an average fs = 0.085. The resulting stoichiometry for PCE and acetate utilization by strain BB1 is shown below.

$$0.125\ CH_3COO^- + 0.212\ H_2O + 0.229\ C_2Cl_4 + 0.125\ H^+ + 0.004\ HCO_3^- + 0.004\ NH_4^+ \rightarrow 0.233\ CO_2 + 0.229\ C_2H_2Cl_2 + 0.458\ HCl + 0.004\ C_5H_7O_2N$$

Bioreactor Experiment. The results of the bioreactor experiment are summarized in Figures 3 and 4. The control reactor lost only 15% of its PCE over the course of the experiment, showing that a relatively constant concentration of PCE can be maintained using Tenax as a source. The maximum growth rate observed in Reactor 1 was 35 μg/l/day. This low rate of growth is most likely due to the low PCE concentration in the reactor (~4 mg/l). Based on the observed chloride increase of 0.78 mM in Reactor 1 and the stoichiometry determined from the batch experiment, we can compare theoretical and experimental results as shown in Table 2.

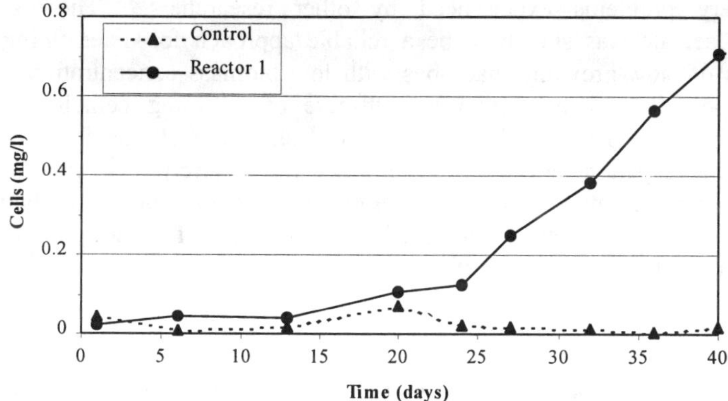

FIGURE 3. BB1 cell yield derived from the utilization of ^{14}C-acetate and PCE in a continuous bioreactor over time.

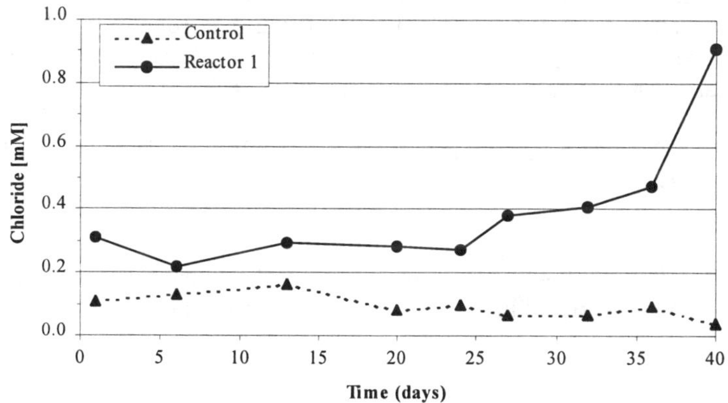

FIGURE 4. Evidence of BB1 dechlorination activity: chloride production in a continuous bioreactor over time.

TABLE 2. Bioreactor experiment: theoretical vs. experimental results

	Theoretical	Observed	% of Theoretical
Cells produced, mg/l	0.770	0.688	89.4
Acetate consumed, mM	0.213	0.257	120.7
DCE produced, mg/l	37.8	15.6	41.3*

**The majority of DCE produced was adsorbed onto the Tenax.*

CONCLUSIONS

Our data confirm that steady-state PCE concentrations can be maintained for several months in halorespiring cultures using Tenax as a source. We have also shown that Tenax adsorbs the byproduct of dechlorination, DCE, before it can accumulate to toxic levels. The use of a Tenax bed may be the solution to VOC delivery problems experienced by other researchers. The use of radiolabeled acetate was shown to be a reliable approach for determining the growth rates of slow-growing anaerobes with low biomass concentrations over short time periods. A maximum growth rate of 0.36 mg cells/l/day was determined for strain BB1 with PCE = 86 mg/l. Using the methods developed in this study, kinetic parameters can be determined for a variety of dechlorinators over a range of chlorinated solvent concentrations. This information can be used to help design more effective bioremediation strategies for the removal of chlorinated solvents from the subsurface.

REFERENCES

Atlas, R.M. 1997. *Handbook of Microbiological Media.* CRC Press: New York.

Bergmann, J.G., and J. Sanik, Jr. 1957. "Determination of Trace Amounts of Chlorine in Napthta." *Anal. Chem.* 29: 241-243.

Löffler, F.E., Li, J., Urbance, J.W., and J.M. Tiedje. 1998. "Characterization of Strain BB1, a Tetrachloroethene (PCE)-Dechlorinating Anaerobe." In *Abstracts of the 98th General Meeting of the American Society for Microbiology 1998*, p. 450. American Society for Microbiology, Washington, D.C.

McCarty, P.L. 1971. "Energetics and Bacterial Growth." In J. Faust and J.V. Hunter (Eds.), *Organic Compounds in Aquatic Environments*, pp. 495-531. Marcel Dekker, Inc., New York, NY.

Pignatello, J.J. 1990. "Slowly Reversible Sorption of Aliphatic Halocarbons in Soils. I. Formation of Residual Fractions." *Environ. Tox. Chem.* 9: 1107-1115.

APPLICATIONS OF A COMPETITIVE HYDROGENOTROPHIC BIOLOGICAL DECHLORINATION TRANSPORT MODEL FOR GROUNDWATER REMEDIATION

Matthew Willis (Ph.D. Candidate, Cornell University, USA)
Christine Shoemaker, James Gossett and Donna Fennell (School of Civil and Environmental Engineering, Cornell University, USA)

ABSTRACT: A groundwater flow and transport model describing competitive hydrogenotrophic anaerobic biological dechlorination of halogenated solvents (such as PCE) is presented. The model incorporates a kinetic description of growth of dechlorinating and competing methanogenic bacteria and the resulting biodegradation of PCE. An application of the model to a field scale model problem demonstrates the potential of the transport model for identification of remediation strategies that minimize competition for hydrogen in the subsurface environment. Additionally, the model may be utilized to determine potential for natural attenuation. Ultimately this model will allow comparison of alternative strategies to predict the effect of engineering decisions (including the type, location, and rate of injection of hydrogen donors) on contaminant movement and biodegradation in the subsurface.

INTRODUCTION

It is the purpose of this study to describe a groundwater model for the anaerobic dechlorination of halogenated solvents such as tetrachloroethene (PCE). Previous research by Fennell et al. (1997, 1998) has indicated that anaerobic bacteria can utilize hydrogen for dechlorination, but that such dechlorination is compromised by the growth of methanogenic bacteria that compete for available hydrogen. For convenience, hydrogen is added in the form of a donor material, such as butyric acid, which is fermented to produce hydrogen. Novel features of the Fennell and Gossett model (1998) include competition for hydrogen, thermodynamic limitations on donor fermentation, and differential biological activity thresholds. For example, organisms that use hydrogen for methanogenesis require higher concentrations of hydrogen than do those facilitating dechlorination. This suggests that such competition for hydrogen may be avoidable.

This paper extends the batch model of Fennell and Gossett (1998) to incorporate groundwater transport of chlorinated ethenes, bacteria, hydrogen and donor. The resultant model is called DECT (dechlorination with competition and transport). The DECT model allows comparison of the efficacy of alternative hydrogen donors and pumping rate strategies to predict the effect of engineering decisions (including the type, location, and rate of injection of hydrogen donors) on contaminant movement and biodegradation in the subsurface. Sample uses of the DECT model are examined.

IMPLEMENTATION

The DECT groundwater transport model was created by extending existing groundwater modeling software whenever possible. Groundwater flow problems are solved using the MODFLOW package, (McDonald and Harbaugh, 1988), a finite difference flow model. Transport is based on the Eulerian-Lagrangian software packages MT3D (Zheng, 1990) and the MT3D-derived package, RT3D (Clement, 1998). Derivative calculations, necessary for solution of the reaction kinetics, were generated using the ADIFOR automatic differentiation package (Bischof et al. 1996).

The DECT model requires a greatly-increased computational effort as compared to the original Fennell-Gossett model. The processes of advection, diffusion and sorption/adhesion to soil are considered for organisms and chemical species, and contribute to the increased computational requirements. Additionally, the model must compute the concentrations at each of many nodes in order to represent the time-varying spatial distribution of these species. By contrast, the Fennell-Gossett model describes reactions at only one point, at which reactants are assumed to be completely mixed.

Application Example. The availability of the DECT groundwater transport and bioremediation model provides the modeler with a tool to evaluate engineering decisions in a practical context. The example case used is a site contaminated with PCE in sandy soil. For modeling purposes it will be assumed to have uniform hydraulic conductivity; the hydraulic characteristics of the site are listed in the Appendix. As shown in Figure 1, the aquifer is polluted with PCE, which forms a plume approximately 230m in diameter, with a maximum aqueous-phase concentration of 110 g/m^3 at the center, representing a total PCE plume of approximately 4×10^6 moles (in sorbed and free phases). For this example, the subsurface environment is assumed to exist at 35°C, a temperature selected to correspond to available experimental lab data, rather than typical subsurface temperatures. It should be understood that reaction rates will be slower for typical groundwater temperatures.

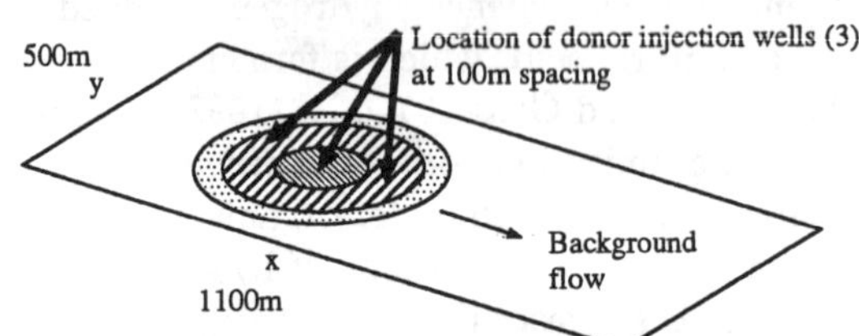

FIGURE 1. Site Geometry for Application Example

The example case is analyzed using three donor injection wells. The hydrogen donor butyrate is injected in the form of butyric acid. Two variables are examined: injection flow rate and injected donor concentration. In this study, the input donor concentrations may be either low or high, and injection rates may be either low or high. Simulation conditions are indicated in Table 1, where "high" concentration denotes 5000 g/m^3, and "low" concentration denotes 500 g/m^3. Pumping rates of 3 m^3/h and 0.3 m^3/h are used for high and low values. Note that Run 2 and Run 3 have the same donor mass loading rate (15 kilograms butyrate per hour).

TABLE 1. Simulations Examined

Run Number	Pumping Rate	Donor Concentration
1	High	High
2	Low	High
3	High	Low
4	Low	Low

Experiments and Results. Competition favors methanogenesis when concentrations of hydrogen are high since methanogenic populations can expand more quickly when hydrogen levels are high. This results in donor waste, an avoidable expense. Above 1.61×10^{-5} g/m^3 dissolved hydrogen, both methanogenesis and dechlorination occur. Between this and the dechlorination threshold, 3.02×10^{-6} g/m^3, is a preferential concentration interval where it is assumed that only dechlorination occurs. Below the lower concentration, neither activity is presumed to occur. Figure 2 shows a plot of hydrogen concentrations during Run 3 after five months of treatment. It gives a snapshot of dynamically-changing hydrogen concentrations, and how this affects competition. Close to the three injection wells, hydrogen concentrations are highest, and the most competition occurs. Outside this region is a region of lower hydrogen concentrations, where only dechlorination occurs. Since background hydrogen concentrations are assumed to be zero, the region with a competitive advantage for dechlorination is narrow. Clearly, regions of competitive advantage are possible, and this knowledge may be manipulated to the advantage of the remediator.

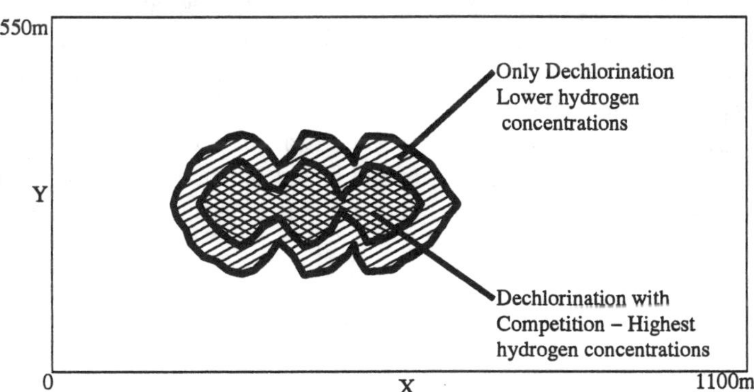

FIGURE 2. Regions with Biological Activity

The growth of methanogens is fastest when the mass rate of donor injection is high, as illustrated in Figure 3, which demonstrates the total methane produced by the competing methanogens. Runs 2 and 3, despite having the same donor mass loading rates (in grams of butyrate/hour), result in different competitive behavior.

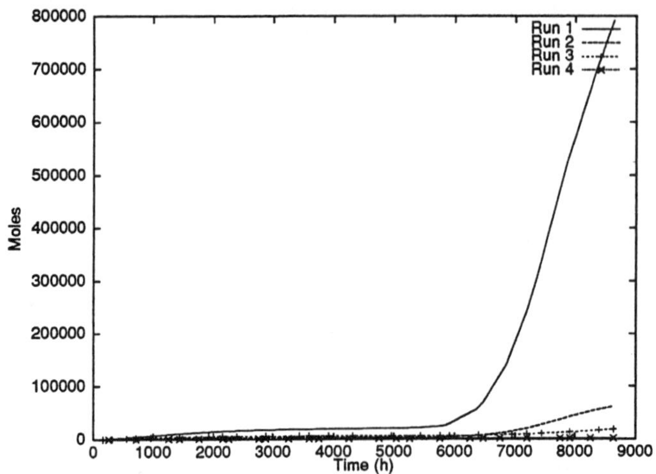

FIGURE 3. Methane produced from Hydrogen Sources

The total moles of PCE converted to other forms, shown in Figure 4, is evidence of dechlorination. Run 1 exhibits the greatest amount of dechlorination, and suggests that the highest donor mass loading rates will also present the highest dechlorination. For situations when the mass of donor added in a period of time is constant, (e.g. Runs 2 and 3), the evidence of dechlorination is greater for the high pumping/low concentration (Run 3) example than for the low pumping/high concentration (Run 2) example. This leads to the conclusion that for a given donor mass loading, higher pumping rates may be more effective in terms of minimizing competition and maximizing degradation of PCE. Note that the flattened profile of Run 1 is not due to total cleanup of PCE (there are approximately 4×10^6 moles PCE). Rather, it is due to the formation of an interception region where competitors utilize hydrogen before it can reach outlying dechlorinators. The rapid acceleration of methanogenesis seen after 6500 h (Figure 3, Run 1) is a result of the formation of a hydrogen interception region surrounding the three injection wells. After its formation, dechlorination essentially ceases.

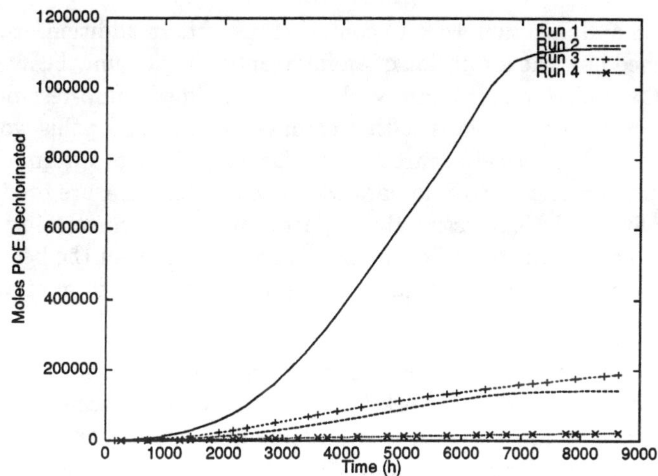

FIGURE 4. Total PCE Converted to Other Forms

DISCUSSION

The results shown in Figures 2–4 demonstrate the use of the DECT model for evaluation of alternative strategies for promoting dechlorination. Increased amounts of donor loading rates (e.g. Run 1) result in the highest dechlorination. Lesser donor loading rates, such as Runs 2 and 3, and Run 4, result in lower dechlorination activity (Figure 4). Runs 2 and 3 are of particular interest, since they represent two ways of applying the same mass of donor. It would appear that, despite the same input mass loading of donor, Run 3 has higher levels of dechlorination. It is theorized that this difference is due to the dynamical nature of the reaction kinetics, and the fact that the initial biological populations grow in response to the injected donor. With a higher pumping rate, a larger area is exposed to donor, and a larger region of dechlorination activity exists. Ultimately, the methanogen populations reach competitive concentrations and stop dechlorination from occurring. It appears that higher pumping rates are beneficial for this reason. Figures 3 and 4 suggest that a donor-injection strategy will be effective until large quantities of methane are observed, which indicate the formation of a hydrogen interception region. In this example, this appears to occur after approximately onc year.

The results given here are for a hypothetical aquifer. However, the DECT model can be adjusted for other field situations using site-specific parameters. It should be noted that the assumed subsurface temperature, 35°C, was chosen to agree with experimental laboratory results, and is unrealistic for groundwater environments in the United States. Real-world dechlorination activity will certainly be slower than predicted here. The effect of temperature is very important and is currently a subject of active research.

We have evaluated the effect of injection pumping rate and donor concentration. This study was by no means exhaustive; we could also examine alternative donors, alternative well locations, natural attenuation, and a mixture of extraction and injection wells to control transport. In addition, it remains to be determined how the formation of an interception region may be avoided. It is theorized that pulsed donor input, with a suitably timed pump-rest-pump cycle, may be effective in this regard. Other extensions planned for this work include improvement in numerical accuracy and efficiency through alternative kinetic formulations, application to field data with realistic temperature profiles, extensive sensitivity analysis to assess the impact of parameter uncertainty on model predictions, and application of optimization analysis to select the best strategies (pumping rates, well locations, donor concentrations, etc.) to minimize costs.

ACKNOWLEDGEMENTS

This research has been funded by a grant to Christine Shoemaker and James Gossett from the New York Center for Advanced Technology (Biotechnology Program) and by a grant to Christine Shoemaker from the IBM Environmental Research Grant program. The author is grateful to T. P. Clement for access to the source code for the RT3D program, and to Christian Bischof and Alan Carle for the use of the ADIFOR automatic differentiation package.

REFERENCES

Bischof C., P. Khademi, A. Mauer, A. Carle. 1996. "Adifor 2.0: Automatic differentiation of Fortran 77 programs" *IEEE Computational Science & Engineering* 3: (3) 18-32.

Clement, T. P. 1998. *RT3D, A Three-Dimensional Numerical Model for Simulation of Multi-species Reactive Transport of Contaminants in Saturated Groundwater Systems*

Fennell, D. E., J. M. Gossett. 1998. "Modeling the Production of and Competition for Hydrogen in a Dechlorinating Culture." *Environmental Science and Technology*, 32: 2450-2460.

Fennell, D. E., J. M. Gossett, and S. H. Zinder. 1997. "Comparison of Butyric Acid, Ethanol, Lactic Acid, and Propionic Acid as Hydrogen Donors for the Reductive Dechlorination of Tetrachloroethene" *Environmental Science and Technology*, 31: 918-926.

McDonald, M. G., A. W. Harbaugh. 1988. *MODFLOW, A Modular Groundwater Flow Model*

Vershureren, K. 1983. *Handbook of Environmental data on organic chemicals, 3ed.* Van Nostrand Reinhold.

Zheng, C. 1990. *MT3D, A Modular Three-Dimensional Transport Model for Simulation of Advection, Dispersion and Chemical Reactions of Contaminants in Groundwater Systems*

APPENDIX

The subsurface conditions for the sample problem are listed in Table 2. Table 3 lists chemical and biological sorption parameters used. Biological parameters used for this study are based on $35°C$ temperature values as used by Fennell and Gossett (1998).

TABLE 2. Hydrogeologic Simulation Parameters

Parameter	Symbol	Value
Porosity	n	0.3
Dry bulk density	ρ_B	1.86 g/ml
Hydraulic conductivity	K	1×10^{-3} cm/s
Aquifer thickness	b	10 m
Aquifer length	X	1100 m
Aquifer width	Y	500 m
Background hydraulic gradient	i	0.023 m/m
Storativity	S	5.0×10^{-4}
Transverse/longitudinal dispersivity	α_T/α_L	0.3
Vertical/longitudinal dispersivity	α_V/α_L	1.0
Molecular diffusion coefficient		0.0
Longitudinal dispersivity	α_L	10 m
Fraction of organic carbon	f_{oc}	0.015
Temperature of subsurface	T	35°C

TABLE 3. Partition Coefficient Values (K_d)

Species	K_d	Note
Butyrate	0 m^3/g	*
Acetate	0 m^3/g	*
PCE	5.46×10^{-6} m^3/g	†
TCE	1.89×10^{-6} m^3/g	†
DCE	0.89×10^{-6} m^3/g	†
VC	0 m^3/g	*
ETH	0 m^3/g	*
H_2	0 m^3/g	*
CH_4	0 m^3/g	*
H_2 methanogens	1.45×10^{-6} m^3/g	**
Acetotrophic methanogens	1.45×10^{-6} m^3/g	**
Dechlorinators	1.45×10^{-6} m^3/g	**
Donor fermenters	1.45×10^{-6} m^3/g	**

*- Sorption ignored (considered a dissolved product in calculations).

†- K_d is calculated using the relation $K_d = f_{oc}k_{oc}$ with $f_{oc} = .015$ and k_{oc} values from Vershuren (1983)

**- equivalent K_d calculated assuming $R = 1 + \rho_B K_d/n$ with a specified retardation $R = 10$, $n = 0.30$ and $\rho_B = 1.86 g/ml$.

CARBON TETRACHLORIDE PRODUCT SHIFT IN A METHANOGENIC, IRON-CONTAINING SYSTEM

E. J. Andrews and P. J. Novak, University of Minnesota, Minneapolis, Minnesota, USA

ABSTRACT: Transformation of carbon tetrachloride (CT) by *Methanosarcina thermophila* can be influenced by the presence of elemental iron (Fe^0). The transformation of CT by *M. thermophila* grown in the presence and absence of Fe^0 (P and A, respectively) was investigated using ^{14}C-CT. In both P and A treatments, the main degradation products were chloroform (CF) and CO_2. However, the P treatment produced greater amounts of CF, but degraded it at a higher rate than the A treatment. The P treatment also produced less $^{14}CO_2$ than the A treatment. The results of ^{14}C-CT degradation experiments using the filtered culture supernatant from *M. thermophila* cultures grown with and without Fe^0 parallel those from the organism-containing systems, with the exception that CO is generated as an intermediate transformation product. The greater amounts of CF and the lesser amounts of CO and CO_2 produced by systems pre-exposed to Fe^0 suggest that Fe^0 creates conditions that lead to more hydrogenolysis of CT than occurs in systems without pre-exposure to Fe^0.

INTRODUCTION

Carbon tetrachloride (CT), once used as a degreaser and dry cleaning solvent (Verschueren, 1996), is now a common groundwater contaminant listed as a priority pollutant by the US EPA. Previous studies showed that transformation of chloroform (CF) occurred more rapidly when methanogens are combined with elemental iron (Fe^0) (Weathers et al., 1997). This enhanced degradation may provide an advantageous strategy for coupling biotic and abiotic remediation technologies. However, an understanding of the influences on product distribution is necessary to develop and optimize such a remediation strategy.

The presence of Fe^0 is one of many factors that can influence product distribution. Research suggests that it may be possible to influence the rate and end products of contaminant degradation. The addition of cyanocobalamin to a methanogenic enrichment culture increased the rate and extent of CF degradation, while simultaneously decreasing the amount of dichloromethane (DCM) and increasing the amount of CO_2 produced (Becker and Freedman, 1994). The addition of short-chain organic acids or alcohols to an anaerobic aquifer microcosm stimulated the degradation of tetrachloroethene (Gibson and Sewell, 1992). In abiotic systems, the end products of CT degradation in the presence of cobalt corrins were dependent on the reducing agent utilized (Lewis et al., 1996).

To determine whether the enhanced transformation of CT in the Fe^0-methanogen system was due to (1) an increase in the rate of transformation, with eventual formation of the same products or (2) an alternative transformation mechanism, resulting in formation of different products, experiments were performed with *Methanosarcina thermophila* and ^{14}C-CT. Products were identified to determine if degradation occurred via hydrogenolysis, generating CF and DCM, or hydrolytic reduction, generating CO. The influence of substrate, cell age, pH, [H_2], and [Fe^{2+}] on the product distribution is being studied.

MATERIALS AND METHODS

Experiments utilized unwashed Fe^0 powder (99.9% pure, Aldrich Chemical Co.), hexanes (98.5% pure, Aldrich Chemical), methanol (HPLC grade, Mallinckrodt Baker Inc.), and ^{14}C-CT (97% pure, DuPont NEN Life Science Products). CH_4, H_2, N_2, CO_2, and CO were supplied by Air Products.

M. thermophila (Deutsche Sammlung von Mikroorganismen) was cultured in the presence (P, 10 g/L) and absence (A) of Fe^0 powder (Novak et al., 1998).

Experimental Procedure. Triplicate bottles were used for all experiments. For experiments with *M. thermophila*, stationary phase cultures were transferred to crimp-top serum bottles in a glovebag. Twenty mL of culture or media was placed into 38-mL serum bottles. In one treatment, Fe^0 powder (10 g/L) was added to bottles containing organisms (referred to as bacteria A + Fe^0) and to bottles containing medium only. Other treatments consisted of organisms only (referred to as bacteria P and bacteria A) and medium only. After bottles were capped in the glovebag, the headspace was flushed with nitrogen. Reactors were incubated at 50°C quiescently for the duration of the experiment.

For experiments with the supernatant from *M. thermophila* cultures, stationary phase bacteria, both P and A, were filtered through a 0.22 μm cellulose acetate bottle top filter (Corning) in a glovebag, and 20 mL of supernatant was transferred to 38-mL serum bottles. Medium and medium containing Fe^0 powder were also filtered and transferred to 38-mL serum bottles. No bacteria or Fe^0 was present in these systems. The bottles were capped, flushed with nitrogen, and incubated quiescently at 50°C for the duration of the experiment.

Gas standards were made by placing volumes of H_2, CH_4, CO_2 and CO into 38-mL serum bottles containing 20 mL water and a N_2 headspace. CT, CF, and DCM calibration standards were made from methanol stock solutions.

Experiments were started by adding 5-8 μM (~0.5 μCi/bottle) ^{14}C-CT. Headspace samples (100 μL) were taken for GC analysis (Hewlett Packard 6890) with a gas-tight locking syringe. CT, CF, DCM, H_2, CO, CO_2 and CH_4 were monitored with a μ-ECD for CT, CF, and DCM measurement and a TCD and gas proportional counter (IN/US) in series for analysis of H_2, CO_2, CH_4, and CO.

After 5 days, a final measurement of CT, CF, and DCM was taken. The reactors were acidified and analyzed for CO_2, CO and CH_4. Fifty μL headspace and 5 μL liquid from each reactor were added to scintillation cocktail for subsequent counting on a liquid scintillation counter (LSC). In experiments with bacteria, each reactor was filtered onto a Whatman glass fiber filter and stored at –69°C until the samples were combusted and the resulting CO_2 analyzed for radioactivity. Empty reactors and stoppers were extracted with 5 mL hexane for 24 hours. Five μL of the extract from each reactor was counted on a LSC.

RESULTS AND DISCUSSION

Effects of Fe^0 on the bacteria. Figures 1a and 1b show the degradation of CT and the formation of CF by *M. thermophila*, respectively. There is only a slight difference in the rate and extent of CT degradation by bacteria P and bacteria A. However, the transformation products differ. The amount of CF formed is greater in the bacteria A + Fe^0 and bacteria P treatments, but is subsequently transformed at a greater rate in the bacteria P treatments than in any of the other systems.

a.

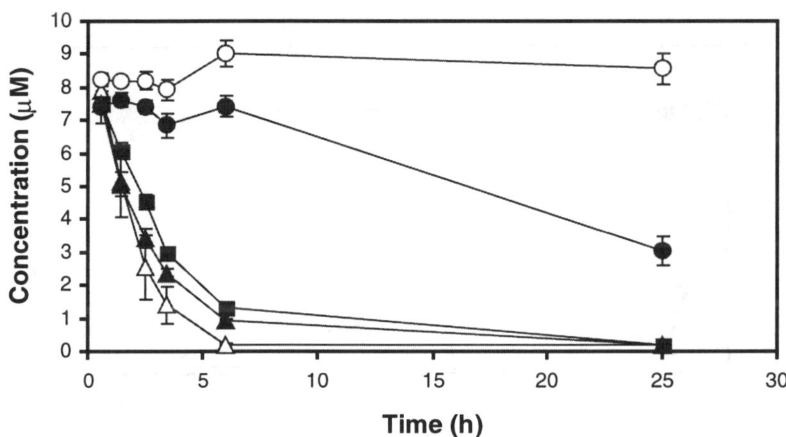

b.

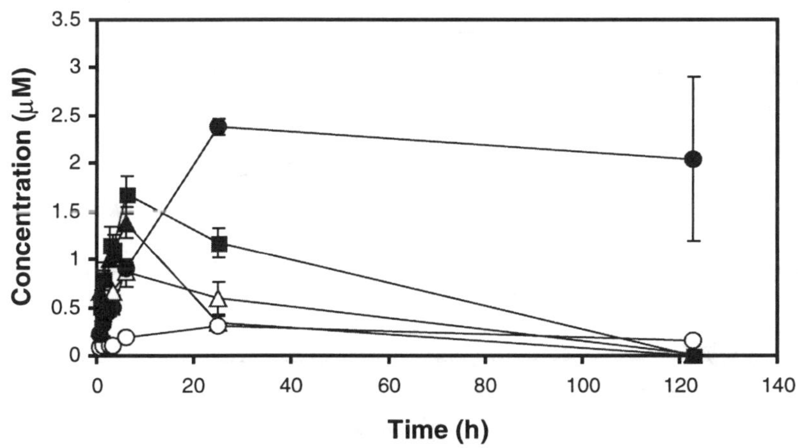

FIGURE 1. CT (a) and CF (b) transformation by *M. thermophila*. Symbols are: bacteria A: Δ, bacteria A + Fe^0: ■, bacteria P: ▲, medium only: O, medium + Fe^0: ●. Error bars are ± a standard deviation.

Bacteria A produced nearly 7 times the amount of $^{14}CO_2$ as the other treatments. All of the cultures produced trace amounts of $^{14}CH_4$. After 7 days, approximately all of the radioactivity could be accounted for (data not shown).

Effects of Fe^0 on the supernatant. The supernatant from *M. thermophila* degrades CT (Novak et al., 1998); therefore, the supernatant was investigated to gain insight into the product distribution observed in the organism-containing systems. Figures 2a and 2b show the transformation profiles for CT and CF in this experiment, respectively.

a.

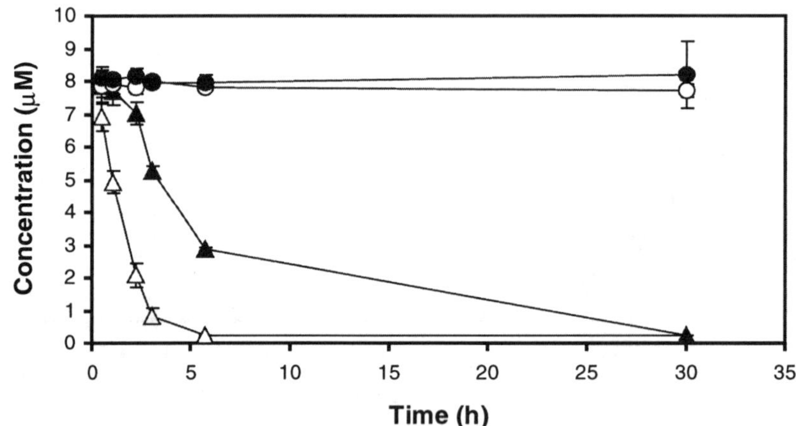

b.

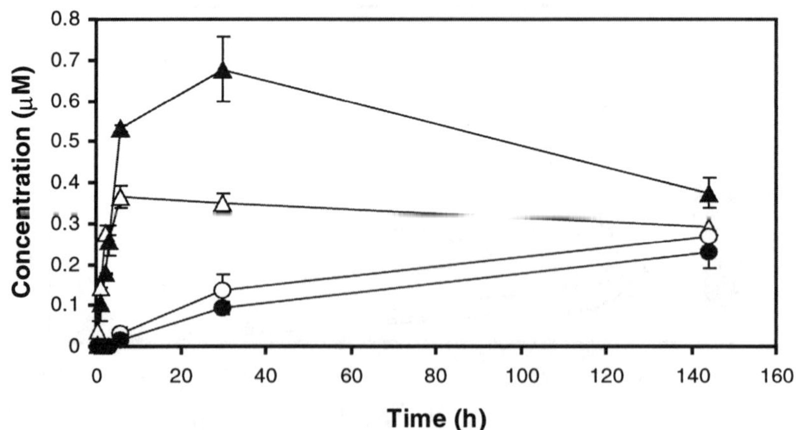

FIGURE 2. CT (a) and CF (b) transformation by *M. thermophila* supernatants. Symbols are: A supernatant: Δ, P supernatant: ▲, medium only: O, medium + Fe^0 (filtered): ●. Error bars are ± a standard deviation.

The transformation trends of CT and CF parallel that seen in the organism-containing systems discussed above. $^{14}CO_2$ levels in the filtered supernatant from bacteria A were again higher than that seen in the filtered supernatant from bacteria P. After 7 days, approximately all of the radioactivity could be accounted for (data not shown).

Supernatant from bacteria P produces significantly less CO, indicating that the presence of Fe^0 leads to conditions that perhaps favor hydrogenolysis over hydrolytic reduction, as suggested by the greater amounts of CF produced. The reaction of microscopic Fe^0 particles (< 0.22 μm) with CT to produce CF could occur, but is an unlikely explanation for the higher amounts of CF observed in the bacteria P treatments. The amount of CF produced in the filtered medium-Fe^0 treatment does not account for the difference in CF formed in the bacteria A and P filtered supernatants. This supports the idea of a biological catalyst excreted by *M. thermophila* in the presence of Fe^0 (Novak et al., 1998).

Another significant observation is ^{14}CO is transformed after CT and CF have been transformed. Figure 3 shows the transformation of CO and CO_2 by the supernatant from bacteria A. The transformation of ^{14}CO corresponds to a subsequent increase in $^{14}CO_2$. This suggests that ^{14}CO is oxidized to $^{14}CO_2$ and $^{14}CO_2$ can be generated by anaerobes in a pathway other than via a trichloromethyl radical (Criddle and McCarty, 1991).

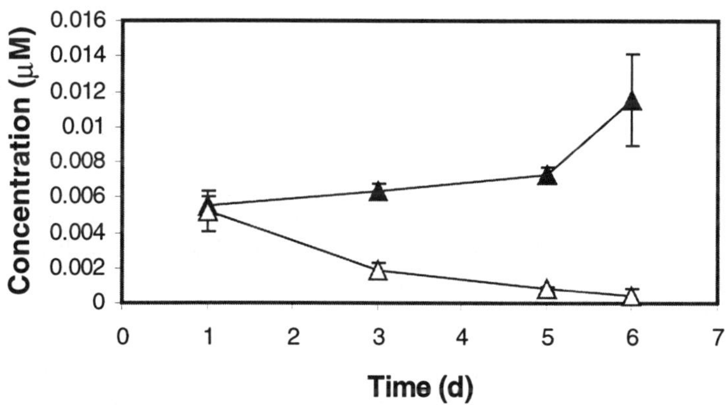

FIGURE 3. Transformations of CO (Δ) and CO_2 (▲) by *M. thermophila* supernatant (A). Error bars are ± a standard deviation.

CONCLUSIONS

Our data suggest that it is possible to influence the pathway used by *M. thermophila* to degrade CT. The presence of Fe^0 seems to reduce the extent to which CT is degraded by hydrolytic reduction. The ability to impact the extent of contaminant mineralization and quantity of toxic intermediates formed is critical to the future use of bioremediation schemes. Knowledge of what factors determine the final product distribution of contaminants provides one with the ability to optimize such processes. Several other factors are currently under investigation that can potentially influence the final product distribution of CT degradation; these include cell age, substrate, pH, and Fe^{2+} concentration.

Insight has also been gained into the generation of CO_2 in anaerobes. Our data suggests that CT is transformed to CO, which is subsequently transformed to CO_2. Although the existence of this mechanism has been hypothesized, to date little experimental evidence has existed to confirm it.

AKNOWLEDGEMENTS

The authors thank C. Just at the University of Iowa for biomass combustion and analysis of $^{14}CO_2$, and B. Koons at the University of Minnesota for developing the GC methods employed in these experiments.

REFERENCES

Becker, J. G. and D. L. Freedman (1994). "Use of a Cyanocobalamin To Enhance Anaerobic Biodegradation of Chloroform." *Environmental Science and Technology 28*: 1942-1949.

Criddle, C. S. and P. L. McCarty (1991). "Electrolytic Model System for Reductive Dehalogenation in Aqueous Environments." *Environmental Science and Technology 25*: 973-978.

Gibson, S. A. and G. W. Sewell (1992). "Stimulation of Reductive Dechlorination of Tetrachloroethene in Anaerobic Aquifer Microcosms by Addition of Short-Chain Organic Acids or Alcohols." *Applied and Environmental Microbiology 58*: 1392-1393.

Lewis, T., A., M. J. Morra, and P. D. Brown. (1996). "Comparative Product Analysis of Carbon Tetrachloride Dehalogenation Catalyzed by Cobalt Corrins in the Presence of Thiol or Titanium(III) Reducing Agents." *Environmental Science and Technology 30*: 292-300.

Novak, P. J., L. Daniels, and G. F. Parkin. (1998). "Rapid Dechlorination of Carbon Tetrachloride and Chloroform by Extracellular Agents in Cultures of Methanosarcina thermophila." *Environmental Science and Technology 32*: 3132-3136.

Verschueren, K. (1996). *Handbook of Environmental Data on Organic Chemicals*. New York, Van Nostrand Reinhold.

Weathers, L. J., G. F. Parkin, and P. J. Alvarez. (1997). "Utilization of cathodic hydrogen as electron donor for chloroform cometabolism by a mixed methanogenic culture." *Environmental Science and Technology 31*: 880-885.

CHLORINATED METHANE TRANSFORMATION BY A METHANOGEN DERIVED BIOMOLECULE

Koons, Brad W. and Novak, Paige J.
University of Minnesota, Minneapolis, Minnesota, USA

ABSTRACT: Contamination of groundwater by chlorinated solvents is a widespread problem in the United States. Previous studies have shown that the combination of Fe^0 and a methanogenic population enhanced both the rate and extent of carbon tetrachloride and chloroform transformation. Further research showed that the methanogen *Methanosarcina thermophila* excreted one or more extracellular biomolecules, which played a role in this enhanced degradation. Work to characterize the extracellular biomolecule(s) is ongoing.

Studies have been conducted to determine the biomolecule activity, defined as the rate of carbon tetrachloride transformation and the ability to transform chloroform, under a variety of pH, reduction potential, and temperature conditions. Examination of the affect of pH on biomolecule activity has indicated that activity is greater under alkaline conditions. At pH 8.5, carbon tetrachloride and chloroform were rapidly degraded by the excreted biomolecule. However, at pH 5.5, chloroform degradation was inhibited and carbon tetrachloride was not degraded significantly faster than in medium controls. When in an oxidized environment, biomolecule activity was completely inhibited. Rereduction of the biomolecule after oxidation returned limited carbon tetrachloride degradation activity, but chloroform transformation was permanently inhibited.

This research could lead to an effective bioremediation approach that utilizes this excreted biomolecule for rapid contaminant destruction.

INTRODUCTION

Carbon tetrachloride (CT) and chloroform (CF) are chlorinated solvents which have been used for decades in cleaning and manufacturing applications. Substantial quantities of these chemicals have been released into the environment and persist in ground and surface waters (Novak et al., 1998b). Because of their widespread contamination and negative impacts on human health, extensive research has been performed to determine strategies for the remediation of these compounds.

Research on CT and CF degradation has centered around the dechlorination of these compounds under anaerobic conditions. Several researchers have reported the ability of mixed cultures and pure strains of methanogenic bacteria to rapidly and effectively transform chlorinated compounds, including CT and CF (Bouwer and McCarty, 1983; Bagley and Gossett, 1995). Additional studies have shown that methanogenic metallo-coenzymes are active in dechlorination, and are most likely responsible for their degradative ability (Krone and Thauer, 1989; Gantzer and Wackett, 1991).

Extracellular biomolecules have been identified that are active in the dechlorination of chlorinated compounds. A study conducted by Dybas and coworkers (1995) found that *Pseudomonas* sp. strain KC was capable of CT dechlorination under denitrifying conditions. This activity was attributed in part to an extracellular factor produced by this bacterium. Previous research has also shown the existence of an extracellular biomolecule excreted from *Methanosarcina thermophila* that is capable of the dechlorination of CT and CF (Novak et al., 1998b). Because this biomolecule facilitates this transformation, further investigation of the environmental conditions that affect its activity is underway, to better understand the nature of this biomolecule and aid in its isolation and identification.

MATERIALS AND METHODS

All chemicals were high-pressure liquid chromatography grade. Unwashed Fe^0 powder (Fisher Scientific) was used in all of the experiments.

M. thermophila (Deutsche Sammlung von Microorganismen) was cultured in the presence (10 g/L) and absence of Fe^0 (Novak et al., 1998b).

Approximately one day after the cultures had reached the stationary phase, as determined by methane concentration, the bottles were uncapped in an anaerobic glovebag (Coy) and filtered (0.22 μm cellulose acetate bottle-top filters (Corning)). No Fe^0 or organisms were present in the experimental samples. Fifeteen-mL volumes were placed into 22-mL headspace autosampler vials and sealed with teflon-lined butyl rubber stoppers and aluminum crimp tops (Wheaton). Medium controls were handled in the same manner.

A saturated aqueous stock of CT was made with approximately 2 mL of neat CT added to a 38-mL serum bottle containing 20 mL milli-Q water. CT and CF calibration stocks were prepared gravimetrically in methanol. Calibration standards were prepared from calibration stocks in 22-mL headspace autosampler vials containing 15 mL milli-Q water.

2.5 μM CT was added to the vials to begin each experiment. CT and CF concentrations were measured in headspace samples via gas chromatography over the course of each experiment. Headspace volumes (250μL) were drawn from vials by a headspace autosampler (Tekmar 7000) and injected into a gas chromatograph (Hewlett-Packard 6890) equipped with a μ–ECD. In all experiments other than that assessing the temperature effects, samples were incubated at 50° C for the duration of the experiment. All experiments were performed with triplicate samples and controls.

Experimental Procedure. For the baseline CT and CF degradation study, culture supernatant from organisms grown in the presence and absence of Fe^0 (referred to as filtrate+Fe^0 and filtrate, respectively) was prepared as described above. Controls consisted of filtered medium only (medium) and filtered medium preexposed to Fe^0 (medium+Fe^0). CT and CF concentrations were monitored with time to determine degradation rates.

To determine the effect of oxygen exposure on CT and CF degradation, the treatments described above were prepared. Two sets of samples were fully oxygenated by exposing the sample to atmospheric gas using a syringe needle attached to a 0.22 μM nylon syringe filter to allow atmospheric air but no airborne organisms to come into contact with the samples. One set of the oxygenated samples was rereduced to anaerobic conditions after one hour of oxygen exposure using 1.67 mg/mL and 5.0 mg/mL anhydrous L-cysteine hydrochloride for the medium samples and Fe^0-containing medium samples, respectively.

The effect of pH on CT and CF degradation was determined using the treatments described above. After filtration, the pH of the filtrates was analyzed with a pH meter and samples were adjusted to the target pH values of 5.5, 7.0, and 8.5 using 1 M HCl solution and 1 M NaOH solution.

The effect of temperature on CT and CF degradation was determined by incubating the treatments described above at 35, 50, and 65°C for the duration of the experiment.

RESULTS AND DISCUSSION

Baseline CT and CF Degradation Study. An experiment was performed in order to establish a baseline as a basis for comparison with subsequent experiments. No significant CT dechlorination was observed for the medium and medium+Fe^0 controls. Rapid CT transformation occurred in the filtrate and filtrate+Fe^0 treatments with first-order dechlorination rates of 0.86/hr±0.01 and 0.36/hr±0.003, respectively. Small CT losses were observed in the medium controls. These losses are attributed to volatilization losses during sampling. Approximately 0.1 μM CF was produced in the filtrate and filtrate+Fe^0; however, no CF degradation was observed.

Effect of Oxygenation on CT and CF Degradation. Oxidation of the excreted biomolecule was investigated to determine if dechlorination could be performed in an oxidizing environment. CT dechlorination was completely inhibited in the oxidized samples, indicating that dechlorination is not possible when the biomolecule is oxidized. As shown in Figure 1, rereduction of the oxidized biomolecule restored some of the CT dechlorination capability; however, dechlorination was much slower than when catalyzed by the culture supernatant that had not been exposed to O_2.

CF dechlorination was also affected by oxidation. CF dechlorination behavior in the fully oxidized species could not be determined because no CF was formed. As shown in Figure 2, CF formation was zero-order in the oxidized/rereduced samples. CF formation rates were 0.051±0.003 μM/hr and 0.091±0.004 μM/hr for the filtrate and filtrate+Fe^0, respectively. The correlating zero-order rates of CT dechlorination in these samples were 0.19±0.01 μM/hr and 0.15±0.02 μM/hr. The higher rate of CF yield from the filtrate and filtrate+Fe^0 treatments could be due to abiotic CT dechlorination by Fe^0 particles <0.22 μm.

Return of partial CT dechlorination ability and apparent lack of CF dechlorination ability after oxidation/rereduction indicates that a portion of the excreted biomolecule is permanently inactivated when exposed to oxygen. This behavior also suggests that the dechlorination could be carried out by two or more biomolecules excreted by *M. thermophila*. Activity of one (or more) biomolecule is permanently inhibited by oxygen, while another performs the limited dechlorination observed in the oxidized/rereduced specimens. This is supported by the change in CT dechlorination behavior from first-order in the unoxidized species to zero-order in the oxidized/rereduced species.

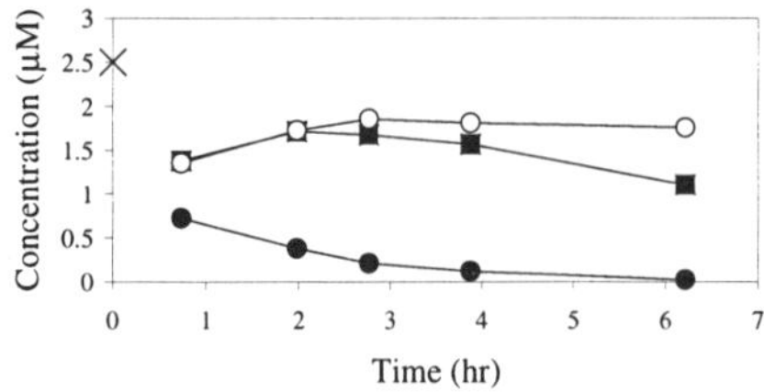

FIGURE 1. CT degradation for oxidation experiment. Symbols are: ●, filtrate+Fe^0; ◆, oxidized/rereduced filtrate+Fe^0; ○, medium+Fe^0; ×, target inoculation concentration

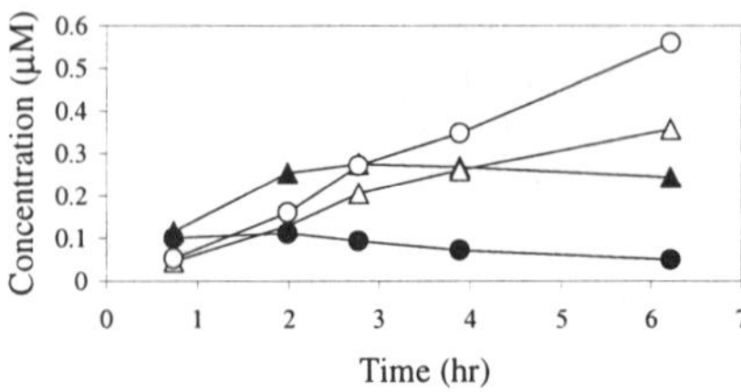

FIGURE 2. CF degradation for oxidation experiment. Symbols are: ▲, filtrate, ●, oxidized/rereduced filtrate+Fe^0, △, oxidized/rereduced filtrate, ○, oxidized/rereduced filtrate+Fe^0

Effect of pH on CT and CF Degradation. As shown in Figure 3, pH has a direct influence on the rate of CT dechlorination. After 4.2 hours of incubation at a pH of 8.5, 95% of the original CT had been degraded. After 4.2 hours at pH values of 7 and 5.5, only 85% and 75% CT degradation had occurred, respectively. As shown in Figure 4, an acidic pH inhibited CF transformation. In addition, total CF formation was greater in the acidic system. The peak CF concentration in the pH 7.0 and 8.5 systems is approximately 0.4 µM CF, whereas the pH 5.5 peak CF concentration is approximately 0.6 µM.

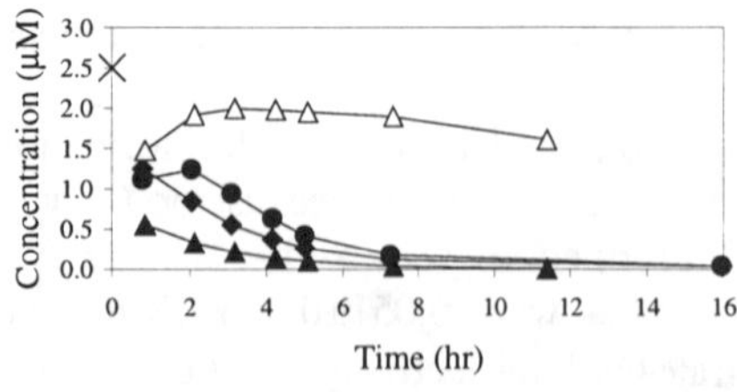

FIGURE 3. CT degradation for pH experiment. Symbols are: ●, pH 5.5 filtrate; ◆, pH 7.0 filtrate; ▲, pH 8.5 filtrate; △, pH 8.5 medium; ×, target inoculation concentration

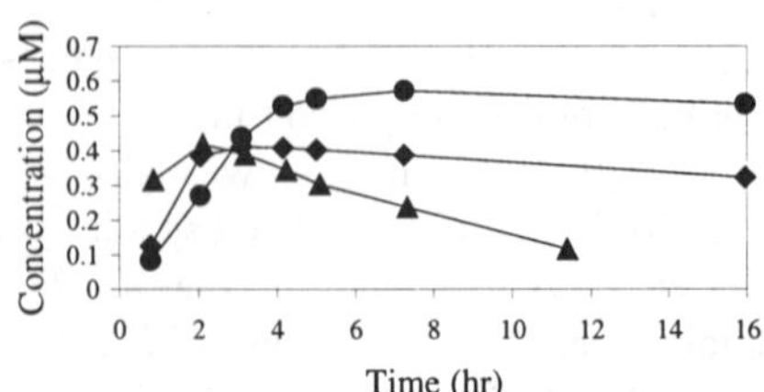

FIGURE 4. CF degradation for pH experiment. Symbols are: ●, pH 5.5 filtrate; ◆, pH 7.0 filtrate; ▲, pH 8.5 filtrate

Earlier studies of CT degradation with filtered *M. thermophila* supernatants indicated that Fe^0 added to the growth medium increased the rate and extent of CT degradation. However, when Fe^0 is added to an anaerobic environment, an increase in pH occurs as the Fe^0 is oxidized to Fe^{+2}. This increase in medium pH could account for the greater activity shift previously attributed to the inclusion of Fe^0 in the growth medium.

Effect of Temperature on CT and CF Degradation. An experiment was performed in order to determine biomolecule activity at temperatures other than 50° C. As seen in Figure 5, temperature changes within the range explored for this study have no significant impact on CT dechlorination rates. The 35° C control was significantly below the targeted starting concentration of 2.5 μM, however, no loss of CT was observed in this control throughout the duration of the experiment. As shown in Figure 6, negligible CF dechlorination was observed at all temperature ranges. The 65° C control showed an increase in CF concentration over time, possibly indicating that an abiotic process for reductive dechlorination of CT in the growth medium is activated in this temperature range.

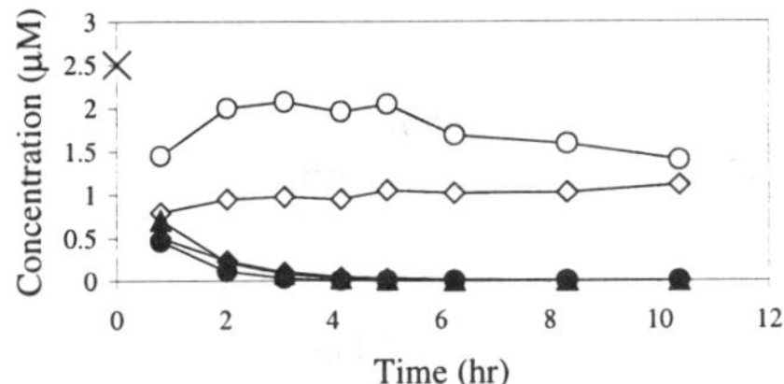

FIGURE 5. Effects of temperature on CT dechlorination activity. Symbols are: ◆, 35°C filtrate; ▲, 50°C filtrate; ●, 65°C filtrate; ◇, 35°C medium; ○, 65°C medium

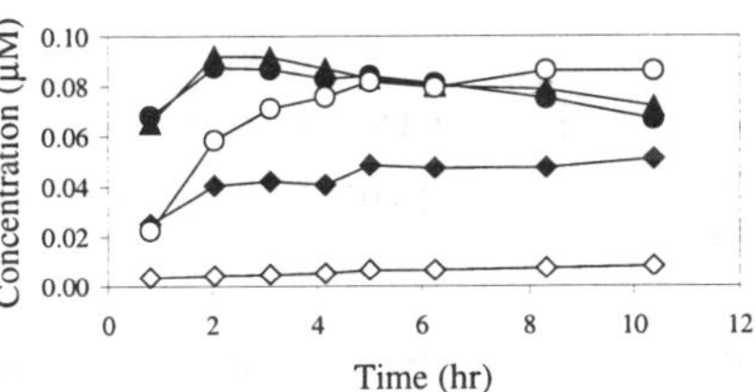

FIGURE 6. Effect of temperature on CF dechlorination activity. Symbols are: ◆, 35°C filtrate; ▲, 50°C filtrate; ●, 65°C filtrate; ◇, 35°C medium; ○, 65°C medium; ×, target inoculation concentration

CONCLUSION

The results obtained in this study indicate that the biomolecule from *M. thermophila* could be used as a catalyst for the remediation of CT and CF contaminated waters. Dechlorination activity in the temperature range tested was rapid; however, additional studies of this phenomenon at lower temperatures would be beneficial to determine if this is always so. As shown, dechlorination is dependant on pH, and knowledge of this effect can assist in designing systems that could take advantage of the greater activity at alkaline pH values. Restriction of this activity to reducing environments is a constraint. The study of this system and others like it could lead to the characterization and use of this and other extracellular biomolecules in the remediation of chlorinated compounds.

REFERENCES

Bagley, D. M., and Gossett, J. M. 1995. "Chloroform Degradation in Methanogenic Methanol Enrichment Cultures and by *Methanosarcina barkeri* 227." *Applied and Environmental Microbiology.* 61(9): 3195-3201.

Bouwer, E. J., and McCarty, P. L. 1983. "Transformations of 1-Carbon and 2-Carbon Halogenated Aliphatic Organic Compounds Under Methanogenic Conditions." *Applied and Environmental Microbiology.* 45(4): 1286-1294.

Dybas, M. J., Tartara, G. M., and Criddle, C. G. 1995. "Localization and Characterization of the Carbon Tetrachloride Transformation Activity of *Pseudomonas* sp. Strain KC." *Applied and Environmental Microbiology.* 61(2): 758-762

Gantzer, C. J., and Wackett, L. P. 1991. "Reductive Dechlorination by Bacterial Transition-Metal Coenzymes." *Environmental Science and Technology.* 25: 715-722.

Krone, U. E., and Thauer, R. K. 1989. "Reductive Dehalogenation of Chlorinated C_1-Hydrocarbons Mediated by Corrinoids." Biochemistry. 28: 4908-4914

Novak, P. J., Daniels, L., and Parkin, G. F. 1998a. "Enhanced Dechlorination of Carbon Tetrachloride and Chloroform in the Presence of Elemental Iron and *Methanosarcina barkeri, Methanosarcina thermophila, or Methanosaeta concilli.*" *Environmental Science and Technology.* 32: 1438-1443.

Novak, P. J., Daniels, L., and Parkin, G. F. 1998b. "Rapid Dechlorination of Carbon Tetrachloride and Chloroform by Extracellular Agents in Cultures of *Methanosarcina themophila.*" *Environmental Science and Technology.* 32: 3132-3136

Wolin, E. A., Wolin, M. J., and Wolfe, R.S. 1963. "Formation of Methane by Bacterial Extracts." *Journal of Biological Chemistry.* 238(8): 2282

C_1 TO C_3 ACIDS IN THE REDUCTIVE DEHALOGENATION OF TETRACHLOROETHYLENE

Robert W. Holden, II, Robert B. Jacko, Loring F. Nies
Purdue University, West Lafayette, Indiana

ABSTRACT: Consumption/conversion of formate, acetate, and lactate by methanogenic enrichment cultures was followed and related to the reductive dehalogenation of (PCE) in a laboratory setting. Experimental results agreed well with the theoretical considerations at the beginning of the experiment and in explaining when dehalogenation could occur. From the results, the conclusion was drawn that no one acid served as an electron donor and provided the electrons for the dehalogenation of the PCE present. The results support the theory that hydrogen production/consumption in relation to the use of the formate, acetate, lactate, and propionate controlled dehalogenation in the samples. The coupling of these short-chained acids to the production/consumption of hydrogen results in a proposed symbiotic relationship in the experimental cultures that is absent in many pure culture experiments. Further, the exploitation of the hydrogen production ability of organic substrates can result in improved dehalogenation abilities in many remedial or treatment schemes.

INTRODUCTION

Tetrachloroethylene (PCE) is a widely employed dry-cleaning solvent, metal degreasing agent, and intermediate in the production of other halogenated chemicals. In 1986, approximately 400 million pounds of PCE were produced in the United States and another 150 million pounds were imported. In a recent study of groundwater contamination, PCE was the tenth most commonly identified groundwater contaminant; and, was rated in the second highest level of concern based on prevalence in groundwater and deleterious health effects (Knox et al., 1996).

To design treatment schemes accordingly the microbial transformation of PCE must be fully understood. Previous research has examined a multitude of differing variables effecting the reaction. The research presented in this discussion is an effort to understand the process by which a bacterial culture uses short-chained organic acids in relation to dehalogenation.

Examination of potential electron donors ability to support dehalogenation has demonstrated that the compound provided as the primary electron donor is not the true electron donor in reductive dehalogenation. Hydrogen can serve as the electron donor (DiStefano et al., 1992). The provided compound serves simply as the primary substrate in generating hydrogen which serves as the electron donor in the dehalogenation process. In coupling the oxidation of the provided substrate with the reduction of water, hydrogen is generated. Recently, a pure culture of bacteria was isolated that is capable of the complete reduction of PCE through ethylene (Maymo-Gatell et al., 1997). This bacteria has been demonstrated to have a continued dehalogenation ability with hydrogen serving as the primary electron donor (Maymo-Gatell et al., 1995). This finding complicates electron donor and acceptor relationships.

Hydrogen generation potential by various substrates selecting for dehalogenating communities has been hypothesized (Fennell et al., 1995; Fennell et al., 1997). Examination of the competition between methanogenic and dehalogenating communities led to the conclusion that the deliberate choice of primary substrates capable of low-level hydrogen release could provide a selective advantage for the dehalogenating communities (Smatlak et al., 1996).

Objective. The objective of this experiment was to define the use of provided organic acids to the sample in relation to the reductive dehalogenation of PCE. Conversion of electron donors provided to different metabolic products may provide a situation where the true electron donor for the dehalogenation process is not the primary substrate provided. The true electron donor may be the metabolic product or a combination of the originally provided substrate and the metabolic product.

MATERIALS AND METHODS

All cultures used throughout the course of research were developed from the anaerobic digestors of the Lafayette and West Lafayette, Indiana wastewater treatment plants. Composition of the final bacterial media used in all liquid culture experimentation was as follows (per liter): NH_4Cl, 100 mg; KH_2PO_4, 1 g; K_2HPO_4, 0.5 g; NaCl, 10 mg; $CaCl_2$, 5 mg; and $MgCl_2$, 10 mg. Additionally, 1 ml of a trace metals solution and 1 mg of a resazurin solution was added. Acetic acid, formic acid, and lactic acid were the original sources of electron donors. After addition of the acids, the pH of the liquid media was adjusted with NaOH pellets to a final value of 7.0. 10^{-4} M sulfide was added to reduce the media.

Analysis of chlorinated ethenes was done using a gas chromatograph equipped with an electron capture detector. A J&W Scientific capillary column (DB-624, 60 m. by 0.25 mm O.D.) was used for separation. Liquid-liquid extraction with pentane was performed to remove the chlorinated compounds from the aqueous sample. Organic acid analysis was done on a Waters High Performance Liquid Chromatograph with a 991 Photodiode Array (PDA) detector. The column used with the HPLC system was a Metachem Inertsil ODS-3 (30 cm x 4.6 mm). Separation of the acids and the water soluble vitamins was accomplished using a 0.1% H_3PO_4 solution as the mobile phase at room temperature.

Experimental Design. Use of formate, acetate, and lactate by two mixed methanogenic cultures was followed in triplicate samples with abiotic controls for a period of six months. Concentrations of chlorinated ethenes and organics acids as well as gas production from the samples were recorded every third day.

RESULTS AND DISCUSSION

The consumption of formate, acetate, and lactate present in the samples varied over time. Dehalogenation in the samples resulted in the limited production of *cis*-1,2 DCE. Over the course of the experiment, three additions of fresh electron donors were made to the samples. These additions are labeled points 1, 2, and 3 in Figures 1 and 2.

Dehalogenation. Dehalogenation in both of the enrichment cultures consisted of the complete conversion of PCE to TCE. Partial conversion of TCE to cis-1,2 DCE was demonstrated after all PCE was converted to TCE. Figure 1 illustrates the recorded concentrations of PCE, TCE, and cis-1,2 DCE with time. Dehalogenation was followed through the addition of three cycles of organic acid consumption, described above. Additionally, a second cycle of dehalogenation was monitored as the samples were respiked with PCE at day 124 of the experiment.

Dehalogenation was first noted in days 6 to 9 when the liquid PCE concentration had fallen to approximately 100 μM due to loss from the sample bottles. Dehalogenation began in the period of time when the initial lactate present in the samples had been consumed and active formate consumption was proceeding. Additional acid amendments to the cultures had no stimulatory effects on the continued dehalogenation of the TCE present in the system. Only after the PCE in the system had been converted to TCE did the sequential conversion of TCE to cis-1,2 DCE occur. During this period of time, acetate and propionate consumption in the samples was occurring.

After the total concentration of halogenated ethenes in the samples had fallen below 50 μM, the samples were respiked with fresh PCE. This second amendment with PCE was done in an effort to demonstrate continued dehalogenation ability. Additionally, the decision was made not to reamend the samples with formate and lactate to determine if the acetate and propionate present in the samples could support dehalogenation. Dehalogenation in the samples after the reamendment was more gradual and resulted in reduced conversion percentages over those demonstrated previously. Dehalogenation was limited to the production of TCE and did not result in the complete conversion of the PCE present. After observing the dehalogenation characteristics of the samples after respiking with PCE for 24 days, the samples were provided 50 mM methanol in an attempt to comment on the consumption of formate and a corresponding conversion to methanol. The point at which the samples were spiked with methanol is labeled point 3 in Figure 1 and 2. The addition of the methanol did not stimulate further dehalogenation.

Organic Acid Consumption. Immediately after inoculation, all of the samples began to consume the lactate and formate present. Within the first 6 days, all of the samples had no lactate remaining. Concurrent with the consumption of lactate was the appearance of propionate and acetate in the samples.

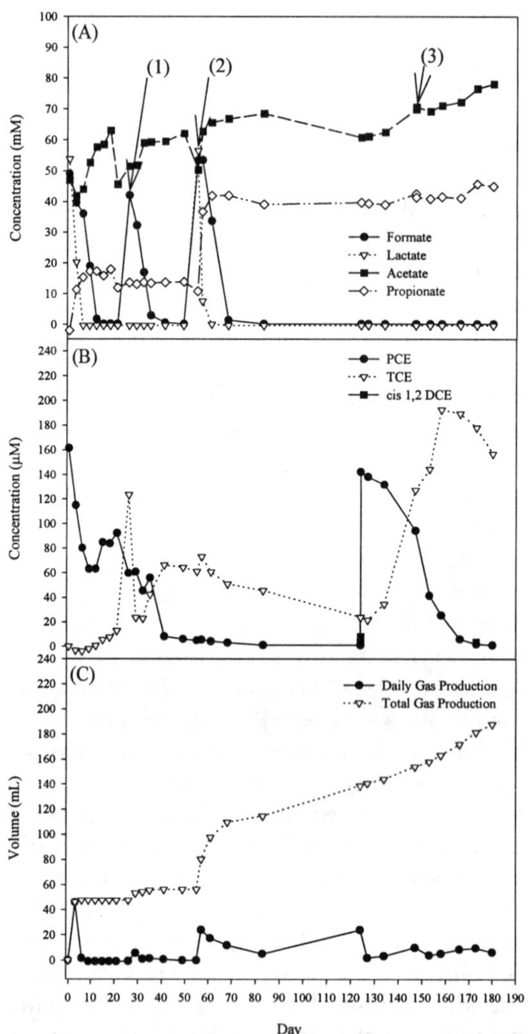

Figure 1 - Experimental results from the LAF4 cultures. Data in the graphs are for (A) organic acid consumption, (B) dehalogenation, and (C) gas production.

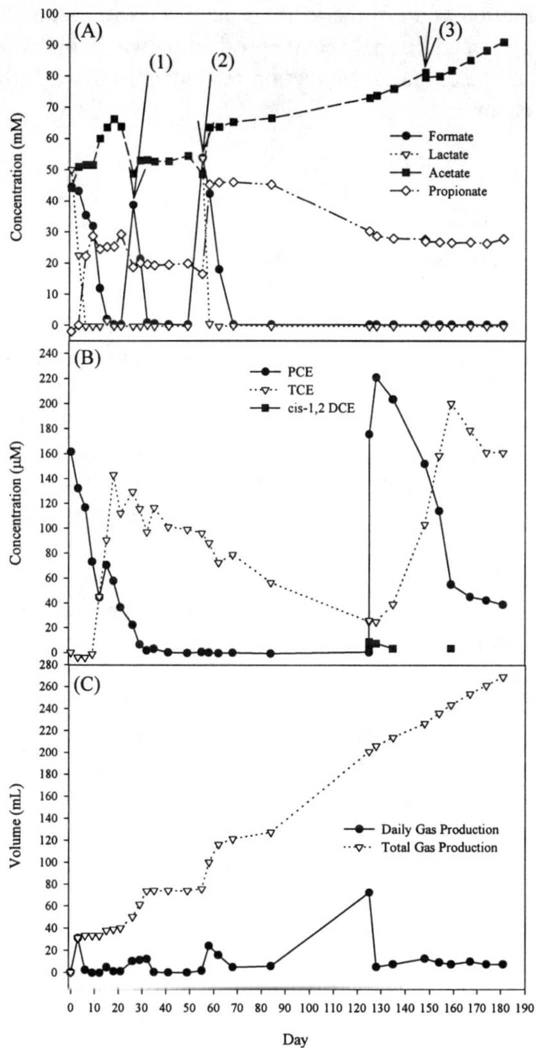

Figure 2 - Experimental results from the WL2 cultures. Data in the graphs are for (A) organic acid consumption, (B) dehalogenation, and (C) gas production.

Additionally, gas production in the samples was noted in significant levels during the consumption of lactate. Formate consumption began from day 3 to day 6 of the experiment. This lag period was simultaneous with the active consumption of lactate. Consumption of all of the formate occurred within the 21 days. During formate consumption additional acetate production was again noted. Equations 1, 2, and 3 represent the oxidation-reduction reactions balanced with hydrogen that were felt to be occur.

$$3\,CH_3CHOHCO_2^- + H_2 \rightarrow 2CH_3CH_2CO_2^- + CH_3CO_2^- + HCO_3^- + H^+ \tag{1}$$

$$HCO_2^- + H_2O \rightarrow HCO_3^- + 2H_2 \tag{2}$$

$$2\,HCO_3^- + H^+ + 4H_2 \rightarrow CH_2CO_2^- + 4H_2O \tag{3}$$

After the first 21 days of the experiment, all of the formate and lactate had been consumed. Acetate and propionate concentrations were constant, and dehalogenation of PCE to TCE had neared a level of completion demonstrated by previous generations of the enrichment cultures. At this point, the decision was made to begin to test whether formate was the electron donor for the dehalogenation and whether the extent of the dehalogenation was limited by the consumption of formate. As a result, the samples were again spiked with formate to bring the cultures back to their original 50 mM concentrations. Consumption of the new formate was complete, began immediately, and resulted in the same limited production of acetate (equations 2 and 3) with no additional dehalogenation being noted. With no additional dehalogenation being noted, the samples were respiked with both lactate and formate to determine if the consumption of the combination of the acids was responsible for the observed dehalogenation. The samples were restored to their original concentration levels of 50 mM formate and lactate. Consumption of lactate after the second amendment was slightly extended over the initial stages of the experiment. Consumption of lactate again resulted in the production of acetate and propionate. Like lactate, formate consumption by the samples at this stage in the experiment was reduced over the first two periods of formate consumption. In this stage, all formate was consumed, but the length of time over which the formate was consumed increased. All of the formate had been used within 28 days after reamendment. Formate consumption again resulted in the production of acetate.

Consumption of the propionate in the samples resulted in the production of acetate and bicarbonate as described in equation 4. Acetate consumption by the cultures was limited throughout the experiment and was only slightly noticeable. Equation 5 was felt to describe acetate consumption best.

$$CH_3CH_2CO_2^- + 3H_2O \rightarrow CH_3CO_2^- + HCO_3^- + H^+ + 3H_2 \tag{4}$$

$$CH_3CO_2^- + H_2O \rightarrow CH_4 + HCO_3^- \tag{5}$$

Conclusions

Examination of the data from the organic acid consumption, dehalogenation, and gas production reveals that the enrichment cultures were using the provided substrates in many different ways. Lactate consumption by the bacteria resulted in the production of acetate and propionate. Formate was more complex in that acetate was produced, but not directly, with the consumption of formate. Acetate consumption was limited in extent as compared to lactate and formate and was present only in the absence of formate and lactate in the samples. Thus, the samples demonstrated acetogenesis and with the confirmation of the presence of methane, methanogenesis as well.

Experimental results from this experiment agree well with the theoretical considerations of free energy at the beginning of the experiment and in explaining when dehalogenation could occur. Hydrogen available to the samples from the atmosphere of the anaerobic chamber used to prepare the samples, to the hydrogen potentially produced from the conversion of the acids supports the hypothesis of hydrogen serving as the true electron donor for dehalogenation. The data presented here leads to the conclusion that no one acid served as an electron donor and provided the electrons for the dehalogenation of the PCE present. Hydrogen production/consumption in relation to the use of the formate, acetate, lactate, and propionate was felt to control dehalogenation in the samples. The hydrogen produced during the conversion of formate to bicarbonate was concluded to be the source of reducing equivalents in the reduction of the initial PCE to TCE. Propionate conversion to acetate and bicarbonate was concluded to be the source of hydrogen for the completion of PCE reduction to TCE and for the continued reduction of TCE to *cis*-1,2 DCE.

References

DiStefano, T.D., J.M. Gossett, and S.H. Zinder 1992. "Hydrogen as an Electron Donor for Dechlorination of Tetrachloroethene by an Anaerobic Mixed Culture." *Applied and Environmental Microbiology.* (*58*):3622-3629.

Fennell, D.E., M.A. Stover, S.H. Zinder, and J.M. Gossett 1995 Comparison of Alternative Electron Donors to Sustain PCE Anaerobic Reductive Dechlorination. In: Bioremediation of Chlorinated Solvents. R.E. Hinchee, A. Leeson and L. Semprini, eds. Battelle Press, Columbus Ohio, pp. 9-16.

Fennell, D.E., J.M. Gossett, and S.H. Zinder 1997. "Comparison of Butyric Acid, Ethanol, Lactic Acid, and Proprionic Acid as Hydrogen Donors for the Reductive Dechlorination of Tetrachloroethylene." *Environmental Science and Technology.* (*31*):918-926.

Knox, R.C. and L.W. Canter 1996. "Prioritization of ground water contaminants and sources." *Water, Air, & Soil Pollution.* (*88*):205-226.

Maymo-Gatell, X., Y. Chien, J.M. Gossett, and S.H. Zinder 1997. "Isolation of a Bacterium That Reductively Dechlorinates Tetrachloroethylene to Ethene." *Science.* (*276*):1568-1571.

Maymo-Gatell, X., V. Tandoi, J.M. Gossett, and S.H. Zinder 1995. "Characterization of an H2-Utilizing Enrichment Culture That Reductively Dechlorinates Tetrachloroethene to Vinyl Chloride and Ethene in the Absence of Methanogenesis and Acetogenesis." *Applied and Environmental Microbiology.* (*61*):3928-3933.

Smatlak, C.R., J.M. Gossett, and S.H. Zinder 1996. "Comparative Kinetics of Hydrogen Utilization for Reductive Dechlorination of Tetrachloroethylene and Methanogenesis in an Anaerobic Enrichment Culture." *Environmental Science and Technology.* (*30*):2850-2858.

ANAEROBIC BIOREMEDIATION OF DNAPL TETRACHLOROETHENE IN SLUDGE MICROCOSM STUDIES

M. D. Lee (Terra Systems Inc., Wilmington DE, USA)
A. A. Biehle (DuPont Company, Houston, TX, USA)
D. E. Ellis (DuPont Company, Wilmington DE, USA)

Abstract: Impoundments at a Gulf Coast site contained sludge and sediments contaminated with tetrachloroethene (PCE) at concentrations near or above its solubility in water which indicated the presence of dense non-aqueous phase liquid (DNAPL). A twenty week microcosm study with sacrificial bottles evaluated the potential for natural attenuation and enhanced anaerobic bioremediation of the PCE and daughter products in the sludge and sediments. The laboratory studies documented the presence of an anaerobic microbial population in the sludge that is capable of degrading DNAPL-level concentrations of PCE under the existing environmental conditions at the site. Microcosms, amended with 250 mg carbon per liter impoundment water of benzoate, acetate, or molasses as substrates to support anaerobic microbial activity, showed extensive degradation of the parent chlorinated solvents. PCE and its chlorinated daughter compounds were not detected at the end of the study in the benzoate or molasses-fed treatments. Detection of higher concentrations of daughter products than the PCE concentrations measured initially demonstrated the removal of the DNAPL tetrachloroethene. High concentrations of sulfate did not inhibit dechlorination by this dechlorinating population. These studies demonstrated the potential for anaerobic bioremediation of chlorinated solvent source areas.

INTRODUCTION

Dense non-aqueous phase liquids (DNAPLs) such as chlorinated solvents are difficult to remediate (Rothmel et al. 1998). Traditional pump-and-treat technologies have little effect on DNAPLs because of the slow rate of dissolution of the chlorinated compounds. Air sparging, vacuum extraction, and surfactant-enhanced aquifer remediation have been used to treat DNAPLs. In situ bioremediation may be another remediation option. Rothmel et al. (1998) demonstrated the use of surfactant foam and bioaugmentation with an adhesion-deficient, aerobic organism to degrade TCE in laboratory simulations. Anaerobic bioremediation is another possibility for treatment of DNAPL chlorinated solvents (Lee et al. 1998). For example, PCE is degraded under anaerobic conditions to trichloroethene (TCE), dichloroethene (DCE), vinyl chloride (VC), and ethene or ethane. Concentrations of PCE as high as 550 μM (91 mg/L) have been degraded to ethene by a mixed microbial consortium isolated from a municipal wastewater sludge fed methanol (Smatlak et al. 1996). Isalou et al. (1998) reported the transformation of 600 μM PCE to about 100 μM VC and 450 μM ethene in column studies. However, other researchers have reported high concentrations of chlorinated ethenes to be inhibitory (Lee et al. 1998).

Three surface impoundments at the Gulf Coast site were used to treat wastewater containing PCE. The primary impoundment has hot spots that contain high concentrations of PCE, TCE, DCE, and VC and high concentrations of sulfate. A microcosm study was initiated to evaluate the potential for anaerobic bioremediation of DNAPL-concentrations of chlorinated contaminants in sludge from the primary impoundment.

MATERIALS AND METHODS

A microcosm study was set up to investigate the rates and extent of PCE, TCE, DCE and VC dechlorination without additional substrates or nutrients (unamended treatment), with sterile sludge autoclaved on successive days for one hour (control treatment), with nitrogen and phosphorus only (nutrient-amended treatment), and at substrate loadings of 250 mg organic carbon per liter of impoundment water (mg C/L) of molasses, sodium benzoate, and sodium acetate under sulfate-reducing conditions. Each substrate-amended microcosm was supplied with 25 mg N, 2.5 mg/L P, and 25 mg C/L yeast extract as a source of trace elements and vitamins. The nutrient-amended treatment received 25 mg N and 2.5 mg/L P, but no substrate. Five mg/L of dissolved PCE was spiked into each microcosm. The sludge contained a mixture of PCE, TCE, DCE, and VC except for the control where these compounds were removed by autoclaving.

Eight hundred gram portions from four samples that had shown the highest concentrations of contaminants in the initial screen were mixed together and used to set up the microcosm studies. The 26 mL microcosms were prepared in an anaerobic chamber with a slurry containing 50% sludge and 50% impoundment water by volume. The microcosms were sealed with Teflon-lined butyl rubber septa and crimp caps. All serum bottles except the Day 0 samples were incubated in the anaerobic chamber upside-down at 22 °C.

Duplicate microcosms from each anaerobic treatment were sacrificed after 0, 2, 4, 6, 8, 10, 12, and 16 weeks. Only one replicate was collected at 20 weeks. The following analyses were conducted on both duplicates of the anaerobic treatments. Volatile contaminants were analyzed in one mL aqueous samples preserved with 0.5 mL methanol by purge and trap gas chromatography/mass spectrometry according to the EPA SW 846 Method 8240 protocols (U.S.E.P.A. 1996). Metabolic endproduct gases (ethene, ethane, and methane) were separated by a headspace sampler-equipped gas chromatograph and analyzed with a flame ionization detector in a modification of EPA SW-846 Method 8015. Dissolved ferric iron was analyzed by inductively coupled plasma emission spectrometry by EPA SW-846 Method 6010B. Anions -- acetate, benzoate, sulfate, nitrate, chloride, and phosphate -- were analyzed by ion chromatography according to APHA-AWWA-WPCF (1985) Method 429. Ammonia was analyzed by ion chromatography according to EPA Method 300.7 (U.S.E.P.A. 1983). Total organic carbon (TOC) measurements were made by a TOC analyzer according to EPA Method 415.2.

RESULTS AND DISCUSSION

Table 1 summarizes the average percent change in concentrations of the chlorinated ethenes, chloride, sulfate, TOC, and fatty acids for each treatment from Week 0 to Week 20. The biodegradation of the chlorinated ethenes is presented graphically for the treatment amended with benzoate and yeast extract (Figure 1). Concentrations of PCE, TCE, DCE, VC, ethene, and ethane are expressed in μM units so that all chloroethenes are expressed on an equivalent basis. Because of the often large variability between replicates of a treatment at a time point due to non-uniform distribution of the DNAPL, the range is presented as error bars.

The microbes found in the impoundments were sulfate-reducers that extensively degraded the substrates. Sulfate concentrations were reduced by 8 to 29 percent in the substrate-amended treatments, but remained in excess of 1,000 mg/L. Very little methane (maximum of 0.45 mg/L) or reduced iron (maximum of 0.23 mg/L) were detected which demonstrated that methanogenesis and iron-reduction were not important electron accepting processes in this study. The microbes reduced the TOC in the substrate-amended microcosms between 35 and 80%. Acetate and benzoate were reduced to below detection limits of 5 mg/L in the treatments amended with substrates.

PCE removals over the twenty week study ranged from 90 percent for the nutrient-amended treatment to greater than 99 percent for the substrate-amended treatments. Initial PCE concentrations were near the aqueous solubility limits of 900 μM or 150 mg/L. The losses of PCE in the control treatment were attributed to incomplete sterilization of the sludge. The absence of ethene in the control treatment and the persistence of PCE in the treatments without substrate addition demonstrated that the removal that was observed in the substrate-amended treatments was biological. Higher DCE concentrations were detected in the substrate-amended treatments than were found in the initial samples. This suggested that residual PCE and TCE in excess of their aqueous solubilities were being degraded. PCE, TCE, DCE, and VC were completely eliminated in the benzoate and molasses treatments. The stoichiometry of the amount of substrate consumed per quantity of chloroethenes degraded was favorable. A maximum stoichiometry of 0.11 mg benzoate/mg PCE can be calculated if all 7,800 μM of VC produced in the benzoate-amended treatment had come from dechlorination of PCE. In comparison, a highly efficient methanol-fed culture required 0.83 mg methanol per mg PCE consumed (Lee et al., 1998). Some residual VC remained in the acetate treatment that suggested that not enough acetate was added to completely degrade all of the chlorocarbons during the 20 week study. Ethene concentrations increased in all treatments except the control. Ethene production accounted for a maximum of about 20 percent of the chloroethenes degraded in the molasses and benzoate amended treatments. Ethane was a minor dechlorination product with a maximum concentration of 1.6 μM.

The maximum first order half-lives were estimated to be 0.084 years for PCE, 0.063 years for TCE, 0.21 years for DCE, and 0.12 years for VC. The calculated intrinsic remediation first order rates for the unamended treatment are very rapid compared to the field rates reported by Lee et al. (1998). Undefined organics in the impoundment water supported this dechlorination activity.

Chloride concentrations increased in all treatments (from 19% for the control to 94% for the benzoate treatment) as the chloroethenes were degraded and possibly as precipitated chloride dissolved in the aqueous phase.

The laboratory studies documented the presence of an anaerobic microbial population in the sludge that is capable of degrading DNAPL-level concentrations of PCE under the existing environmental conditions at the site. Dissolved organics in the unamended and nutrient-amended treatments supported limited degradation of PCE with an increase in the daughter products, which suggested the sludges have capability for intrinsic bioremediation of the chlorinated ethenes at relatively rapid rates compared to other sites. Addition of nitrogen and phosphorus did not enhance the extent of intrinsic dechlorination over the unamended treatment. All three of the substrates tested supported the anaerobic biodegradation of chloroethenes. Benzoate and molasses supported more complete degradation of the chloroethenes than acetate during the twenty week study. High concentrations of sulfate did not inhibit dechlorination by this dechlorinating population. These studies demonstrated the potential for anaerobic bioremediation of chlorinated solvent source areas where an acclimated microbial population was present.

REFERENCES

APHA-AWWA-WPCF. 1985. *Standard Methods for the Examination of Water and Wastewater.* Washington, DC.

Isalou, M., B. E. Sleep, and S. N. Liss. 1998. "Biodegradation of High Concentrations of Tetrachloroethene in a Continuous Flow Column System." *Environ. Sci. Technol.* 32(22): 3579-3585.

Lee, M. D., J. M. Odom, and R. J. Buchanan, Jr. 1998. "New Perspectives on Microbial Dehalogenation of Chlorinated Solvents. Insights from the Field." *Ann. Rev. Microbiol.* 52: 423-452.

Rothmel, R. K., R. W. Peters, E. St. Martin, and M. F. DeFlaun. 1998. "Surfactant Foam/Bioaugmentation Technology for In Situ Treatment of TCE-DNAPLS." *Environ. Sci. Technol.* 32(11): 1667-1675.

Smatlak, C., J. M. Gossett, and S. H. Zinder. 1996. "Comparative Kinetics of Hydrogen Utilization for Reductive Dechlorination of Tetrachloroethene and Methanogenesis in an Anaerobic Enrichment Culture. *Environ. Sci. Technol.* 30(9): 2850-2858

U.S. Environmental Protection Agency. 1983. *Methods for Chemical Analysis of Water and Wastes.* EPA 600/4-79-020. Washington, DC.

U.S. Environmental Protection Agency. 1996. *Test Methods for Evaluating Solid Waste, Physical/Chemical Methods.* SW-846. Washington, DC.

TABLE 1. Impoundment Microcosm Study Initial Concentrations and Percent Change Over 20 Weeks.

Treatment	Unamended Initial	% Change	Control Initial	% Change	Nutrient-Amended Initial	% Change
PCE (μM)	810	-98	20	-96	1120	-90
TCE (μM)	150	-6.7	<0.57	0	107	1080
DCE (μM)	665	147	<0.77	0	380	150
VC (μM)	102	166	<1.2	0	86	388
Sum CE (μM)	1727	20	20	-96	1693	62
Ethene (μM)	38	23	<0.89	0	26	47
Ethane (μM)	1.1	-22	<0.83	0	0.77	31
Chloride (mg/L)	900	33	605	19	875	37
Sulfate (mg/L)	1500	20	2200	-9.1	1850	8
TOC (mg/L)	62	-50	110	0	96	-66
Fatty Acid (mg/L)	<5 Ace		<5 Ace		<5 Ace	

Treatment	Benzoate Initial	% Change	Molasses Initial	% Change	Acetate Initial	% Change
PCE (μM)	1300	>-99	875	>-99	815	>-99
TCE (μM)	74	>-92	66	>-91	56	>-90
DCE (μM)	355	>-98	310	>-98	350	>-98
VC (μM)	68	>-82	63	>-81	70	84
Sum CE (μM)	1797	>-98	1543	>-98	1292	>-90
Ethene (μM)	30	457	40	1900	34	635
Ethane (μM)	0.77	7.8	1.1	>-21	0.88	>-5.7
Chloride (mg/L)	980	94	1040	35	960	15
Sulfate (mg/L)	1950	-8	2250	-20	2250	-29
TOC (mg/L)	109	-35	115	-69	155	-80
Fatty Acid (mg/L)	830 Benz	>-99 Benz	50 Ace	>-90 Ace	370 Ace	>-99 Ace

Abbreviations: >- = removal to below detection limit. The detection limit was used to calculate the extent of removal. Ace = acetate. Benz = benzoate. TOC = total organic carbon. CE = Chloroethenes.

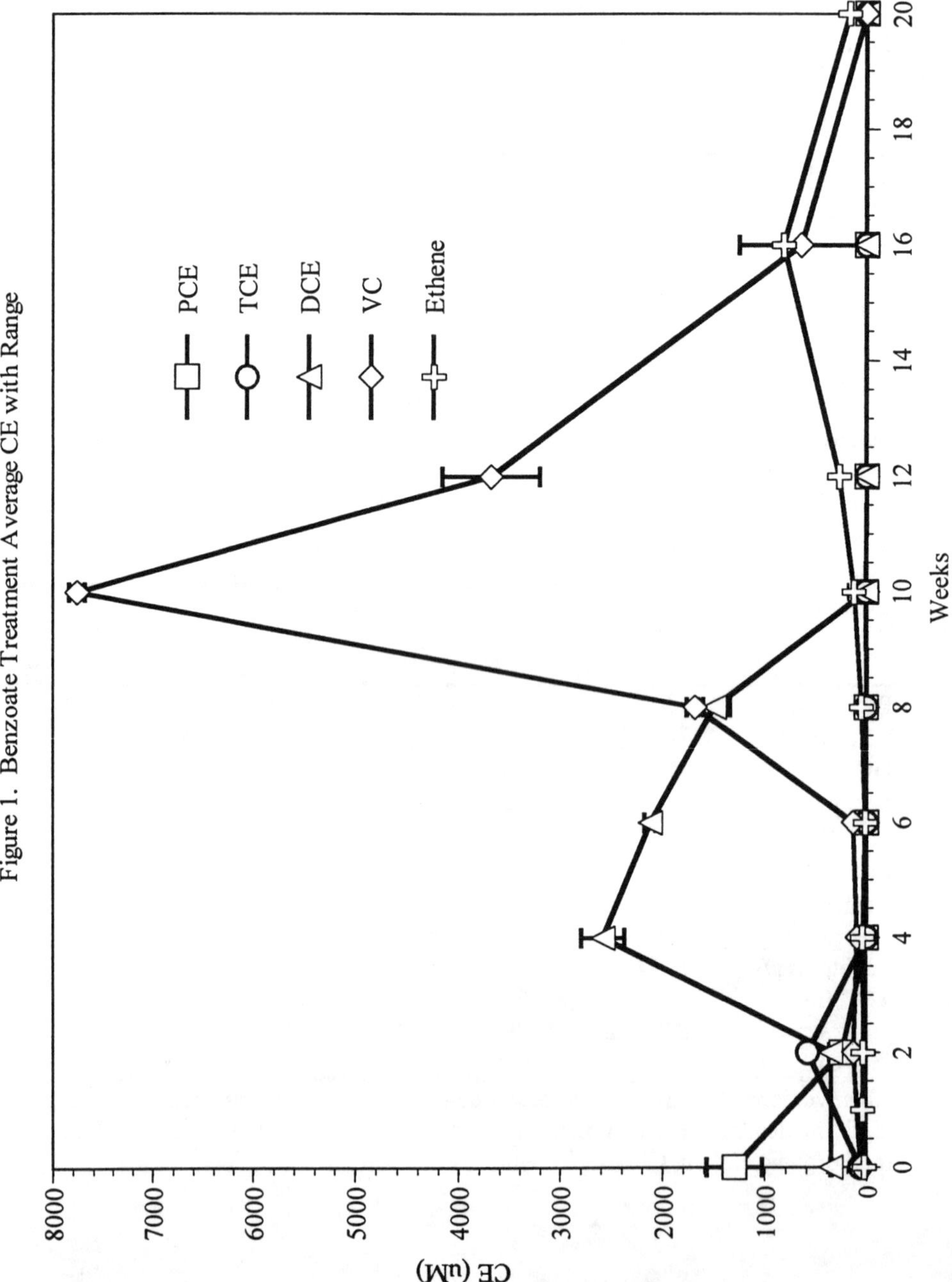

Figure 1. Benzoate Treatment Average CE with Range

FIELD APPLICATION OF A LACTIC ACID ESTER FOR PCE BIOREMEDIATION

John K. Sheldon (Montgomery Watson, Des Moines, IA, USA)
Stephen S. Koenigsberg (Regenesis, San Juan Capistrano, CA, USA)
Kenneth J. Quinn (Montgomery Watson, Madison, WI, USA)
Craig A. Sandefur (Regenesis, San Juan Capistrano, CA, USA)

ABSTRACT: As part of a strategy to enhance reductive dechlorination with a timed release of hydrogen, an application of glycerol polylactate (GPL) was made to a dry cleaner site contaminated with perchloroethylene (PCE). GPL is one of several proprietary lactic acid esters sold commercially as Hydrogen Release Compound (HRC™). When HRC contacts the saturated contaminated zone, the lactic acid moiety is released slowly by hydrolysis and is eventually degraded through pyruvic acid to acetic acid producing hydrogen at each step. Under the conditions of this test the GPL-HRC lasted between four and five months; other formulations in development could last considerably longer.

PCE degradation rates in the HRC injected zone were 11.5 times faster than background rates at Day 70 and 4.9 times faster at Day 120. Overall, 80% of the mass of PCE was removed in 253 days with good mass balance relative to TCE and DCE. HRC injection technology is conceptually applicable to enhancing the reductive dechlorination of any of the chlorinated aliphatic hydrocarbons (CAHs) and possibly any of the other chlorinated compounds that are anaerobically degradable.

INTRODUCTION

Chlorinated hydrocarbon contamination in the environment has been a difficult and persistent problem requiring and stimulating innovative solutions. Monitored Natural Attenuation (MNA), offered as a remedy in other less complicated petroleum hydrocarbon venues, is not immediately practical for many sites impacted with chlorinated hydrocarbons. Consequently, the enhancement of Monitored Natural Attenuation is a valid and important focus. To this end, we have successfully demonstrated the field application of Hydrogen Release Compound (HRC™); specifically glycerol polylactate ester (GPL). GPL is one of a number of proprietary HRC compounds that provide a time-release source of hydrogen which, under the right conditions, can promote enhanced rates of reductive dechlorination.

MATERIALS AND METHODS

A dry cleaning site impacted with PCE was chosen for the field study, because in exhibited the recognized biogeochemical signs of natural attenuation. A zone about 60 square feet was chosen for the study. It was injected by Geoprobe® with 240 pounds of a slightly flowable formulation of GPL-HRC; at 20 pounds per hole in 12 holes. Figure 1 shows the combination of existing monitoring wells and newly installed Geoprobe wells that were used to monitor upgradient, downgradient and within the injection area at intervals of two to eight weeks for eight months; it also shows some hydrogen data to be discussed later.

Samples were analyzed for the parameters deemed important in a number of standard manuals on MNA. The monitoring of hydrogen was made once on Day 149. The monitoring well network was constructed of 1-in. diameter PVC. The geology consists of a saturated sand layer of 1.8 ft. to 5 feet thick. A thick clay layer is present below the sand. Weathered silt and clay is present above the sand layer. Groundwater flow is almost due South and traveled at 0.01 – 0.1 ft./day.

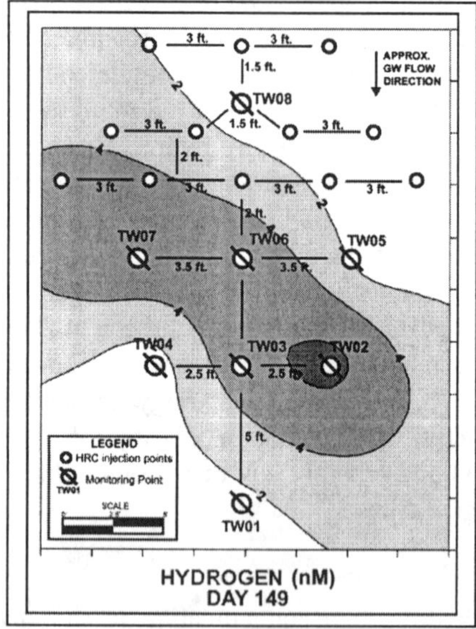

FIGURE 1. HRC injection diagram with hydrogen concentration at Day 149

RESULTS

The first effect of the HRC on the aquifer was to further depress the redox potential. Nitrate, sulfate and iron levels responded accordingly in the classical patterns and there was an immediate reduction of PCE accompanied by the expected increases in the daughter products - trichloroethylene (TCE) and dichloroethylene (DCE). In latter points in time over the 253 days test period, the degradation of daughter products began and vinyl chloride began to appear. Laboratory studies on saturated soil from the site, spiked with TCE, displayed this pattern in an accelerated fashion as shown in Table 1.

TABLE 1. TCE test tube experiments data summary.

Days	TCE	DCE	VC	TCE	DCE	VC
0	10	0	0	25	0	0
9	5.27	2.19	0	8.06	1.98	0
15	2.32	0.29	0	3.6	0.22	0
21	2.13	0.32	0.49	2.24	0.25	0.17
29	0.9	0	0	1.5	0.08	0.06

The remainder of the Tables and Figures give a comprehensive picture of the enhanced degradation dynamics in this aquifer. Figure 2 presents a series of contour plots and total mass calculations for PCE, TCE and DCE at key points in time as indicated. Also, returning to Figure 1, the injection array map, we see the results of the single measurement of hydrogen made on Day 149. It shows that a zone of hydrogen exists in the range that is associated with reductive dechlorination; considered to be 2-10 nM. The individual wells and their hydrogen measurements are as follows: TW01 (.8), TW02 (6.73), TW03 (5.42), TW04 (.8), TW05 (.8), TW06 (5.56), TW07 (5.86) and TW08 (.8). If these data are considered in conjunction with degradation rate data as presented in Tables 2 and 3, it is clear that hydrogen had an immediate impact where it was applied and then migrated readily out of the injection area to impact CAHs downgradient.

Table 2 summarizes changes in total mass in the system over different points in time showing both the percent change in mass for each CAH constituent and a mass balance between parent and daughter products. Mass balances in excess of 25% are considered to be evidence for biodegradation as a dominant mechanism in PCE loss; a mass balance of 60% would be considered ideal in studies of this nature.

Table 3 enables one to provide some quantification of the term enhanced natural attenuation. PCE data from Table 2 is abstracted and used to calculate various decay rates at different points in time for different elements in the system. The elements chosen are 1) total system mass as derived from the contouring effort in Figure 2, 2) the mass in TW08 - the most reactive well with clear evidence of HRC impact based on loss of PCE and the concomitant appearance of daughter products and 3) the mass in TW01 - the least reactive well in which there was minimal if any impact of HRC based on minimal loss of PCE and an actual decline of daughter products.

After the half lives are calculated by the first order rate equation, one can look at various ratios between the elements and quantify natural attenuation. Higher degradation rates are not occurring uniformly throughout the entire system so the total system mass loss is lower than for active wells like TW08. Rates in the HRC injected zone were 11.5 times faster than background at Day 70 and 4.9 times faster at Day 120; which was near the practical extinction point of the HRC. If one compares the total system rate to the least impacted point (TW01) the enhanced natural attenuation rates are about half; 5.11 times faster at Day 70 and 2.66 times faster at Day 120. These data also provide evidence of residual HRC effects that are still apparent at Day 253.

Figure 3 provides some other general information including the full spectrum of mass changes over time for all the CAHs, lactic acid and its derivatives and key electron acceptors. On the matter of vinyl chloride (VC) - it was never detected at the site before the HRC application, however, one can see it begins to emerge between Day 191 and Day 253. Its ability to form was also accelerated in the microcosm study (Table 1) which demonstrates microbially degradation competency in the aquifer. Lactic acid is readily released and is converted to both propionic acid and pyruvic acid, the latter being quite ephemeral leading to a more immediate registration of acetic acid. Propionic acid will essentially revert to back lactic acid and eventually move through to acetic acid. This creates a fortuitous "reserve mechanism" as propionic acid is slower to degrade. The practial endpoint for the dissolution of GPL-HRC is 4 months, but through propionic acid one sees persistence through 6 months and signs of CAH degradation through 8 months.

CONCLUSION

Injection of HRC at the site has resulted in a more rapid and significant degradation rate of PCE throughout and downgradient of the injection zone and has begun to impact the daughter products. Ultimately plume size for the daughter products will be reduced relative to their potential spread under conditions of monitored natural attenuation. An enhanced rate of natural attenuation was clearly demonstrated and shown to be persistent based on a simple cost effective application of HRC.

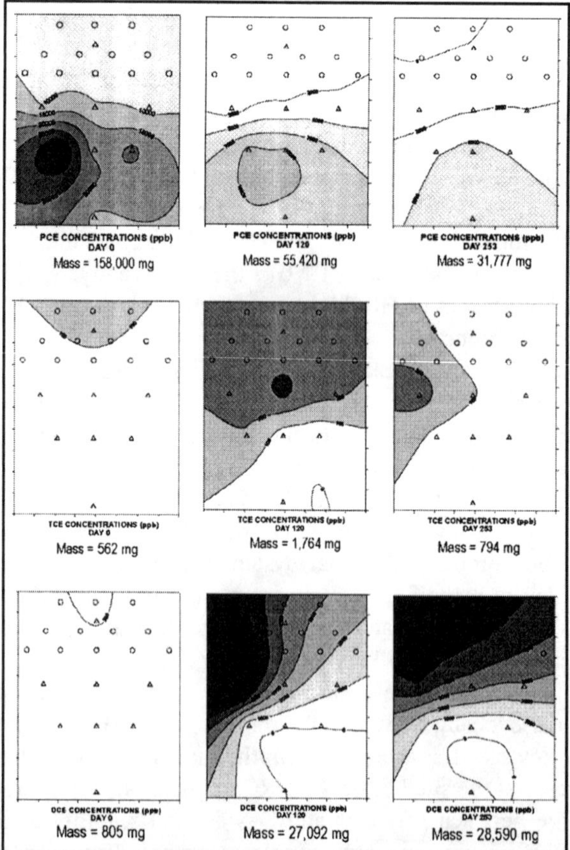

FIGURE 2. PCE, TCE, and cis-1,2-DCE concentration contour maps at days 0, 120, and 253.

TABLE 2. PCE, TCE, cis-1,2-DCE, and VC mass summary.

Mass Summary in Grams - TW01-TW08 (**Days 0-70**)				
	PCE	**TCE**	**DCE**	**VC**
Day 0	158	0.56	0.81	ND
Day 70	75.9	1.38	13.94	ND
Grams Loss/(Gain)	82.1	(0.82)	(13.13)	-
% Decrease/(Increase)	52%	(246%)	(1,721%)	-
Moles Loss/(Gain)	0.501	(0.006)	(0.139)	-
Mass Balance (DP/PP)*		29%		

Mass Summary in Grams - TW01-TW08 (**Days 0-120**)				
	PCE	**TCE**	**DCE**	**VC**
Day 0	158	0.56	0.81	ND
Day 120	55.4	1.76	27.09	ND
Grams Loss/(Gain)	102.6	(1.2)	(26.28)	-
% Decrease/(Increase)	65%	(314%)	(3,344%)	-
Moles Loss/(Gain)	0.626	(0.009)	(0.279)	-
Mass Balance (DP/PP)*		46%		

Mass Summary in Grams - TW01-TW08 (**Days 0-149**)				
	PCE	**TCE**	**DCE**	**VC**
Day 0	158	0.56	0.81	ND
Day 149	47.0	1.10	17.66	ND
Grams Loss/(Gain)	111.0	(0.54)	(16.85)	-
% Decrease/(Increase)	70%	(196%)	(2,180%)	-
Moles Loss/(Gain)	0.677	(0.004)	(0.179)	-
Mass Balance (DP/PP)*		27%		

Mass Summary in Grams - TW01-TW08 (**Days 0-191**)				
	PCE	**TCE**	**DCE**	**VC**
Day 0	158	0.56	0.81	ND
Day 191	56.2	1.41	17.82	ND
Grams Loss/(Gain)	101.8	(0.85)	(17.01)	-
% Decrease/(Increase)	64%	(252%)	(2,200%)	-
Moles Loss/(Gain)	0.606	(0.006)	(0.174)	-
Mass Balance (DP/PP)*		30%		

Mass Summary in Grams - TW01-TW08 (**Days 0-253**)				
	PCE	**TCE**	**DCE**	**VC**
Day 0	158	0.56	0.81	ND
Day 253	31.8	0.79	28.59	0.021
Grams Loss/(Gain)	126.2	(0.23)	(27.78)	(0.021)
% Decrease/(Increase)	80%	(141%)	(3,530%)	(210%)
Moles Loss/(Gain)	0.751	(0.002)	(0.283)	(0.00033)
Mass Balance (DP/PP)*		38%		

$$\text{*Mass Balance} = \frac{\text{Moles Daughter Products (DP)}}{\text{Moles Parent Products (PP)}} = \frac{\text{Moles TCE+DCE+VC}}{\text{Moles PCE}}$$

TABLE 3. PCE degradation rates.

Days	Total Mass (grams)	Half Life (d) Total Mass	Half Life (d) Well TW01	Half Life (d) Well TW08	Ratio TW01/Total Mass	Ratio TW01/TW08
70	158	66	338	29	5.11	11.50
120	75.9	79	211	43	2.66	4.94
149	55.4	85	256	53	3.00	4.84
191	47.0	128	309	45	2.41	6.80
253	56.2	109	205	38	1.87	5.41

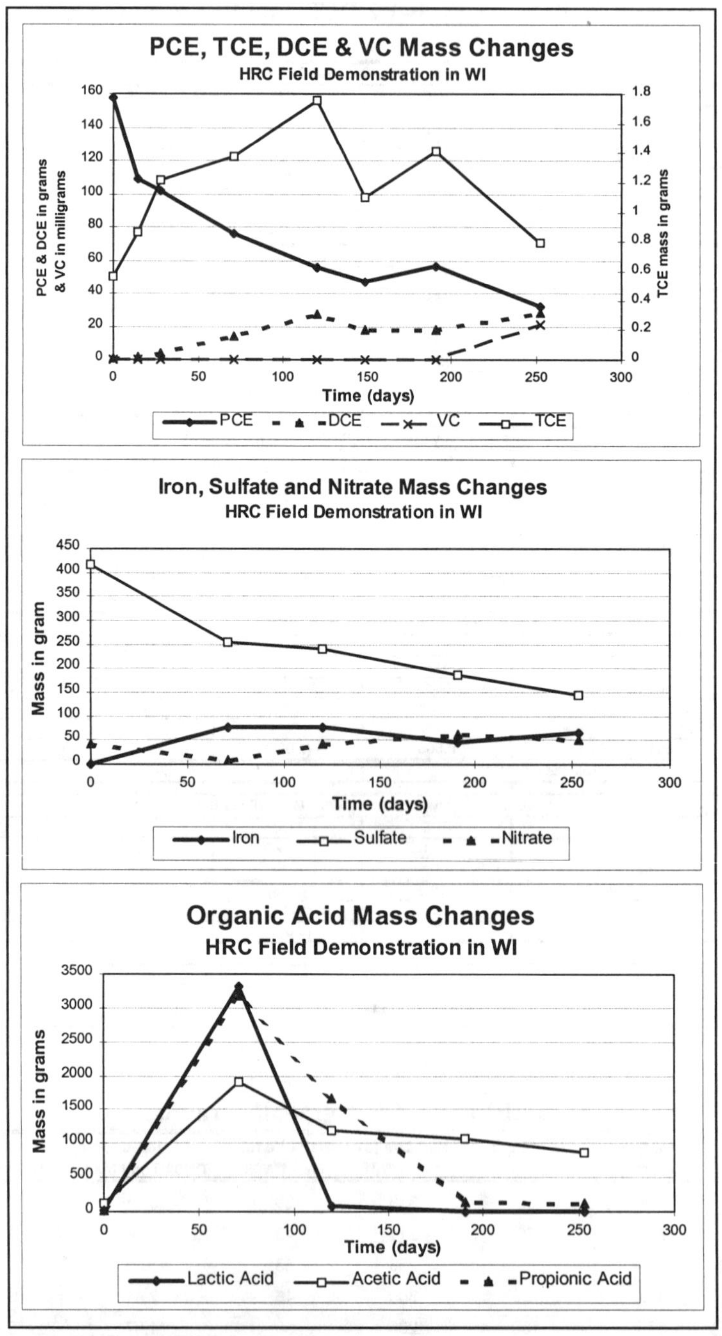

FIGURE 3. Mass changes in chlorinateds, iron, sulfate, nitrate and organic acids.

THE USE OF HYDROGEN RELEASE COMPOUND (HRC™) FOR CAH BIOREMEDIATION

Stephen S. Koenigsberg (Regenesis, San Juan Capistrano, California)
William A. Farone (Applied Power Concepts, Orange, CA)

ABSTRACT: Hydrogen Release Compound (HRC™) is a simple, passive, low-cost and long-term option for the anaerobic bioremediation of chlorinated aliphatic hydrocarbons (CAHs) via a reductive dehalogenation pathway. Applications to other classes of chlorinated compounds that are anaerobically degradable by this mechanism are under investigation. HRC should be viewed as a tool for the enhancement of natural attenuation at sites that would typically require high levels of capital investment and operating expense.

HRC is a proprietary, food grade, polylactate ester that, upon being deposited into the subsurface, slowly degrades to lactic acid. Lactic acid is then metabolized to hydrogen, which in turn drives the reductive dechlorination of CAHs. This has been demonstrated effectively in the laboratory and in the field. HRC is a moderately flowable, injectable material, that facilitates localized treatment and passive barrier designs, for the remediation of dissolved phase plumes and the associated contaminant that is hydrophobically sorbed. The use of HRC is contraindicated for free-phase DNAPL unless the total mass to be remediated is within the scope of economic feasibility in comparison to alternative treatments.

Evidence suggests there is competition between reductive dehalogenators and methanogens in which the methanogens compete for the use of hydrogen in the conversion of carbon dioxide to methane. It is believed that a low concentration of hydrogen favors the reductive dehalogenators and starves out the methanogens. The objective, therefore, is to keep hydrogen concentrations low. The time release feature of HRC, which is based on the hydrolysis rate of lactic acid from the ester and the subsequent lag time to hydrogen conversion, facilitates this objective. HRC, therefore, becomes a passive form of enhanced natural attenuation in contrast to the more capital and management intensive alternatives now available. Laboratory and field results will be presented that will expand on the first uses of HRC by various members of the engineering and consulting firm community.

INTRODUCTION

Hydrogen Release Compound (HRC) offers a passive, low-cost treatment option for in-situ anaerobic bioremediation of chlorinated aliphatic hydrocarbons. HRC is a proprietary, environmentally safe, food quality, polylactate ester specially formulated for the slow release of lactic acid upon hydration. Bioremediation with HRC is a multi-step process. Indigenous anaerobic microbes metabolize the lactic acid generated by HRC and produce hydrogen. The resulting hydrogen can be used by reductive dehalogenators which are capable of dechlorinating CAHs. Major target compounds in this group include PCE, TCE, TCA and their derivatives.

HRC is a high viscosity, flowable liquid that can be pressure injected using various direct push technologies. By providing a long-lasting, time-released hydrogen source, HRC can enhance anaerobic reductive dechlorination of CAHs. The following are some of the key advantages of HRC.

1) Low maintenance and low cost - unlike actively engineered systems, continuous mechanical operation and maintenance is eliminated, dramatically reducing overall operations and maintenance costs.

2) Constant and persistent hydrogen source - HRC is a semi-solid material that will remain where emplaced and generate highly diffusable hydrogen slowly over time. Since CAH plumes are difficult to locate, a continuous, highly diffusable hydrogen source increases the effectiveness of contact, containment, and remediation,

3) Enhance desorption of CAHs - The continuous hydrogen source provided by HRC can reduce dissolved phase CAH concentrations. This creates a larger concentration gradient which in turn facilitates desorption of CAHs from the soil matrix and

4) Favored reductive dechlorination over possible competing methanogenic activity - Results from several university studies suggest that there is competition for hydrogen between the reductive dechlorinators and methanogens While methanogen survival is favored under elevated hydrogen conditions, reductive dechlorinators are best supported in conditions of more moderate hydrogen concentration.

METHODS, RESULTS AND DISCUSSION

HRC Microcosm Studies with TCE. The use of HRC for remediation of TCE was studied in 200 ml test tube experiments in which the release of lactic acid from HRC was measured as a function of bacterial concentration and HRC concentration. In the experiments, 10 grams of sterilized sand was added to each test tube followed by a solution of TCE with a concentration up to 140 mg/L. Various quantities of bacteria, capable of metabolizing TCE, were then added. Finally, 0.5 or 1.5 grams of HRC was added to each test tube. Each day, 6 ml samples were taken and analyzed for TCE and lactic acid.

Results indicated that TCE was remediated under all conditions. Results from one representative experiment are presented in Figure 1, which shows that reduction in TCE follows an increase in lactic acid release. It is important to note that most of the initial drop in TCE (within the first hour) was due to adsorption of TCE on the sand. This TCE eventually desorbed from the soil as the dissolved phase TCE was remediated during the progression of the experiment.

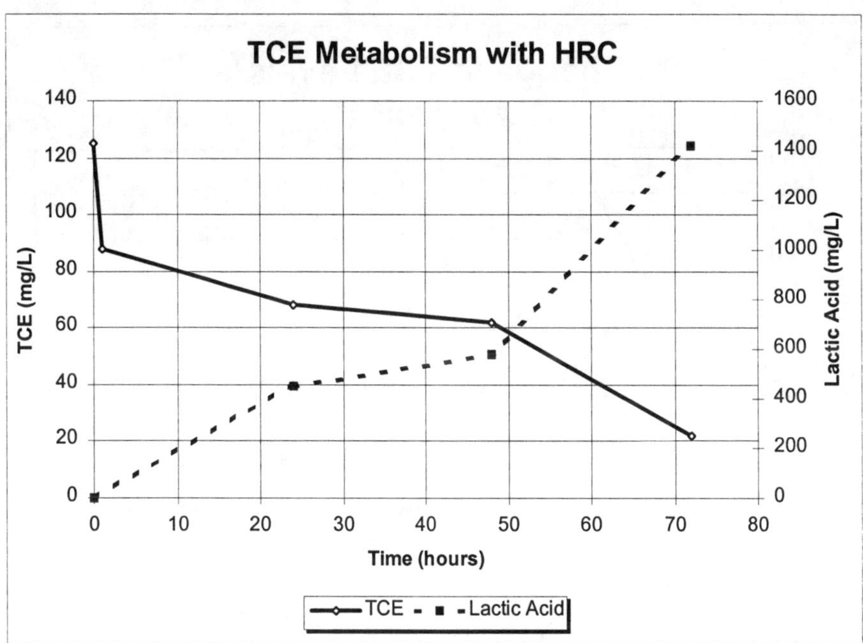

FIGURE 1. HRC microcosm study results comparing rate of release of lactic acid from HRC to rate of TCE degradation.

Aquifer Simulation Vessel (ASV) Studies. The Aquifer Simulation Vessel (ASV) is used to establish the influence of important field-scale parameters on the efficacy of HRC. The ASV consists of a horizontal six inch diameter/six foot length pipe. The ASVs are designed to allow measurement at six inch intervals along the pipe. Each pipe is packed with actual contaminated soil from the field. HRC is placed in the system at the "CAH-water" inlet side, such that the flowing water will pass through the HRC and then move through the pipe. The water can be added with various levels of CAHs and remediation rates measured. The distribution of lactic acid and its breakdown products can also be measured.

In the initial studies, the ability of HRC to facilitate the reductive dechlorination of TCE was measured. In the experiments, an ASV was filled with soil. Then, the TCE was added to the soil at the CAH-water inlet side at a concentration of approximately 6 mg/L. The ASV was allowed to acclimate over a period of 6 days during which time baseline TCE concentration profiles were developed. Finally, a "slug" of HRC was added to the inlet side and the system was run at a flow rate of 0.5 ft/day for a period of 9 days. Results from one experiment, in which TCE levels were measured at days 1, 6, and 9 at each six inch interval along the ASV, are presented in Figure 2.

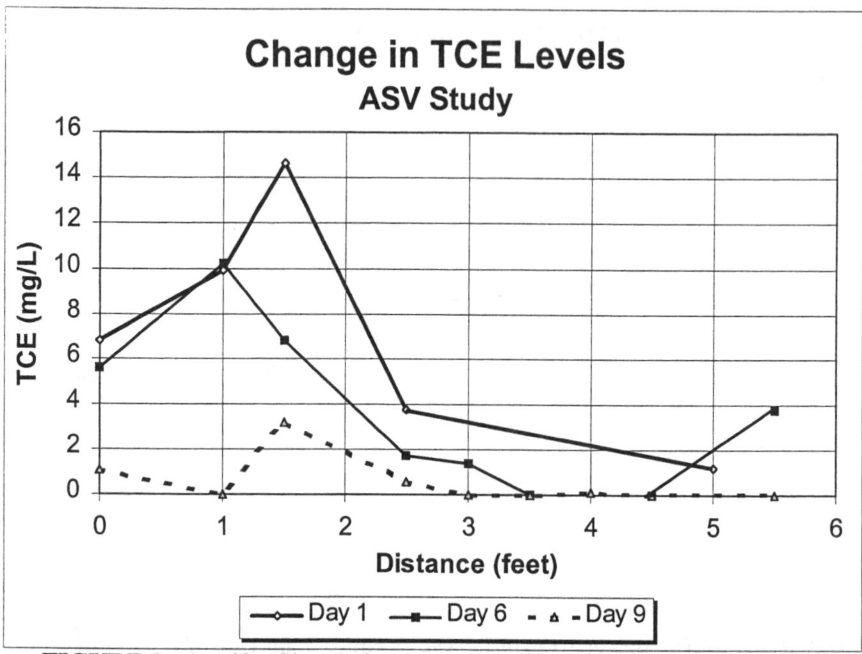

FIGURE 2. Aquifer Simulation Vessel (ASV) TCE degradation study.

Field Study - Single Well Application. The effects of HRC-containing canisters in a single well were studied at a site in Florida. Eleven, four foot long canisters were placed in a five inch monitoring well containing moderate concentrations of TCE, cis-1,2-DCE, and VC. The contaminant plume, contained in a fine to medium grained sand aquifer, measured 120 feet in length by 60 feet in width. Due to a flat gradient, groundwater velocity is estimated to be less than 0.1 foot per day. Reductions in contaminant concentrations in the well are presented in Figure 3. Following three months of treatment, reductions in TCE, cis-1,2-DCE, and VC were 96%, 98%, and 99%, respectively. Absence of DO and highly negative redox levels confirmed the existance of a highly reduced environment.

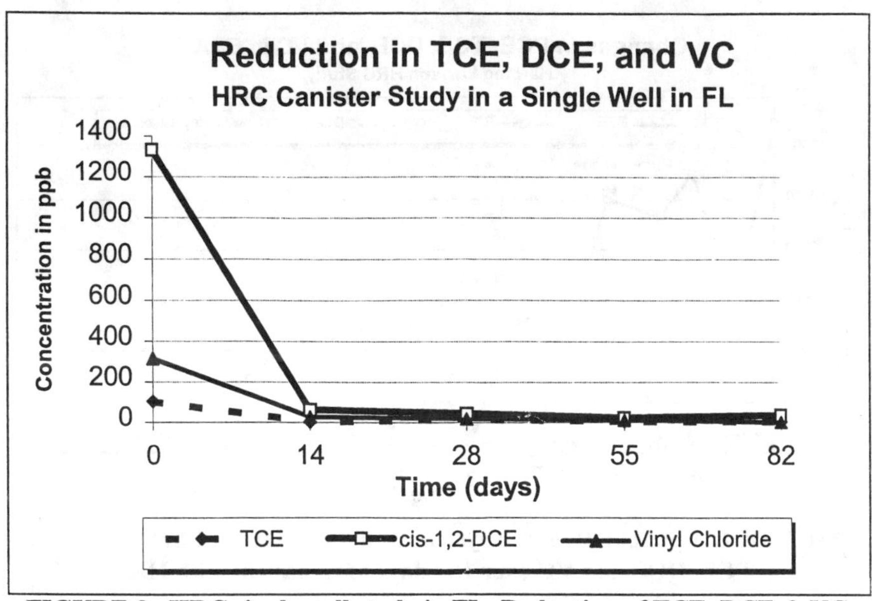

FIGURE 3. HRC single well study in FL -Reduction of TCE, DCE & VC.

Field Study - Recirculating Well System. As part of the USEPA SITE program, Harding Lawson Associates tested the efficacy of HRC in remediating CAHs in a recirculating well system. HRC canisters were inserted into injection wells; circulation was maintained between extraction and re-injection wells. Weekly monitoring over 189 days indicated changes in PCE, TCE, DCE, and VC concentrations as represented in Figure 4. Harding Lawson Associates concluded: 1) HRC can reduce redox conditions in wells containing HRC and along the downgradient flow path, 2) TCE biodegradation was demonstrated, 3) biodegradation products cis-DCE, VC, and ethene were detected.

Field Study - Full Scale Injection. The direct-push injection of HRC is currently being studied with Montgomery Watson at a site contaminated with high levels of PCE. 240 pounds of HRC were injected at 12 delivery points in a 60 square foot area. Approximately 8 months following the installation of HRC, PCE mass was reduced 126 grams, representing a reduction of 80%. PCE degradation rates in the HRC injected zone were 11.5 times faster than background rates at Day 70 and 4.9 times faster at Day 120.

Concurrent increases in PCE-degradation daughter products, TCE, DCE and VC, were also documented as was further sequential degradation through some of the daughter products themselves at different times. These total changes in mass over time for all the CAHs is presented in Figure 5. The continued effects of a single cost-effective application of HRC are demonstrated to be present for at least 253 days.

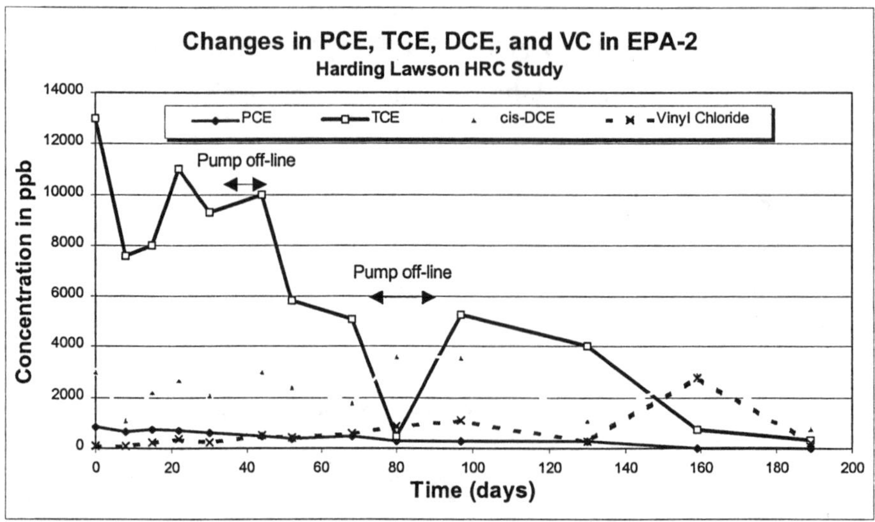

FIGURE 4. Recirculating well system study with HLA. Changes in PCE, TCE, DCE and VC over 189 days of treatment with HRC.

Although not shown, there were mass balances between parent and daughter products of between 27% and 46%; an important indicator that the HRC injections facilitated contaminant removal by biodegradation. There were several other indicators that the proper conditions for facilitating reductive dechlorination were established, including: highly negative redox levels, significant reductions in sulfate and nitrate, formation of acetic acid, increased dissolved hydrogen levels, and a substantial increase in total plate counts for anaerobic bacteria.

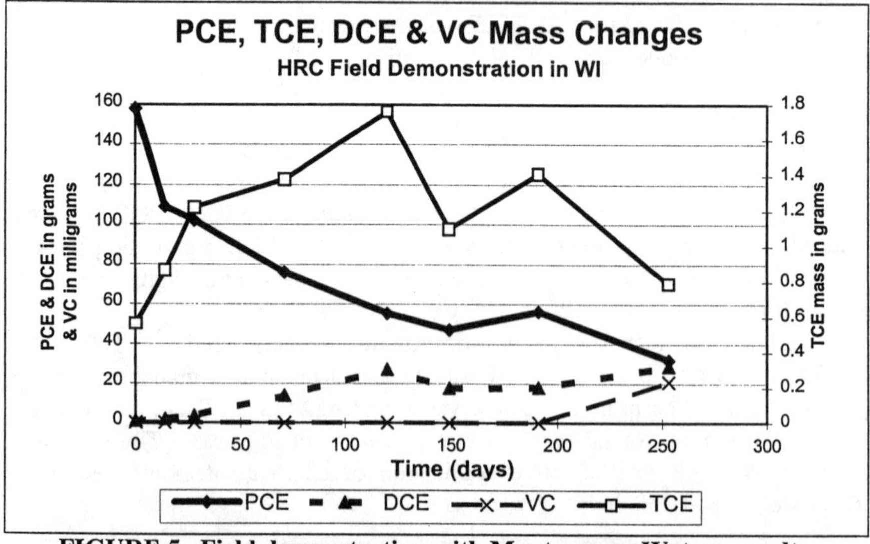

FIGURE 5. Field demonstration with Montgomery Watson results.

ASSESSMENT AND MONITORING OF 1,2-DICHLOROETHANE DECHLORINATION

Jan Gerritse, Arjan Borger, Erwin van Heiningen, Huub H.M. Rijnaarts and Tom N.P. Bosma (TNO Institute of Environmental Sciences, Energy Research, and Process Innovation, Apeldoorn, The Netherlands)
Jan Taat (Delft Geotechnics, Delft, The Netherlands)
Bas van Winden (Arcadis, Waalwijk, The Netherlands)
John Dijk and Jan A.M. de Bont (Wageningen Agricultural University, Wageningen, The Netherlands)

ABSTRACT: Biotransformation of 1,2-dichloroethane (1,2-DCA) in anaerobic groundwater was studied on an industrially polluted site and in the laboratory. Analysis of groundwater samples indicated that in the field 1,2-DCA was intrinsically transformed to mainly ethene. In batch microcosms containing groundwater and sediment from the site "fermentative" transformation of 1,2-DCA occurred to ethene and CO_2. Addition of electron donor substrates strongly enhanced reductive dechlorination of 1,2-DCA completely to ethene under both sulphate and carbon-dioxide reducing conditions. The presence of $Fe(OH)_3$, MnO_2, NO_3^- or oxygen inhibited this dihaloelimination. However, under nitrate or oxygen reducing conditions 1,2-DCA appeared to be degraded oxidatively. To our knowledge this paper presents the first evidence for nitrate-coupled 1,2-DCA oxidation. Maximum rates of 1,2-DCA transformation obtained under different redox conditions were: 18.5 μM/day for non-stimulated fermentation; 300 μM/day, formate stimulated dihaloelimination; 925 μM/day, nitrate coupled oxidation; and 1260 μM/day for aerobic oxidation.

It is concluded that: i) reductive, fermentative as well as oxidative pathways can be involved in natural attenuation of 1,2-DCA, depending on the actual redox conditions *in situ*, and ii) stimulation of reductive or oxidative 1,2-DCA transformation may be obtained in anaerobic aquifers through infiltration of electron donor substrates or nitrate, respectively.

INTRODUCTION

1,2-Dichloroethane, also known as ethylene-dichloride (EDC), is used as bulk chemical intermediate for the production of vinylchloride (VC) and other chlorinated chemicals, as lead scavenger, fumigant and as solvent for pharmaceutical products. 1,2-DCA dissolves relatively well in water and moves rapidly in soil. Transformation of 1,2-DCA occurs under both aerobic and anaerobic conditions. 1,2-DCA is used as a growth substrate by aerobic bacteria, such as *Xanthobacter*, *Ancylobacter* and *Pseudomonas* strains, which completely oxidize 1,2-DCA to CO_2, H_2O and chloride (Van den Wijngaard et al., 1992). Alternatively, methanotrophic bacteria can co-metabolise 1,2-DCA (Lazarone and

McCarty, 1990). Aerobic mineralization of 1,2-DCA is applied in various bioreactor systems for wastewater and groundwater treatment (Freitas dos Santos and Livingston, 1994; Stucki and Thüer, 1995). Under acetogenic, sulphate reducing or methanogenic conditions 1,2-DCA is reductively dehalogenated to mainly ethene (Egli et al., 1987; Holliger et al., 1990; Wild et al., 1995). This dihaloelimination is believed to be a co-metabolic transformation by sulfate reducing and methanogenic bacteria. VC, chloroethane and ethane may be formed as by-products. Anaerobes capable of halorespiration using 1,2-DCA as electron acceptor have not yet been identified. Although oxidative microbial degradation with nitrate, oxidised manganese or iron has been demonstrated for mono- and dichloro-aliphatics, information on such transformations of 1,2-DCA is lacking. Abiotic transformation of 1,2-DCA, including alkaline hydrolysis to VC and hydrolytic substitution yielding ethylene glycol, also occurs under anaerobic conditions. Compared with microbial transformation of 1,2-DCA these processes are relatively slow with half-lives in the order of 10 to 70 years (Barbash and Reinhard, 1989; Jeffers et al., 1987)

Objective. The goal of the present study is to determine biotransformation pathways of 1,2-DCA possible under different redox conditions in anaerobic groundwater systems. This information is used i) to assess and monitor the process of natural attenuation of 1,2-DCA in aquifers, and ii) to explore possibilities to develop microbiological means for control or remediation of 1,2-DCA in the subsurface.

MATERIALS AND METHODS

Field site and sampling procedures. Studies were focused on a 1,2-DCA polluted industrial site of a vinylchloride producing factory in the Rotterdam harbour area in The Netherlands. At this site the major fraction of the 1,2-DCA contamination is present in a confined anaerobic aquifer located between 20 and 30 metres depth below surface (Bosma et al., 1998). The redox conditions range from iron reducing to methanogenic. The redox potential of the groundwater is -120 to -180 mV and the pH is around neutral. Spreading of the contamination plume is controlled using pump-and-treat methods, during which more than 300 tons of 1,2-DCA have already been removed from the aquifer.

Soil samples were obtained in cores from bore holes made to install monitoring wells near the core of the contamination at 23, 26 and 29 metres depth, respectively. From these wells groundwater was collected in gas-tight bottles using a peristaltic pump. The bottles were filled completely to prevent the introduction of oxygen. For the analysis of volatile aliphatic compounds groundwater was sampled using flow-through bottles which could be sealed instantaneously during sampling thus minimising losses through degassing of the groundwater. Soil material and groundwater samples were cooled on ice and transported the same day to the laboratory for analyses and preparation of batches and soil columns.

Batch experiments. Batches were prepared in an anaerobic glove-box under a N_2 atmosphere in 120 ml bottles crimp-sealed with butyl rubber stoppers. Soil microcosm bottles contained 50 ml of groundwater, amended with a NaK-phosphate buffer (50 mM, pH 7) and approximately 20 grams of sediment. From these microcosms transfers were made in batch cultures containing phosphate buffered growth medium composed of nutrient minerals, trace elements, resazurin, yeast extract (100 mg/l) and a vitamin solution. Electron donors were added from separately autoclaved stock solutions to final concentrations of 10 - 50 mM. Electron acceptors were added to final concentrations of 5 mM (Na_2SO_4, $NaNO_3$ or MnO_2) or 10 mM ($Fe(OH)_3$). Amorphous $Fe(OH)_3$ (ferrihydrite) was synthesised by neutralising a $FeCl_2$ solution with KOH. Manganese oxide was prepared by mixing equal amounts of $KMnO_4$ and $MnCl_2$ and adjusting pH to 10. Before use the solutions were washed with milli-Q water by centrifugation. All solutions were prepared under a N_2 atmosphere. Batches were routinely incubated at 20°C.

Analyses. Volatile aliphatic compounds were determined by gas chromatography based on headspace analysis. Ions in groundwater were determined using ion chromatography and by ICP-AES analysis.

RESULTS AND DISCUSSION

Field analyses. Field research was done on an industrial site heavily polluted with 1,2-DCA. Analysis of groundwater samples obtained from 24 monitoring wells installed at 23, 26 and 29 metres below surface revealed 1,2-DCA concentrations up to 3.5 mM (Figure 1). Highest concentrations were present in the lower layers of the aquifer. Ethene was detected to above 3 mM indicating that dihaloelimination is a major pathway for intrinsic transformation of 1,2-DCA in the aquifer. In a previous study the half-life for 1,2-DCA in this aquifer was estimated to range from 1 to 30 years (Bosma et al,. 1998). Ethane and VC -other known transformation products of 1,2-DCA- were also found, but their concentrations remained below 40 μM and 90 μM, respectively. Other chlorinated aliphatics detected included: cis-1,2-dichloroethene (<1 mM), trichloroethene (<30 μM) tetrachloroethene (<4 μM) and dichloromethane (<4 μM). Hence, it is feasible that reduction of chloroethenes is the major source of the risk compound VC. Groundwater contained low concentrations of sulphate (55 μM), nitrate (<10 μM) and ferrous iron (2-3 μM). Methane was present in similar concentrations (up to 3.3 mM) as 1,2-DCA and ethene, indicating that methanogenic conditions prevail in the aquifer. Methanogenic bacteria are known to be capable of transformation of 1,2-DCA to ethene. However, methanogens are probably not important for 1,2-DCA transformation on this location since 1,2-DCA dechlorination by these bacteria is relatively inefficient yielding less than 1.2 mmol of ethene per mol of methane produced (Holliger et al., 1990).

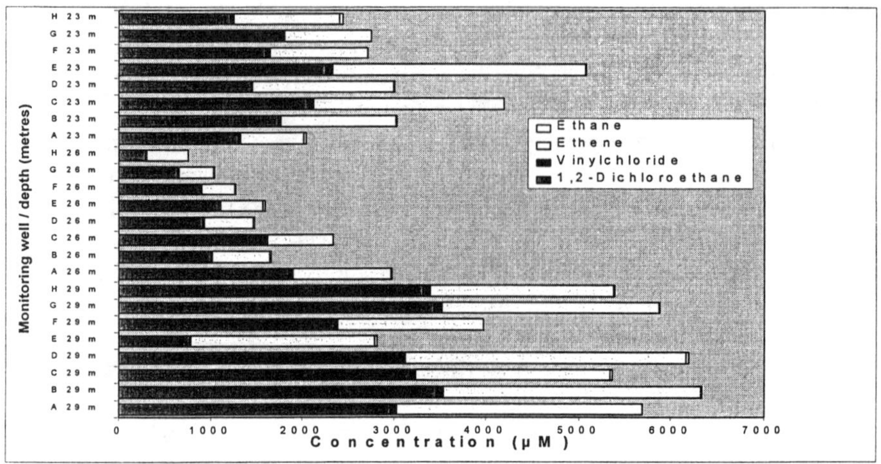

FIGURE 1: 1,2-Dichloroethane and its transformation products in monitoring wells.

Laboratory tests for intrinsic transformation of 1,2-dichloroethane. In batch and column experiments with aquifer material from the monitoring wells 1,2-DCA was completely removed without additional carbon sources and electron acceptors supplied (Bosma et al., 1998). This "spontaneous" 1,2-DCA transformation was a biologically mediated process with an optimum at 20°C. Initially, ethene the yield varied between 10% to 50% of the 1,2-DCA removed, suggesting that, in addition to reductive dechlorinating bacteria, anaerobes were present capable of 1,2-DCA degradation though oxidative and/or fermentative pathways. Upon repeated addition of 1,2-DCA, its removal rate increased from 1.1 μM/day to 6.6 μM/day and 18.5 μM/day, indicating the enrichment of 1,2-DCA transforming bacteria. In parallel, the amount of ethene produced increased to about 90% of the 1,2-DCA dechlorinated (Figure 2A). Mass-balances determined with transfers in anaerobic growth medium suggested that 1,2-DCA was transformed "fermentatively" : $6C_2H_4Cl_2 + 4H_2O \rightarrow 5C_2H_4 + 2CO_2 + 12HCl$.

Influence of electron donors and acceptors on 1,2-DCA transformation. When a carbon and electron source (either yeast extract, methanol, ethanol, formate, acetate or molasses) was added to aquifer material an immediate and stoichiometric transformation of 1,2-DCA to ethene was found at rates up to 300 μM/day (Figure 2B). This indicates the potential for rapid 1,2-DCA degradation through dihaloelimination. Traces of VC and chloroethane were occasionally detected but their concentrations remained below 1% of 1,2-DCA reduced. Transfers in medium with 25 mM formate dehalogenated up to 13 mM 1,2-DCA.

In such cultures significant methane production did not occur. These observations strongly suggest that dihaloelimination of 1,2-DCA was not a co-metabolic but halorespiration-like process in which 1,2-DCA acts as electron acceptor: $C_2H_4Cl_2 + CHOOH \rightarrow C_2H_4 + CO_2 + 2HCl$. Until now halorespiration is described for bacteria dehalogenating chlorinated benzoic acids, chlorinated phenols or chlorinated ethenes, but not for reduction of 1,2-DCA.

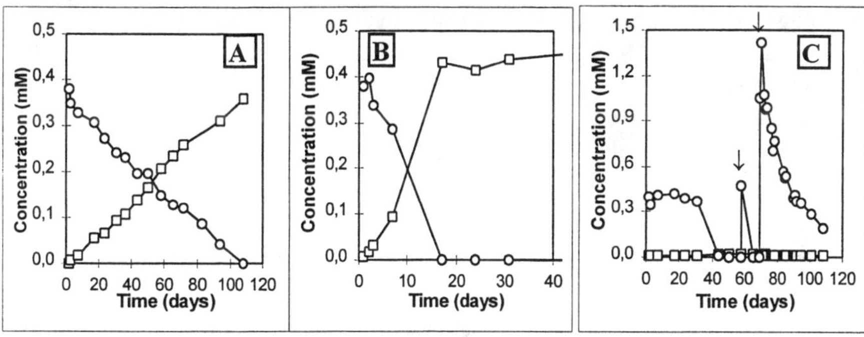

FIGURE 2. Biotransformation of 1,2-dichloroethane (○) to ethene (□) in groundwater / sediment microcosms. A. Without additional electron donor or acceptor, B. with 20 mM formate, or C. with 5 mM nitrate. ↓ 1,2-DCA addition.

1,2-DCA transformation was also studied in the presence of different electron acceptors. Sulphate did not influence 1,2-DCA reduction to ethene. In contrast, in microcosms with $Fe(OH)_3$ or MnO_2 as alternative electron acceptor, 1,2-DCA biotransformation was strongly inhibited, both in the absence or presence of additional electron donors (transformation rates < 0.1 μM/day). Interestingly, in the presence of nitrate 1,2-DCA was degraded completely without significant accumulation of ethene or VC, suggesting that 1,2-DCA was used as an electron donor and mineralized oxidatively by denitrifying bacteria (Figure 2C). Indeed, transfers in anaerobic nitrate-containing media yielded cultures using 1,2-DCA as the sole carbon and electron source. In these batches 1,2-DCA degradation proceeded at rates of 250 to 925 μM/day. Batches incubated under an air atmosphere oxidised 1,2-DCA at similarly high rates of 250 to 1260 μM/day.

Conclusions. Field data indicate intrinsic transformation of 1,2-DCA to mainly ethene in the anaerobic methanogenic aquifer. In laboratory scale batch experiments 1,2-DCA was transformed via different pathways depending on the redox conditions imposed (Figure 3). Under "natural" non-stimulated conditions 1,2-DCA was slowly converted to ethene and CO_2. This suggests that fermentative pathways may be important for natural attenuation of 1,2-DCA in anaerobic groundwater systems low in organic material and alternative electron acceptors. Electron donor substrates strongly stimulated reductive dechlorination of 1,2-

DCA to ethene, an observation also reported by Klečka et al. (1998) for 1,2-DCA reduction in Louisiana sand/water microcosms. Stimulation of reductive dechlorination is therefore a promising tool for the enhancement of 1,2-DCA transformation in the subsurface. Anaerobic oxidation by nitrate reducing bacteria may be a feasible alternative for enhanced 1,2-DCA removal. On this particular site, however, nitrate may inhibit dechlorination of other pollutants such as chlorinated ethenes. Further research is aimed at i) revealing the identity and physiology of anaerobic micro-organisms involved in reductive and oxidative dechlorination of 1,2-DCA, and ii) testing different remediation concepts in the field by creating a "biological activated zone" in the subsurface where the microbial activity is stimulated through infiltration of electron donor and/or electron acceptor substrates (Bosma et al., 1998).

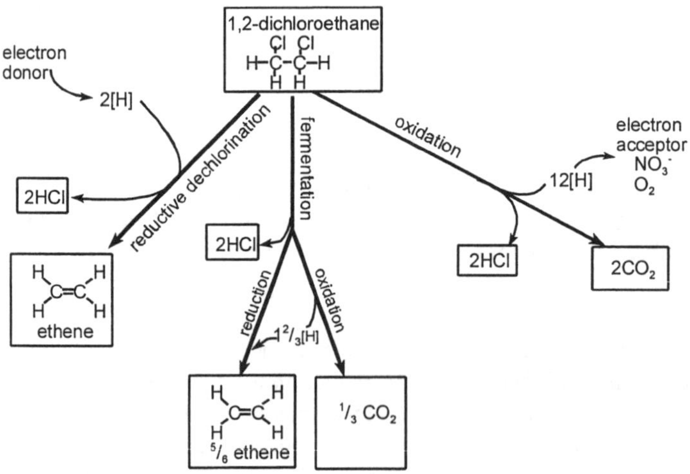

FIGURE 3. 1,2-Dichloroethane biotransformation pathways

Acknowledgement. This work is partially performed within the TNO/WAU Center for Soil and Sediment Remediation Research, and is financially supported by the Netherlands Research Program for Biotechnological In-Situ Remediation (NOBIS) and by the owner of the site.

REFERENCES

Belay, N. and L. Daniels. 1987. "Production of ethane, ethylene and acetylene form halogenated hydrocarbons by methanogenic bacteria." *Appl. Environ. Microbiol. 53*: 1604-1610.

Barbash, J.E. and M. Reinhard. 1989. "Abiotic dehalogenation of 1,2-dichloroethane and 1,2-dibromoethane in aqueous solution containing hydrogen sulfide." *Environ. Sci. Technol. 23*(11):1349-1358.

Bosma, T.N.P., M. van Aalst-van Leeuwen, J. Gerritse, E. van Heiningen, J. Taat, and M. Pruijn, 1998. "Intrinsic dechlorination of 1,2-dichloroethane at an industrial site." *Contaminated Soil '98*, Thomas Telford Publishing, London, pp. 197-202.

Egli, C., R. Scholtz, Cook, A. and T. Leisinger. 1987. "Anaerobic dechlorination of tetrachloromethane and 1,2-dichloroethane to degradable products by pure cultures of *Desulfobacterium* sp. and *Methanobacterium* sp." *FEMS Microbiol. Lett. 43*: 257-261.

Freitas dos Santos L.M. and A.G. Livingston. 1994. "Extraction and biodegradation of a toxic volatile organic compound (1,2-dichloroethane) from waste-water in a membrane bioreactor." *Appl. Microbiol. Biotechnol. 42*: 421-431.

Holliger C., G. Schraa, A.J.M. Stams and A.J.B. Zehnder. 1990. "Reductive dechlorination of 1,2-dichloroethane and chloroethane by cell suspensions of methanogenic bacteria." *Biodegradation 1*(4): 253-261.

Jeffers, P.M., L.M. Ward, L.M. Woytowitch and N.L. Wolfe. 1989. "Homogenous hydrolysis rate constants for selected chlorinated methanes, ethanes, ethenes and propanes." *Environ. Sci. Technol. 23*(8): 965-969.

Klečka, G.M., C.L. Carpenter and S.J. Gonsior. 1998. "Biological transformations of 1,2-dichloroethane in subsurface soils and groundwater." *J. Contam. Hydrol. 34*: 139-154.

Lazarone, N.A. and P.L. McCarty. 1990. "Column studies on methanotrophic degradation of trichloroethene and 1,2-dichloroethane." *Ground Water 28*(6):910-919.

Stucki G. and M. Thüer. 1995. "Experiences of a large-scale application of 1,2-dichloroethane degrading microorganisms for groundwater treatment." *Environ. Sci. Technol. 29*(9): 2339-2345.

Wijngaard van den, A.J., K.W.H.J. van der Kamp, J. van der Ploeg, F. Pries, B. Kazemier, and D.B. Janssen. 1992. "Degradation of 1,2-dichloroethane by Ancylobacter aquaticus and other facultative methylotrophs." *Appl. Environ. Microbiol. 58*(3): 976-983.

Wild A.P., W. Winkelbauer and T. Leisinger. 1995. "Anaerobic dechlorination of trichloroethene, tetrachloroethene and 1,2-dichloroethane by an acetogenic mixed culture in a fixed-bed reactor." *Biodegradation 6*: 309-318.

AEROBIC BIODEGRADATION OF VINYL CHLORIDE AND CIS-1,2-DICHLOROETHYLENE: LABORATORY AND FIELD STUDIES

Liselotte Ludvigsen & Ole Kiilerich (HOH Water Technology, Greve, Denmark)
Kim Broholm (VKI, Hørsholm, Denmark)
Lars Deigaard (Scanrail Consult, Copenhagen, Denmark)

ABSTRACT: An anaerobic shallow aquifer at a site in Copenhagen, Denmark, is polluted with vinyl chloride (VC) and cis-1,2-dichloroethylene (cis-1,2-DCE). In order to investigate the biodegradability of these compounds when changing the redox conditions to aerobic, laboratory microcosms experiments were performed. Biodegradation of vinyl chloride (VC) and cis-1,2-dichloroethylene (cis-1,2-DCE) was demonstrated in the microcosms experiments amended with oxygen and containing sediment and groundwater from the anaerobic aquifer. Addition of both oxygen and methane to the laboratory microcosms stimulated growth of methanotrophic bacteria and resulted in complete degradation of VC within 84 days and 90 % degradation of cis-1,2-DCE within approximately 200 days. These results indicate that stimulation of growth of methanotrophic bacteria by addition of oxygen and methane to the anaerobic aquifer may be a feasible in situ bioremediation technique for removal of VC and cis-1,2-DCE at the polluted site. Therefore, a field scale bioremediation programme involving injection of methane and air to the anaerobic plume was initiated in July 1998. Currently, the plume is being monitored.

INTRODUCTION

Numerous laboratory studies have shown biodegradation of vinyl chloride (VC) and cis-1,2-dichloroethylene (cis-1,2-DCE) under aerobic conditions in the presence of a primary energy source like methane, propane, ammonia, or toluene (Tsien et al., 1989, Wacket et al., 1989, Vannelli et al., 1990, Wackett and Gibson, 1988). Biodegradation of VC may also occur without the presence of additional primary substrates (Davis and Carpenter, 1990). However, only few field studies have demonstrated biodegradation of VC and cis-1,2-DCE.

A shallow anaerobic aquifer in Copenhagen is polluted with VC and cis-1,2-DCE probably caused by anaerobic dechlorination of the original contaminant source of TCE. The aim of this study is to investigate the degradability of VC and cis-1,2-DCE by stimulating growth of methanotrophic bacteria when altering the anaerobic aquifer into an aerobic aquifer by addition of oxygen and methane. For that purpose, firstly laboratory degradation experiments are used to evaluate the potential of this remediation technique on the removal of VC and cis-1,2-DCE using aquifer material from the site. Secondly, a pilot scale in situ biodegradation experiment is performed at the site using injection of methane and oxygen into the anaerobic aquifer.

MATERIALS AND METHODS

Site Description. The site is located in Copenhagen, Denmark, and contains a workshop for locomotive engines as well as various stocks of spare parts, lubricating oil etc. The geology consists of 1-2 m of sandy fill material with high organic carbon content, followed by 2 m of clayey till overlaying 2-3 m of sandy till. The sandy till is followed by 1-2 m of clay that hydraulically separates the sandy till from an underlying sand layer. The lower sand layer is about 8 m thick and rests on Danien limestone. The limestone and the lower sand layer represent the primary groundwater aquifer, whereas the sandy till serves as a secondary groundwater aquifer. The secondary aquifer is confined as the groundwater piezometric head is about 2.5 m below ground level. The groundwater flow direction in the secondary aquifer is towards the south. The hydraulic conductivity is estimated to $1.7 \cdot 10^{-3}$ m^2/s.

The secondary aquifer is anaerobic and contains VC and cis-1,2-DCE at concentrations of up to 270 μg/L and 150 μg/L, respectively. The contaminated aquifer is characterized by concentrations of total organic carbon (NVOC) of 30-180 mg/L, methane of maximally 1 mg/L, sulfate of 70-240 mg/L, dissolved iron of 0.5-14 mg/L, and ammonia of 10-15 mg/L. No nitrate is present in the contaminated area. Furthermore, other contaminant components in terms of isomers of glycol ethers were observed in concentrations typically at mg/L level.

Laboratory degradation experiments. The laboratory set-up consisted of 2.3-L glass bottles equipped with a glass valve used for sampling as described in Nielsen et al. (1995). Each batch microcosm contained 700 g (wet weight) sandy sediment and 1.8 L groundwater. The sediment was sampled from the sandy aquifer 3.5-5.5 m below surface and the groundwater was collected at the same depth. The batch microcosms were spiked with VC and cis-1,2-DCE at resulting concentrations of 330 μg/L and 170 μg/L, respectively, and incubated aerobically at 21 °C during a period of 220 days. To a triplicate set of microcosms methane were added in concentrations of 2.5 vol.-%. Oxygen and methane were added several times during the incubation to insure oxygen and/or methane containing conditions. For control experiments sodium azide (2 g/L) was added to microcosms. Groundwater was sampled from the microcosms by forcing a slight over-pressure of air (supplied with methane in the bottles containing methane) into the bottle, pushing the water sample out of the microcosm. Samples were analyzed for VC, cis-1,2-DCE, methane, and oxygen during incubation. The headspace in the bottles increased over time caused by the water sampling and the concentrations of the volatile compounds were corrected for the loss of mass to the headspace according to Henrys law.

Methane oxidation bioassays. Rates of methane oxidation were examined using 118-mL serum glass bottles containing 35 g of sediment (dry weight) and 57 mL of groundwater. In order to investigate the stimulating effect of methane and oxygen addition on the growth of methanotrophic bacteria, the rates of methane oxidation were estimated using both sediment not exposed to methane previously and sediment from the laboratory batch experiment which had been exposed to methane during the incubation period of 200 days. Sodium azide killed bioassays served as controls. Methane was added to the bioassays in concentrations of 3 mg/L. The bioassays were incubated aerobically for 12 days and water samples were obtained during this period for analysis of methane.

Field scale study. In July 1998 a number of 20 air injection wells were established on a horizontal line across the pollution plume. Air containing 2 vol.-% methane was injected at a flow rate of approximately 20 m^3/h in pulses of 1 minute each 4 hours. The width of the injection cross-section is 20 m. Injection occurs at 5.5 to 6 m below ground surface in the secondary aquifer.

Three lines each consisting of four monitoring wells were established: one upgradient of the biological wall, one about 4 m downgradient, and one about 23 m downgradient of the wall. To supplement the net of monitoring wells several single wells were established upgradient (10-30 m) of the injection wells and downgradient (7 and 45 m) of the injection wells. The monitoring consists of quarterly measurements of groundwater level and sampling of all monitoring wells. The samples are analyzed for chlorinated solvents, glycol ethers and redox parameters.

RESULTS AND DISCUSSION

Laboratory experiments. The results of the laboratory study show that the addition of either oxygen or a combination of methane and oxygen stimulated transformation of VC and cis-1,2-DCE. The relative concentrations of VC and cis-1,2-DCE (relative to the initial concentration) as a function of time in duplicate microcosms are presented in Figure 1. The highest percentages of biodegradation of VC and cis-1,2-DCE occurred in the microcosms amended with methane. Complete degradation of VC was observed after an incubation period of 84 days. In microcosms amended only with oxygen, complete degradation of VC was observed after approximately 200 days. More than 90 % of cis-1,2-DCE was degraded after an incubation period of 200 days in microcosms amended with methane, whereas approximately 50 % of cis-1,2-DCE was degraded in microcosms amended with oxygen only.

These studies indicate that methane served as a primary substrate for methanotrophic bacteria in the transformation of VC and cis-1,2-DCE. After 200 days of stimulating the biomass with oxygen and methane in the laboratory microcosms, a methane oxidation rate was estimated to approximately 0.020 mg CH_4/g dw/day based on the bioassay. For comparisons a rate of 0.015 mg CH_4/g dw/day was observed for the bioassay with a biomass not preexposed to methane. No methane consumption was observed in the sterile control.

The fact that VC and partly cis-1,2-DCE were degraded in microcosms without addition of methane, indicated that also other compounds present in the groundwater (e.g. ammonia, glycol ethers etc.) may serve as primary substrate. However, VC and cis-1,2-DCE may also be degraded by bacteria that do not need primary substrate. The glycol ethers present in the groundwater (initial concentration of approximately 850 µg/L) were degraded completely within 84 days of incubation and only a minor part was removed abiotically in the controls (Table 1). This shows that glycol ethers may stimulate biological activity. However, the degradation of VC and cis-1,2-DCE continued after 84 days in microcosms showing that the biodegradation of VC and cis-1,2-DCE was not dependent of glycol ethers.

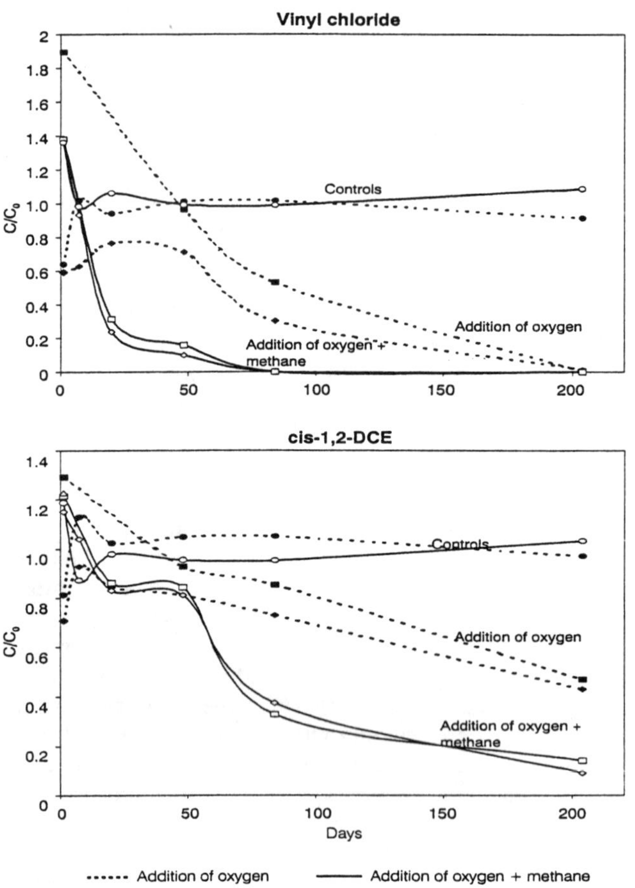

FIGURE 1. The relative concentration of VC and cis-1,2-DCE as a function of time in duplicate microcosms amended with either oxygen or oxygen + methane.

Unfortunately, the conversion of ammonia was not monitored during the experiment. However, it is not likely that the concentration of ammonia has been high enough during the whole experimental period of 200 days to serve as primary substrate. Thus, at least for the last part of the experiment, VC and cis-1,2-DCE may have been degraded by bacteria that did not need additional primary co-substrates.

In conclusion, the laboratory degradation experiment demonstrated a potential for degradation of VC and cis-1,2-DCE present in the groundwater by changing the redox conditions from anaerobic to aerobic. Addition of methane would furthermore speed up the degradation processes.

TABLE 1. Removal of glycol ethers in microcosms of sediment and groundwater after 84 days of incubation.

Additions	Percentage conversion of glycol ethers[a]	
	Glycols, total	Dipropyleneglycol-monomethylethers
Oxygen	100	100
Oxygen + methane	100	100
Control (oxygen + methane)	18	29

a) Initial concentration of total content of glycols: 850 μg/L, initial concentration of dipropyleneglycolmonomethylethers: 38 μg/L.

Field study. Based upon the promising laboratory results the field scale bioremediation study was initiated. The injection of air and methane has now been carried out for 5 months in order to introduce oxygen and a primary substrate into the aquifer and hereby stimulate biodegradation of VC and cis-1,2-DCE.

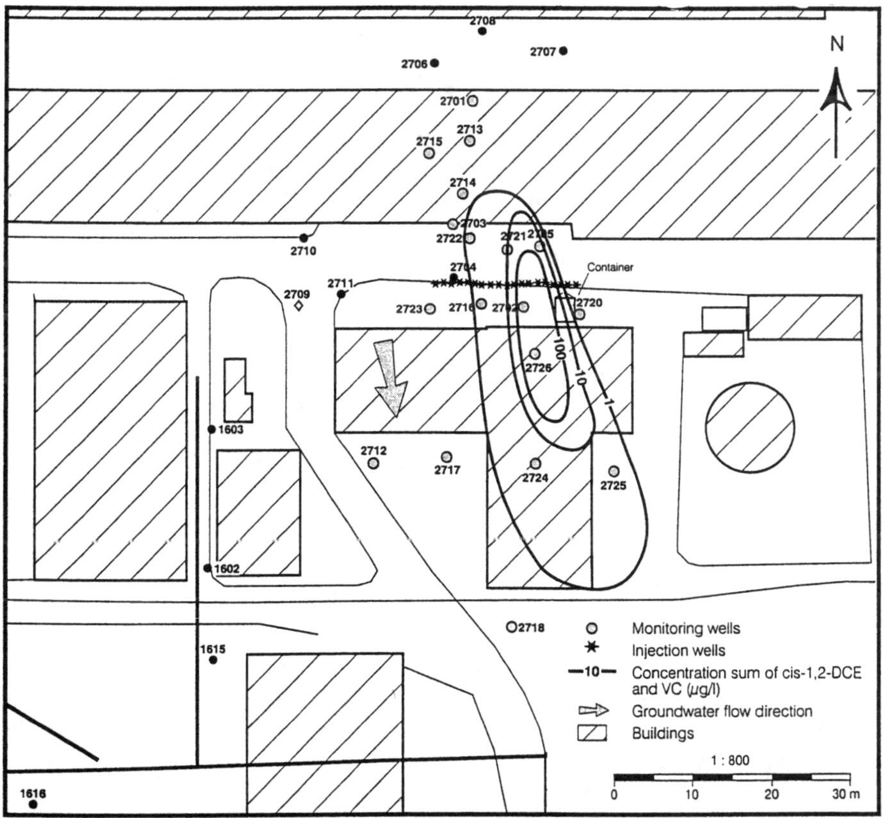

FIGURE 2. Distribution of VC and cis-1,2-DCE in the groundwater at the Copenhagen freightyard site.

Evaluation of the preliminary results shows that the injected amount of oxygen has been too low to permanently establish aerobic conditions. The measured concentrations of oxygen in the plume 3 m downgradient of the injection have not yet exceeded 1 mg/L. The low oxygen concentrations observed in the aquifer is presumably partly caused by the oxidation of reduced species (e.g. Fe(II), reduced sulfur species, ammonia, and organic carbon) in the aquifer. Since only the soluble reduced species – and not the solid species attached to the sediment - have been quantified, the exact oxygen demand needed for oxidation of the reduced species is difficult to estimate.

No significant reduction in the concentration of VC and cis-1,2-DCE has been observed yet. This may be due to the low oxygen concentrations in the aquifer or due to low degradation rates caused by a low number of methanotrophic bacteria since they are slow growing. The distribution of chlorinated compounds in terms of the sum of VC and cis-1,2-DCE based on the latest monitoring results is shown in Figure 2.

To examine the effect of the oxygen concentration the air injection rate has been increased and future results will show if this has an effect on the degradation of VC and cis-1,2-DCE.

ACKNOWLEDGMENTS

The Danish National Railway Agency has initiated a development project centred on environmental/economic evaluation and optimizing of in situ techniques. The project contents following main activities: 1) Development of a decision-making model for optimal restoration of soil and groundwater including preparation of guidelines for an all-inclusive environmental accounting and, 2) Demonstration and optimization of in situ remedial methods on five railroad localities in Denmark. Aerobic biodegradation of halogenated solvents is used as remediation technique at one of the demonstration projects, which is presented in this paper.

This development project was funded by the LIFE-program (no. 96ENV/DK/0016) from the European Economic Community, the Environmental Protection Agency of Denmark and the National Danish Railway Agency. The total budget amounts to approximately 2,3 mill. USD. The project was initiated in the spring 1997 and the final results are expected in year 2000.

REFERENCES

Davis, J. W., and C. L. Carpenter. 1990. "Aerobic degradation of vinyl chloride in groundwater samples". *Appl. Environ. Microbiol. 55*: 3878-3880.

Nielsen, P. H., H.-J. Albrechtsen, G. Heron, and T. H. Christensen. 1995. "In situ and laboratory studies on the fate of specific organic compounds in an anaerobic landfill leachate plume, 1. Experimental conditions and fate of phenolic compounds". *J. Contam. Hydrol. 20*: 27-50.

Tsien, H. C., G. A. Brusseau, R. S. Hanson, and Waclett, L. P. 1989. "Biodegradation of trichloroethylene by Methylosinus trichosporium OB3b". *Appl. Environ. Microbiol. 55*: 3155-3161.

Vannelli, T., H. Logan, D. M. Arciero, and A. B. Hooper. 1990. "Degradation of halogenated aliphatic compounds by the ammonium-oxidizing bacteria Nitrosomonas europaea". *Appl. Environ. Microbiol. 56*: 1169-1171.

Wackett, L. P., and D. T. Gibson. 1988. "Degradation of trichloroethylene by toluene-dioxygenase in whole-cell studies with Pseudomonas putida F1". *Appl. Environ. Microbiol. 54*: 1703-1708.

Wackett, L. P.,G. A. Brusseau, S. R. Householder, and R. S. Hanson. 1989. "Survey of microbial oxygenases: Trichloroethylene degradation by propane-oxidizing bacteria". *Appl. Environ. Microbiol.* 55: 2960-2964.

DEGRADATION OF TRICHLOROETHYLENE BY *MYCOBACTERIUM* SP. TA27

Akiko Hashimoto (CREST, Japan Science and Technology Corporation, Tsukuba, Ibaraki, JAPAN)
Kazuhiro Iwasaki, Naou Nakasugi and Osami Yagi
(National Institute for Environmental Studies, Tsukuba, Ibaraki, JAPAN)

ABSTRACT: We isolated from the soil an ethane-utilizing bacterium, strain TA27, which can degrade trichloroethylene (TCE) and 1,1,1-trichloroethane (TCA). Strain TA27 was identified as a *Mycobacterium* sp. by its physiological characteristics and by 16S rRNA gene analysis. To evaluate the usefulness of strain TA27 for cleaning up contaminated soil and groundwater, we studied its TCE degradation characteristics. Under a head-space gas containing 3% ethane, strain TA27 degraded more than 95% of TCE at an initial concentration of 1 $mg{\cdot}l^{-1}$ within 3 days. At initial TCE concentrations of 10 and 30 $mg{\cdot}l^{-1}$, 40 and 30% of the TCE were degraded within 7 days, respectively. With an initial pH 7 and 12% ethane in the head-space, good growth and 50% degradation of TCE at an initial concentration of 50 $mg{\cdot}l^{-1}$ were observed within 14 days. We suggest that strain TA27 is useful in the bioremediation of contaminated soil and groundwater.

INTRODUCTION

TCE and TCA have been widely used as solvents and degreasing agents, and spillage and inappropriate disposal of these compounds have frequently resulted in contamination of groundwater. Contamination with chlorinated aliphatic compounds such as TCE and TCA has led to the development of treatment methods such as air stripping and activated carbon adsorption. These physical and chemical remediation methods are expensive, and there is a need for a cheaper but complete pollutant destruction technology.

Bioremediation is one of the most promising new technologies for cleaning up groundwater contamination, because of its low cost and its complete destruction of pollutants. There are many reports of the use of degradation by aerobic bacteria in bioremediation. These bacteria include, methane-oxidizing bacteria (Oldenhuis et al., 1989; Uchiyama et al., 1989), ammonia-oxidizing bacteria (Vannelli et al., 1990), propane-oxidizing bacteria (Vanderberg et al., 1995), toluene- and phenol- degrading bacteria (Nelson et al., 1987) and propylene-oxidizing bacteria (Ensign et al., 1992).

We have already isolated the TCA-degrading bacterium, strain TA27, which can utilize ethane. In this report, we describe the TCE degradation characteristics of strain TA27.

MATERIALS AND METHODS

Bacterium and Culture Conditions. The ethane-utilizing gram-positive bacterium *Mycobacterium* sp. strain TA27, isolated from dry cleaning factory soils in Japan, was used in this investigation. A minimal medium (MM medium) was used throughout the degradation experiments. The composition of the MM medium was as follows ($mg{\cdot}l^{-1}$): NH_4Cl, 2,140; K_2HPO_4, 1,170; KH_2PO_4, 450; $MgSO_4{\cdot}7H_2O$, 120; $FeSO_4{\cdot}7H_2O$, 28; $Ca(NO_3)_2{\cdot}4H_2O$, 4.8; H_3BO_3, 0.005; $MnSO_4{\cdot}4\text{-}6H_2O$, 0.06; $Na_2MoO_4{\cdot}2H_2O$, 0.001; $Co(NO_3)_2{\cdot}6H_2O$, 0.06; H_2SeO_4, 0.004; $NiSO_4{\cdot}7H_2O$, 0.006; $CuSO_4{\cdot}5H_2O$, 0.006; $ZnSO_4{\cdot}7H_2O$, 0.01; the final pH was 6.8. The bacterium was cultured in a 155-ml serum flask containing 30 ml of MM medium. The flask was crimped with butyl rubber stopper and aluminum cap. The culture was incubated at 30°C with a rotary shaker at 120 rpm. Growth was determined by monitoring the optical density (OD) at 660 nm.

Analysis of 16S rRNA Gene. We extracted the genomic DNA of strain TA27 and amplified the 16S rRNA gene. The polymerase chain reaction (PCR) product was sequenced with a Dye Terminator and a Big Dye Terminator cycle sequence kit (Applied Biosystems, Perkin-Elmer). The sequence reaction mixture was electrophoresed using an Applied Biosystems model 310 genetic analyzer. The 16S rRNA gene sequence of strain TA27 was compared with the 16S rRNA gene database of the DNA Data Bank of Japan.

Degradation Experiments. For the TCE biodegradation experiments, we used 69-ml serum bottles, each containing 15 ml of MM medium and crimped with a butyl rubber stopper and an aluminum cap. TCE and ethane gas were added to the serum bottles, and the cultures were incubated under aerobic conditions at 30°C with shaking at 120 rpm. The TCE degradation was monitored by measuring the concentration in the head-space gas with gas chromatography.

Gas Chromatograph Conditions. A gas chromatograph (model GC-14B, Shimadzu Co., Kyoto) equipped with Flame Ionization Detector (FID) was used for quantitative determination of TCE and ethane gas. The glass column (0.3 X 200 cm) was packed with Silicone DC550 (GL Science Co., Tokyo). The injection, oven and detector temperatures were 250, 120 and 250°C, respectively. Nitrogen was used as the carrier gas. The oxygen concentration in the head-space gas was also determined, using a gas chromatograph equipped with a glass column (0.3 X 200 cm) packed with Molecular Sieve 5A (GL Science Co., Tokyo). Analytical data were obtained with a computing integrator (model C-R6A; Shimadzu Co., Kyoto).

RESULTS AND DISCUSSION

Phylogenetic Analysis of Strain TA27. Table 1 shows the results of a comparison of the 16S rRNA gene sequence of strain TA27 with the DNA database. Strain TA27 shows the highest level of 16S rRNA gene similarity with

Mycobacterium gilvum (level of similarity, 99.2%). Strain TA27 shared a short helix at positions 451 to 482 (*Escherichia coli* numbering) with the fast-growing mycobacteria (Stahl and Urbance, 1990). This result, together with the physiological characteristics of the bacterium confirmed that strain TA27 belonged to the *Mycobacterium* genus.

TABLE 1. Homology of strain TA27

Name	Homology of strain TA27	Slow/fast growing
M. gilvum	[99.2% / 1394 bp]	fast
M. parafortuitum	[98.8% / 1398 bp]	fast
M. chlorophenolicum	[98.7% / 1392 bp]	fast
M. aichiense	[98.4% / 1420 bp]	fast
M. sphagni	[98.2% / 1420 bp]	fast
M. fortuitum	[98.2% / 1390 bp]	fast
M. aurum	[97.0% / 1420 bp]	fast
M. vaccae	[96.7% / 1422 bp]	fast
M. tuberculosis	[95.4% / 1404 bp]	slow
R. rhodochrous	[94.7% / 1476 bp]	

Influence of Initial pH. We studied the growth and TCE degradation characteristics of strain TA27. By the effect of the initial pH on the degradation of TCE at 1 $mg{\cdot}l^{-1}$, we found that the optimum initial pH was 5.7 (data not shown).

Influence of Ethane Gas Concentration. Figure 1 shows the effect of the ethane concentration on TCE degradation at an initial pH 5.7. Under a head-space gas containing 3% ethane, strain TA27 degraded about 95% of the TCE within 3 days. Strain TA27 can't utilize TCE as the sole carbon source without ethane gas. TCE degradation was inhibited at an ethane concentration of 12%. Therefore, it seemed that TCE degradation was competitively inhibited by ethane gas.

Influence of TCE Concentration. The degradation of TCE by strain TA27 under 3% ethane and with an initial pH 5.7 was affected by the TCE concentration (Figure 2). Strain TA27 degraded 95% of 1 $mg{\cdot}l^{-1}$ TCE within 7 days. At initial concentrations of 10 and 30 $mg{\cdot}l^{-1}$, 40 and 30% of the TCE was degraded within 7 days, respectively. At a TCE concentration of more than 50 $mg{\cdot}l^{-1}$, the growth was inhibited.

Influence of Ethane Concentration at a High Initial TCE Level and Neutral pH. The growth of strain TA27 appeared to be influenced by the ethane concentration at a neutral pH and 50 $mg{\cdot}l^{-1}$ TCE (Figure 3). At an initial pH 7 and 12% ethane concentration, strain TA27 grew well within 14 days (Figure 3B), and degraded 50% of the TCE (Figure 3A). The high ethane concentration

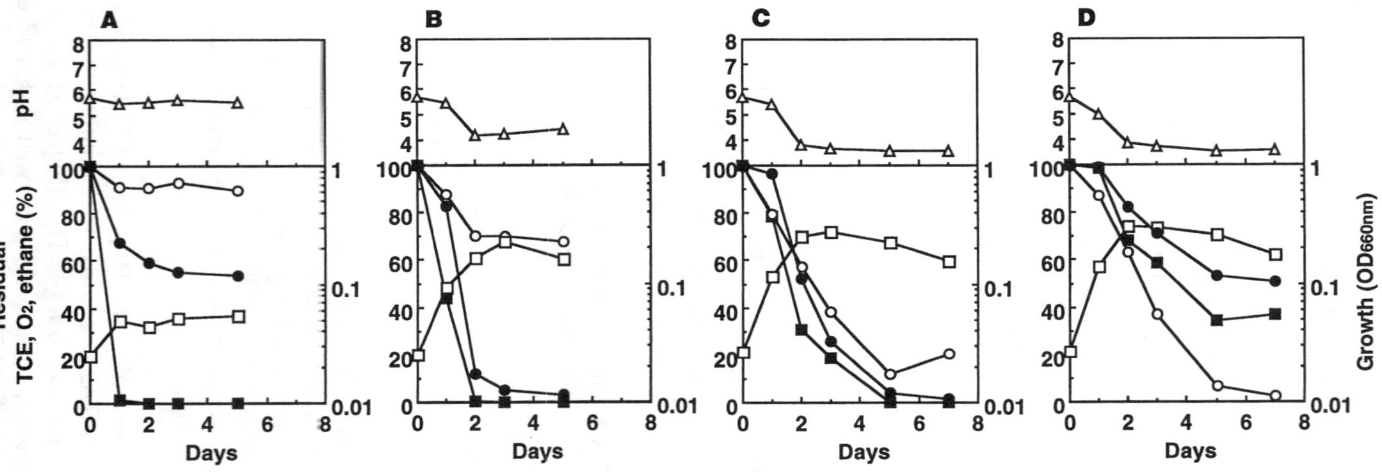

FIGURE 1. Effect of ethane concentration on TCE degradation by strain TA27.
TCE concentration: 1 mg·l^{-1}; initial pH: 5.7; temperature: 30°C; ethane concentration: A, 1%; B, 3%; C, 6%; D, 12%. Symbols: ●, residual TCE (%); ○, O2 (%); ■, ethane (%); □, growth of strain TA27 (OD 660 nm); △, pH.

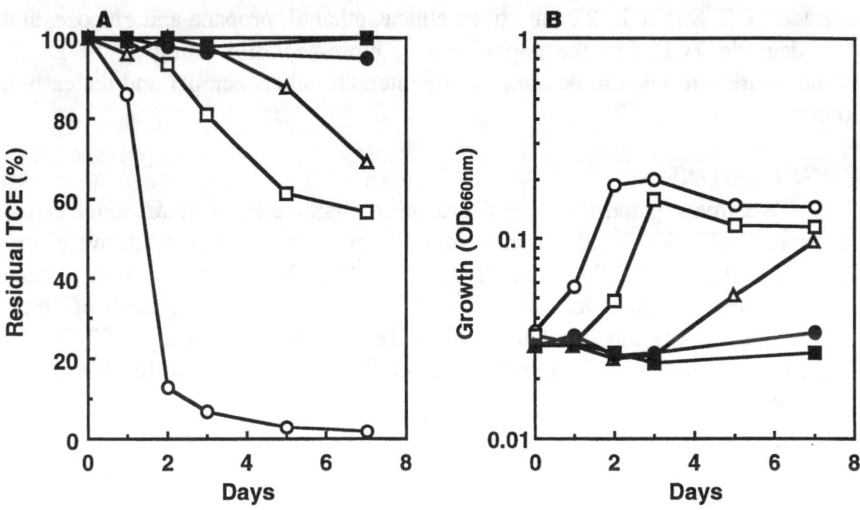

FIGURE 2. Effect of TCE concentration on TCE degradation (A) and growth of strain TA27 (B). Initial pH: 5.7; ethane concentration: 3%. TCE concentration (mg·l^{-1}): ○, 1 ; □, 10 ; △, 30 ; ●, 50 ; ■, 70.

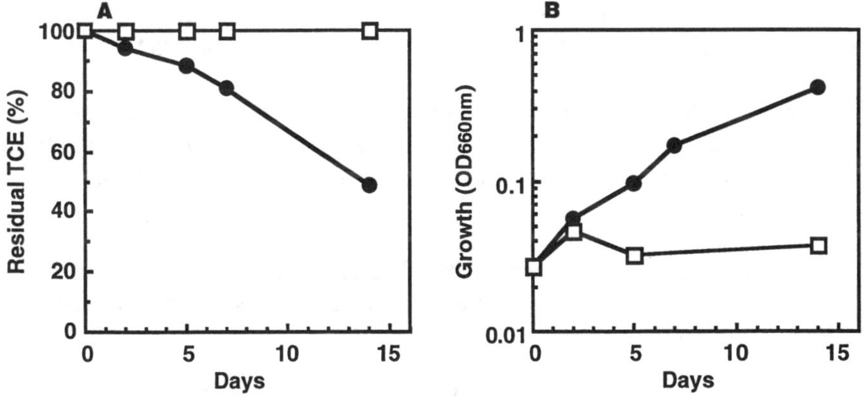

FIGURE 3. Effect of ethane concentration on TCE degradation (A) and growth of strain TA27 (B). TCE concentration: 50 mg·l^{-1}; initial pH: 7. Symbols: □, 3% ethane concentration; ●, 12% ethane concentration.

appeared to mitigate the toxicity of the TCE.

Degradation of 1 mg·l^{-1} TCE by strain TA27 was optimized at a relatively low initial pH 5.7 and a low ethane concentration of 3%, but degradation of 50 mg·l^{-1} TCE was optimized at an initial pH 7 and a high ethane concentration of 12%. Therefore, in our study the initial pH and ethane concentration were very important factors in the degradation of TCE by strain TA27. This may be due to the existence of two kinds of ethane monooxygenase that have the capacity to degrade TCE. Strain TA27 can utilize ethane, ethanol, propane and glucose, and then degrade TCE. For the application of bioremediation using strain TA27, further work will need to be done on the interaction between pH and the carbon source.

CONCLUSIONS

We investigated the TCE degradation characteristics of *Mycobacterium* sp. strain TA27. We found that the initial pH and ethane concentration were very important factors in the degradation of TCE. Moreover, a high ethane concentration mitigated the toxicity of $50 mg \cdot l^{-1}$ TCE on the growth of strain TA27. Strain TA27 can simultaneously degrade high concentrations of TCE and TCA, and is useful for bioremediation on TCE- and TCA- polluted soil and groundwater.

REFERENCES

Ensign, S. A., M. R. Hyman, and D. J. Arp. 1992. "Cometabolic degradation of chlorinated alkenes by alkene monooxygenase in a propylene-grown *Xanthobacter* strain." *Appl. Environ. Microbiol.* 58(9): 3038-3046.

Nelson, M. J. K., S. O. Montgomery, W. R. Mahaffey, and Pritchard, P. H. 1987. "Biodegradation of trichloroethylene and involvement of an aromatic biodegradative pathway. " *Appl. Environ. Microbiol.* 53(5): 949-954.

Oldenhuis, R., R. L. Vink, D. B. Janssen, and B. Witholt. 1989. "Degradation of chlorinated aliphatic hydrocarbons by *Methylosinus trichosporium* OB3b expressing soluble methane monooxygenase." *Appl. Environ. Microbiol.* 55(11): 2819-2826.

Stahl D. A., and J. W. Urbance. 1990. "The division between fast- and slow-growing species corresponds to natural relationships among the mycobacteria." *J. Bacteriol.* 172(1): 116-124.

Uchiyama, H., T. Nakajima, O. Yagi, and T. Tabuchi. 1989. "Aerobic degradation of trichloroethylene by a new type II methane-utilizing bacterium, strain M." *Agric. Biol. Chem.* 53(11): 2903-2907.

Vanderberg, L. A., B. L. Burback, and J. J. Perry. 1995. "Biodegradation of trichloroethylene by *Mycobacterium vaccae*." *Can. J. Microbiol.* 41(3): 298-301.

Vannelli, T., M. Logan, D. M. Arciero, and A. B. Hooper. 1990. "Degradation of halogenated aliphatic compounds by the ammonia- oxidizing bacterium *Nitrosomonas europaea*." *Appl. Environ. Microbiol.* 56(4): 1169-1171.

BIODEGRADATION OF TRICHLOROETHYLENE BY A PROPANE-UTILIZING BACTERIUM *MYCOBACTERIUM* SP. TCE28

Satoshi IMANO, Takashi OHASHI
(Asano Engineering Co. Ltd, Chuo-ku, Tokyo, Japan)
Kazuhiro IWASAKI, ***Osami YAGI*** (National Institute for Environmental Studies and CREST, Tsukuba, Ibaraki, Japan)

ABSTRACT: We studied the application of bioremediation technology in cleaning up soil and groundwater contaminated with trichloroethylene (TCE). We isolated a propane-oxidizing bacterium, strain TCE28, that could degrade TCE. Strain TCE28 was identified as a *Mycobacterium* sp. by its physiological characteristics and by 16S rDNA analysis. Strain TCE28 was not able to utilize TCE as sole carbon source, but we found that it could utilize ethane, butane and *n*-butanol as well as propane, and could thus degrade TCE cometabolically. With propane as the sole carbon source, this strain completely degraded TCE at an initial concentration of 1 mg·l^{-1} in liquid culture within 5 days, and 95% of TCE at an initial concentration of 20 mg·l^{-1} within 14 days. The optimum propane concentration, pH and temperature for the growth and TCE degradation were found to be 5%, 7–8 and 30℃, respectively. Strain TCE28 could also degrade *cis*-1,2-dichloroethylene, 1,1-dichloroethane, 1,2-dichloroethane, dichloromethane and chloroform. Resting cells of this strain degraded 30% of TCE28 at an initial concentration of 5 mg·l^{-1} in 24 hours. Strain TCE28 appears to be useful for cleaning up TCE-contaminated groundwater.

INTRODUCTION

Volatile chlorinated aliphatic compounds such as trichloroethylene (TCE), tetrachloroethylene (PCE) and 1,1,1-trichloroethane (TCA) have been detected in soil and groundwater. These compounds are suspected to be carcinogenic, so it is very important to clean up sites contaminated with them. Various methods have been used to clean up pollution in the soil and groundwater: air-stripping and carbon adsorption, vapor extraction and excavation are very common technologies. However, these processes do not degrade the chemicals; they merely transfer them to other media. In contrast, bioremediation is an effective and low-cost way of cleaning up soil and groundwater pollution.

Several studies have examined TCE degradation by methane-utilizing bacteria (McDonald et al. 1997), propane-utilizing bacteria (Vanderberg and Perry 1994), phenol utilizing bacteria (Sun and Wood 1996), and ammonia-oxidizing bacteria (Vannelli et al. 1990). However, the aromatic hydrocarbons and ammonia that act as carbon sources for bacteria can in themselves be secondary contaminants. We isolated a propane-utilizing bacterium, TCE28, that could degrade TCE. This paper describes the TCE-degradation characteristics of strain TCE28.

MATERIALS AND METHODS

Isolation and Identification of TCE-degrading Bacterium. As a first step we attempted to isolate TCE-degrading bacteria from various soils under aerobic conditions, using an enrichment-culture method. Small amounts of soil were added to 69-ml serum bottles, each containing 15 ml of a mineral-salt medium. The composition of the mineral-salt medium was as follows (in $mg{\cdot}l^{-1}$): NH_4Cl, 2,140; K_2HPO_4, 1,170; K_2HPO_4, 450; $MgSO_4{\cdot}7H_2O$, 120; $FeSO_4{\cdot}7H_2O$, 28; Ca $(NO_3)_2{\cdot}4H_2O$, 4.8; $MnSO_4{\cdot}4\text{–}6H_2O$, 0.06; H_3BO_3, 0.005; $ZnSO_4O{\cdot}7H_2O$, 0.01; $Na_2MoO_4{\cdot}2H_2O$, 0.001; Co $(NO_3)_2{\cdot}6H_2O$, 0.06; $NiSO_4{\cdot}7H_2O$, 0.006; $CuSO_4{\cdot}5H_2O$, 0.006; and H_2SeO_4, 0.004. The final pH of the medium was 6.8. The serum bottles were sealed with butyl rubber caps and crimped-on aluminum rings. Three milliliters of air was withdrawn from each bottle with a syringe, and 3 ml of propane was injected as a carbon source. Aqueous TCE was added to make an initial concentration of 1 $mg{\cdot}l^{-1}$. The cultures were incubated at 30℃ on a shaker. At the end of each week, 0.3 ml of culture broth were transferred to 15 ml of fresh medium. After 3 transfers, the culture broth was spread on to agar plates and incubated in the presence of 10% propane at 30℃. An isolated colony was then picked up and re-spread on the agar plate. This procedure was repeated to make a pure culture. The cells used to determine the physical and chemical characteristics of the mycobacterium were grown on the mineral-salt medium in a 10% propane atmosphere.

Degradation Experiments in Liquid Culture. The TCE-degradation experiments to determine the effects of various carbon sources, and to determine the effects of various initial propane and TCE concentrations, were prepared in a similar way to the enrichment-culture preparation described above. That is, TCE28 was added to 69-ml serum bottles with 15 ml of mineral-salt medium, and the bottles were sealed with butyl rubber caps and aluminum rings. TCE and propane or various other carbon sources were added to the serum bottles, and the culture was incubated under aerobic conditions at 30℃. If the carbon source to be added was gaseous, air was withdrawn from each bottle with a syringe, and 3 ml of the gas to be evaluated were injected to replace the air. Liquid and solid substrates were added at a concentration of 100 $mg{\cdot}l^{-1}$. The cultures were incubated at 30℃ and pH 7, while being shaken at 120 rpm. Bacterial growth was determined by monitoring the optical density at 660 nm. The amount of TCE degradation was determined by measuring the TCE concentration in the head-space gas using gas chromatography.

Degradation Experiments in Soil Microcosms. The degradation of various initial concentrations of TCE in soil microcosms without the addition of other carbon sources was determined as follows. Soil solution (containing 10 g of wet soil), resting cells of strain TCE28, and TCE in 20 mM of phosphate buffer were added to each 69-ml serum bottle to make a volume of 15 ml. The initial concentrations of TCE varied from 1 to 20 mg·l^{-1}. The serum bottles were sealed with teflon-lined butyl rubber caps and aluminum rings. The TCE-degradation experiments in the soil microcosms were carried out in duplicate.

Resting cells were prepared as follows: strain TCE28 was cultured with propane as a carbon source in the mineral-salt medium and harvested by centrifugation at the exponential growth phase, then suspended in 20 mM phosphate buffer at pH 7. The resting cells were added to each soil microcosm to give a final optical density (OD) of 660 nm.

Gas chromatography analysis. A gas chromatograph equipped with a flame-ionization detector was used for quantitative determination of the TCE and propane. The glass column (0.3 × 200 cm) was packed with Silicone DC550. The injection port, oven and detector temperatures were 250℃, 120℃ and 250℃, respectively. The carrier gas was nitrogen (30 ml·min^{-1}).

RESULTS AND DISCUSSION

Characteristics of Strain TCE28. The strain TCE28 bacteria were found to be non-motile, gram-positive, catalase-positive and oxidase-negative rods. The 16S rDNA sequence showed that the highest level of homology (97%) was with *Mycobacterium chlorophenolicum*. Thus the results showed that strain TCE28 belonged to the *Mycobacterium* genus.

TABLE 1. Effect of various carbon sources on TCE degradation in liquid culture

Compounds		Degradation (%)	Growth (OD660nm)	pH
Aliphatic hydrocarbon	Methane	33.2	0.020	6.4
	Ethane	90.4	0.115	6.7
	Propane	94.3	0.225	6.2
	Butane	60.4	0.050	6.6
	Ethylene	-	0	6.8
	Acethylene	-	0.005	6.8
Aromatic hydrocarbon	Benzene	-	0.005	6.8
	Toluene	-	0	6.8
	Xylene	-	0	6.8
	Phenol	3.3	0.010	6.8
Ketone	Acetone	-	0	6.9
Alcohol	Methanol	8.3	0.055	6.6
	Ethanol	3.2	0.045	6.6
	Propanol	15.4	0.060	6.6
	Butanol	80.1	0.050	6.8

TCE: 1 mg·l^{-1}; volume of aliphatic hydrocarbons added: 3 ml; concentration of other (solid and liquid) hydrocarbons added: 100 mg·l^{-1}; –: No degradation

Effects of Various Carbon Sources on TCE Degradation in Liquid Culture. Table 1 shows the effects of various carbon sources on TCE degradation by TCE28. TCE28 could utilize not only propane, but also methane, ethane, butane, methanol, ethanol, 1-propanol and *n*-butanol. Strain TCE28 had a wide assimilation spectrum for aliphatic hydrocarbons and alcohol, but could not assimilate aromatic hydrocarbons and organic acids. Whenever TCE28 grew in the presence of the carbon sources, it could degrade TCE cometabolically.

Effect of Propane Concentration on TCE Degradation in Liquid Culture. Figure 1 shows the effect of the propane concentration on the degradation of TCE at an initial concentration of 1 mg·l^{-1}. Under a head-space gas of 5% propane, strain TCE28 degraded more than 90% of the TCE within 7 days. TCE degradation had almost stopped at 7 days because of lack of propane; strain TCE28 could not utilize TCE as a sole carbon source, that is, without the propane gas. TCE degradation was not inhibited at a propane concentration of 20%. Therefore, it seemed that TCE degradation was not competitively inhibited by propane gas. The pH decreased with the growth of strain TCE28. It appeared that

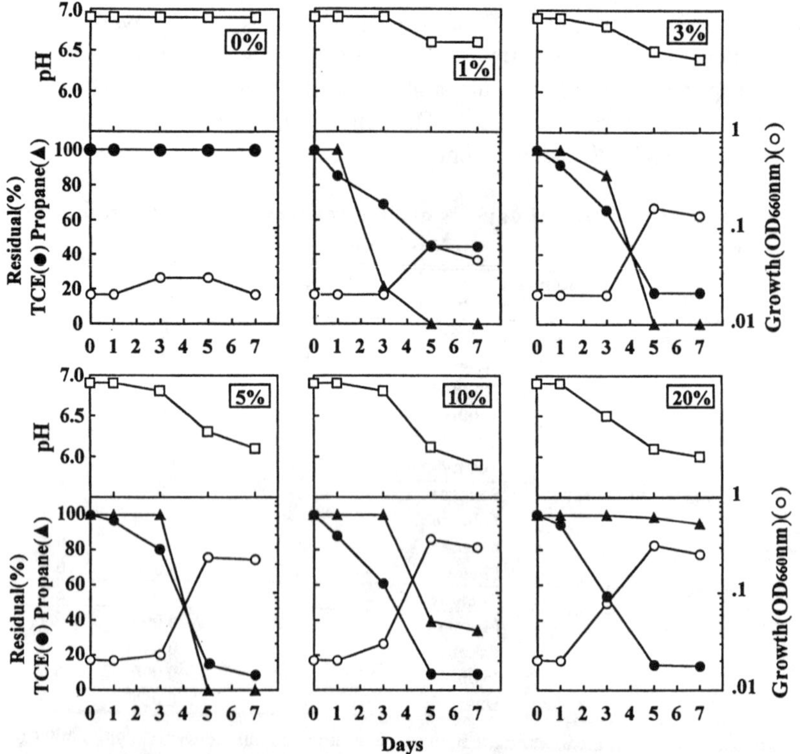

FIGURE 1. Effect of initial propane concentration on TCE degradation by strain TCE28 in liquid culture.

ensuring that enough of the carbon source was present and that the pH was kept at neutral might maintain optimum activity.

Effect of Initial TCE Concentration on TCE Degradation in Liquid Culture. The degradation of TCE under 5% propane was affected by the initial TCE concentration (Figure 2). Strain TCE28 degraded 95% of 5 mg · l⁻¹ TCE in 7 days. With TCE at initial concentrations of 20 and 50 mg · l⁻¹, 90% and 15% of the TCE had been degraded by 30 days.

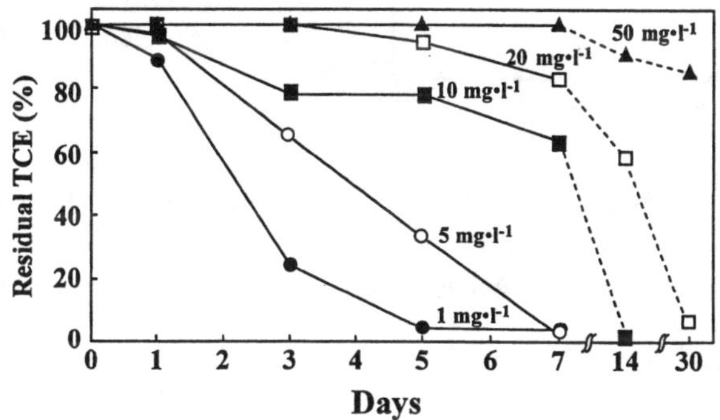

FIGURE 2. Degradation of TCE at various initial concentration under 5% propane gas in liquid culture.

Effect of *n*-butanol on TCE Degradation in Liquid Culture. Figure 3 shows the effect of the *n*-butanol concentration on TCE degradation and growth of TCE28 after 4 days. The TCE and *n*-butanol concentrations used in this experiment were 1 mg · l⁻¹ and 100 to 1000 mg · l⁻¹, respectively. The cell growth increased with the n-butanol concentration. At an initial *n*-butanol concentration of 500 mg · l⁻¹, good growth was observed, and the TCE was completely degraded in 4 days. At higher initial *n*-butanol concentrations, the high cell density was maintained, but TCE degradation was competitively inhibited.

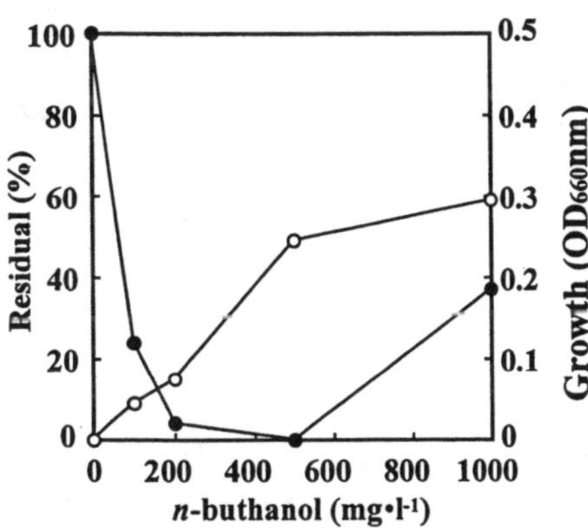

FIGURE 3. Effect of *n*-butanol concentration on degradation of TCE (initial concentration 1 mg•l⁻¹) by strain TCE28 after 4 days in liquid culture.
○: growth, ●: residual TCE

Effect of Initial TCE Concentration on TCE Degradation in Soil Microcosms.
Figure 4 shows the effect of the initial TCE concentration on TCE degradation in the soil microcosms

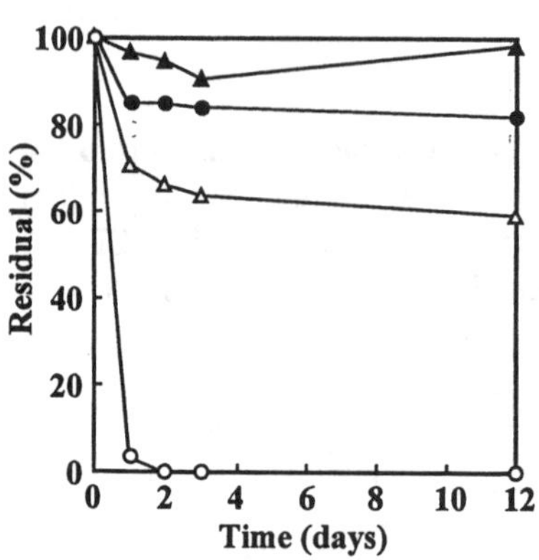

FIGURE 4. Effect of initial TCE concentration on TCE degradation in soil microcosms (no other added carbon sources).
○, △, ●, ▲: 1, 5, 10, 20 mg•l^{-1}of TCE, respectivery.

without any other added carbon source. At an initial TCE concentration of 1 mg·l^{-1} the TCE was completely degraded in 2 days. At an initial TCE concentration of 5 mg·l^{-1}, 30% of the TCE was degraded in 1 day; after that, the degradation rate decreased, and by 12 days only 40% of the TCE had been degraded.

CONCLUSIONS.

We studied the TCE-degradation characteristics of *Mycobacterium* sp. strain TCE28. In liquid culture, strain TCE28 could cometabolically degrade TCE with various carbon sources, and resting cells of this strain could degrade TCE in soil without other carbon sources. TCE28 appeared to be useful for the bioremediation of TCE-contaminated soil and groundwater.

REFERENCES

McDonald, I. R., H. Uchiyama, S. Kambe, O. Yagi, and J. C. Murrell. 1997. "The soluble methane monooxygenase gene cluster of the trichroloethylene–degrading methanotroph *Methylocystis* sp. strain M." *Appl. Environ. Microbiol.* 63(5): 1898-1904.

Sun, A. K., and T. K. Wood. 1996. "Trichloroethylene degradation and mineralization by *Pseudomonads* and *Methylosinus trichosporium* OB3b." *Appl. Microbiol. Biotechnol.* 45(1-2): 248-256.

Vanderberg, L. A., and J. J. Perry. 1994. "Dehalogenation by *Mycobacterium vaccae* JOB-5: role of the propane monooxygenase." *Can. J.Microbiol.* 40(3): 169-172.

Vannelli, T., M. Logan., D. M. Arciero., and A. B. Hooper. 1990. "Degradation of halogenated aliphatic compounds by the ammonia-oxidizing bacterium *Nitrosomonas europaea.*" *Appl. Environ. Microbiol.* 56(4): 1169-1171.

DEGRADATION OF TRICHLOROETHYLENE IN SOIL COLUMNS BY *METHYLOCYSTIS* SP. M

Katsuyuki KUBOTA, Manabu HASHIMOTO and Hiroshi GOHDA
(TOWA KAGAKU Co. Ltd., Naka-Ku, Hiroshima, JAPAN)
Kazuhiro IWASAKI and ***Osami YAGI***
(National Institute for Environmental Studies and CREST,
Tsukuba, Ibaraki, JAPAN)

ABSTRACT: We studied the application of bioaugmentation in cleaning up soil and groundwater contaminated with trichloroethylene (TCE). A TCE-degrading bacterium, *Methylocystis* sp. strain M, which can utilize methane or methanol as a sole carbon source, was used in this experiment. We examined the TCE-degradation characteristics of strain M in serum bottles and in soil columns. In the serum bottles experiments, strain M almost completely degraded TCE at an initial concentration of 10 $mg{\cdot}l^{-1}$ in 1 day. In the soil column experiments, 280 ml stainless-steel columns filled with soil as models for the groundwater environment. TCE, nitrogen-, phosphorus-, oxygen- and methane- amended groundwater flowed through the columns with a 48 h detention time. The effects of the initial strain M concentration on TCE degradation were determined. A total of 1.4×10^9, 7.0×10^9 or 2.8×10^{10} strain M cells was added to each column (initial concentrations: 5.0×10^6, 2.5×10^7 and 1.0×10^8 $cells{\cdot}ml^{-1}$, respectively). In the column to which 2.8×10^{10} strain M cells had been added, about 40% of the TCE had been degraded by 16 days. It appears that strain M amendment is useful for cleaning up TCE-contaminated groundwater.

INTRODUCTION

Volatile aliphatic chlorinated compounds such as TCE and tetrachloroethylene have been detected in groundwater throughout Japan. Various soil and groundwater clean-up technologies are now being developed. Soil vapor-extraction and pumping-up and treatment are very common approaches. Bioaugmentation is one of the most promising new technologies for cleaning up groundwater contamination, because of its low cost and complete destruction of pollutants. We studied the application of bioremediation for cleaning up TCE-contaminated soil and groundwater. To this end, we first isolated a TCE-degrading bacterium, which was named strain M (Uchiyama et al., 1989; Nakajima et al., 1992). Strain M was found to have the ability to utilize methane or methanol as a sole carbon source (Yagi et al., 1994; McDonald et al., 1997). We then studied the TCE-degradation characteristics of strain M in serum bottles and in soil columns.

MATERIALS AND METHODS

Chemicals. TCE was purchased from Wako Pure Chem. Co. (Osaka, Japan). Methane and oxygen gas were obtained from Sumitomo Seika Co. (Tokyo, Japan).

Bacterium and culture conditions. The methane-utilizing gram-negative bacterium *Methylocystis* sp. strain M, which we had isolated from soil, was used in this investigation. A mineral salt medium was used throughout the degradation experiments. The composition of the mineral salt medium was as follows (in $mg \cdot l^{-1}$): NH_4Cl, 2,140; K_2HPO_4, 1,170; KH_2PO_4, 450; $MgSO_4 \cdot 7H_2O$, 120; $FeSO_4 \cdot 7H_2O$, 28; $Ca(NO_3)_2 \cdot 4H_2O$, 4.8; H_3BO_3, 0.005; $MnSO_4 \cdot 4\text{-}6H_2O$, 0.06; $Na_2MoO_4 \cdot 2H_2O$, 0.001; $Co(NO_3)_2 \cdot 6H2O$, 0.06; $ZnSO_4 \cdot 7H_2O$, 0.01; $NiSO_4 \cdot 7H_2O$, 0.006; $CuSO_4 \cdot 5H_2O$, 0.006; H_2SeO_4, 0.004. The final pH of the medium was 6.8. The bacterial culture was prepared in 155 ml serum bottles, each containing 30 ml of mineral-salt medium and 5% methane in the head-space gas. The bottles were sealed with butyl rubber stoppers and crimped-on aluminum caps. The culture was incubated at 30°C in a rotary shaker at 120 rpm. Growth was determined by monitoring the optical density at 580 nm.

Serum bottle experiments. To determine the rates at which varying initial concentrations of strain M could degrade TCE, we prepared 69 ml serum bottles. We added 50 ml of a soil - groundwater mixture to each bottle; the 50 ml mixture contained 30 g of soil (wet weight). A spiking solution containing TCE at 1100 $mg \cdot l^{-1}$ in distilled water was prepared and added to each bottle, to make an initial TCE concentration of 10 $mg \cdot l^{-1}$. The serum bottles were sealed with teflon-lined butyl rubber caps and incubated in darkness at 20°C. These bottles acted as the controls. The experimental bottles were prepared in the same way, but with the addition of varying numbers of strain M cells ranging from 2.5×10^8 to 2.5×10^{10} (to give initial concentrations ranging from 5.0×10^6 to 5.0×10^8 $cells \cdot ml^{-1}$). TCE at initial concentrations ranging from 10 to 50 $mg \cdot l^{-1}$ was added to serum bottles prepared as above.

TABLE 1. Experimental conditions in each soil column.

Column conditions	Column No.			
	1	2	3	4
Strain M (cells)	—	1.4×10^9	7.0×10^9	2.8×10^{10}
K_2HPO_4 ($mg\text{-}P \cdot l^{-1}$)	50	50	50	50
NH_4Cl ($mg\text{-}N \cdot l^{-1}$)	10	10	10	10
Methane	add	add	add	add
Oxygen	add	add	add	add
TCE ($mg \cdot l^{-1}$)	1	1	1	1
Flow rate ($ml \cdot day^{-1}$)	70	70	70	70

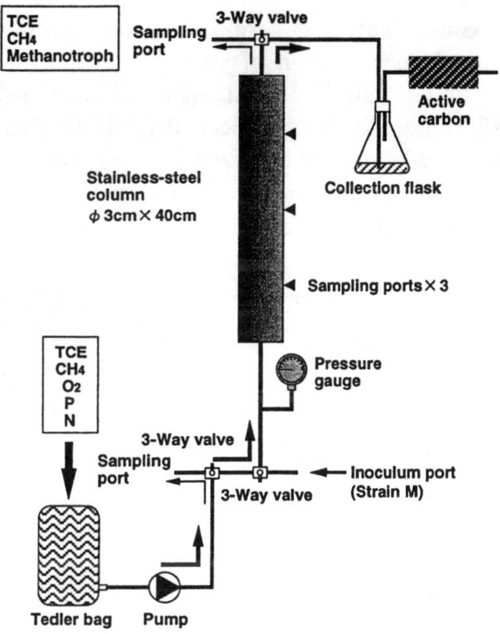

FIGURE 1. The soil column.

Soil column experiments. Stainless-steel columns (Munakata-Marr et al., 1996), each filled with 410 g of soil (dry weight), were used as models for the groundwater environment (Figure 1). TCE-, nitrogen-, phosphorus-, oxygen- and methane- amended groundwater flowed through the soil columns with a detention time of about 48 h. The system was operated at 20°C. The experimental conditions in each of the 4 columns are shown in Table 1. No strain M cells were added to the control column. A total of either 1.4×10^9 or 7.0×10^9 or 2.8×10^{10} cells was added to each of the 3 experimental columns (giving final concentrations of 5.0×10^6, 2.5×10^7 or 1.0×10^8 cells·ml^{-1}). The numbers of methanotrophs and the TCE and methane concentrations in the influent and effluent were determined. The final numbers of strain M cells were measured by the most probable number (MPN) culture method, using methane as a sole carbon source.

GC analysis. The concentration of TCE and methane in the head-space gasas periodically determined by gas chromatography (model GC-14A, Shimadzu Co., Tokyo) using a flame-ionization detector and a glass column (3 mm by 3 m) with 15% Silicon DC550 on 60/80-mesh Uniport HP (GL Science Co., Tokyo). The injection, oven and detector temperatures were 200°C, 100°C and 200°C, respectively. Nitrogen gas was used as the carrier gas.

RESULTS AND DISCUSSION

Effects of strain M concentration in serum bottles. About 60% and 100% of the TCE from an initial concentration of 10 $mg \cdot l^{-1}$ in the serum bottles was degraded in 2 days at initial strain M concentrations of 2.5×10^8 and 5.0×10^8 $cells \cdot ml^{-1}$, respectively (Figure 2). No significant degradation was observed at 5.0×10^6 $cells \cdot ml^{-1}$. No degradation was observed at 3 days without strain M. This result showed that strain M was able to degrade TCE in the soil and groundwater; high cell density was associated with high TCE-degradation.

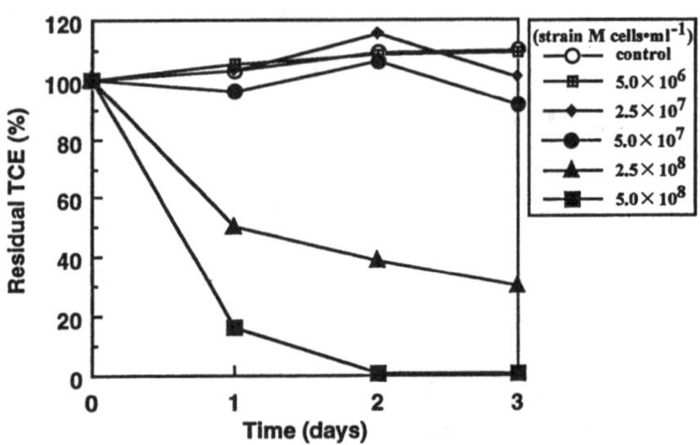

FIGURE 2. Effects of strain M concentration on degradation of TCE in serum bottles.

Effect of TCE concentration in serum bottles. TCE degradation was significantly accelerated by the addition of strain M at 5.0×10^8 $cells \cdot ml^{-1}$. With an initial TCE concentration of 10 $mg \cdot l^{-1}$, all the TCE was degraded in 2 days, whereas with an initial TCE concentration of 50 $mg \cdot l^{-1}$, only 20% and 40% of the TCE were degraded in 1 and 3 days, respectively (Figure 3).

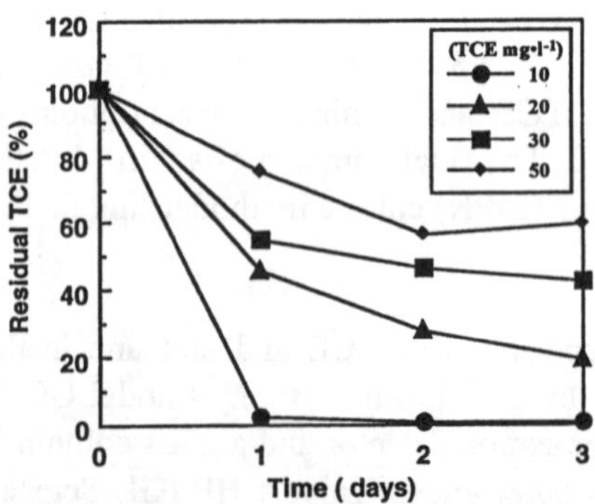

FIGURE 3. Effects of TCE concentration on TCE-degradation in serum bottles.

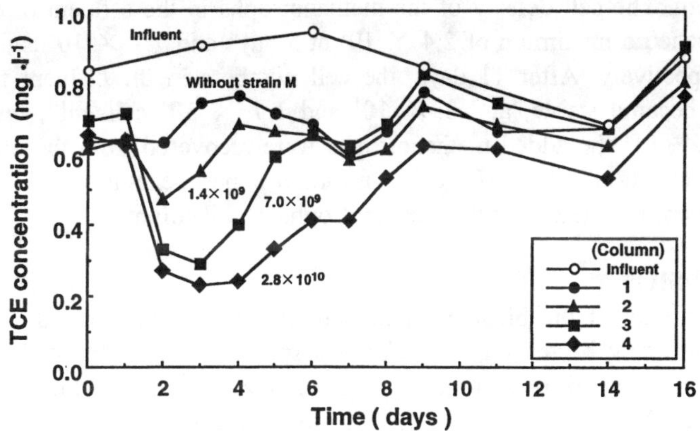

FIGURE 4. Periodic measurements of TCE concentration in the influent and effluent of each soil column.

TCE degradation in soil columns. The TCE concentration in the influent and effluent was determined periodically. The TCE concentration in the influent fluctuated between 0.68 and 0.94 mg·l^{-1}; the average value was 0.84 mg·l^{-1}. TCE degradation activity in columns 2 (total number of strain M cells, 1.4 × 10^{9}), 3 (7.0 × 10^{9} cells) and 4 (2.8 × 10^{10} cells) was maintained for 4, 6 and 16 days, respectively. At the highest cell concentration of 2.8 × 10^{10} cells, the lowest TCE concentration in the effluent was 0.2 mg·l^{-1} and about 40% of the added TCE was degraded over 16 days (Figure 4).

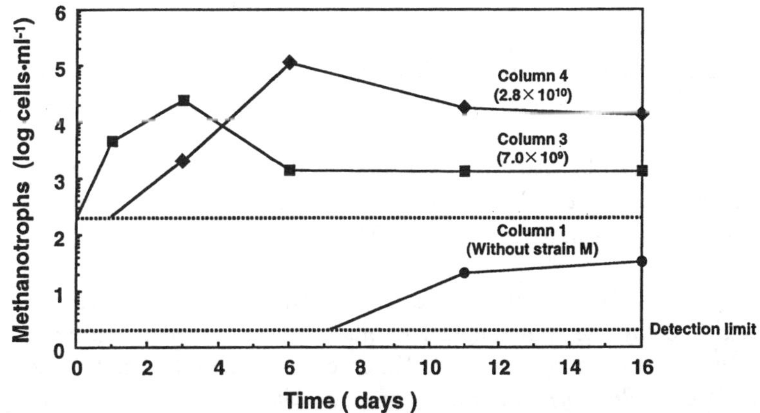

FIGURE 5. Changes in methanotroph numbers in the effluent of each column over 16 days.

We also determined the numbers of methanotrophs in the effluent (Figure 5). The number of effluent methanotrophs in column 1 (without added strain M) increased by a factor of 10 after 11 days. Methane showed a stimulatory effect on cell growth. The cell density of the methanotrophs in the effluent of columns 3 and 4 reached a maximum of 2.4×10^4 at 3 days and 1.1×10^5 cells•ml^{-1} at 6 days, respectively. After 11 days, the cell numbers in the effluent from these columns became stable, at 1.3×10^3 and 1.7×10^4 cells•ml^{-1}, respectively. About 0.2% of the added methanotrophs were recovered from the effluent at a total cell density of 2.8×10^{10} cells. In further experiments it would be desirable also to determine the fate of the methanotrophs in soil columns.

CONCLUSION

We studied the bioaugmentation of the TCE-contaminated soil using a methane-utilizing bacterium, *Methylocystis* sp. M. About 40% of the TCE had been degraded by 16 days in the column at 2.8×10^{10} strain M cells. It appears that strain M amendment is useful for cleaning up TCE-contaminated groundwater.

REFERENCES

McDonald I. R., H. Uchiyama, S. Kambe, O. Yagi, and J. C. Murrell. 1997. "The soluble methane monooxygenase gene cluster of the trichloroethylene-degrading methanotroph *Methylocystis* sp. strain M." *Appl. Environ. Microbiol.* 63(5): 1898-1904.

Munakata-Marr J., P. L. McCarty, M. S. Shields, M. Reagin, and S.C. Francesconi. 1996. "Enhancement of trichloroethylene degradation in aquifer microcosms bioaugmented with wild type and genetically altered *Burkholderia* (*Pseudomonas*) *cepacia* G4 and PR1." *Environ. Sci. Technol.* 30(6): 2045-2052.

Nakajima, T., H. Uchiyama, O. Yagi, and T. Nakahara. 1992. "Novel metabolite of trichloroethylene in a methanotrophic bacterium, *Methylocystis* sp. M, and hypothetical degradation pathway." *Biosci. Biotech. Biochem.* 56(3): 486-489.

Uchiyama H., T. Nakajima, O. Yagi, and T. Tabuchi. 1989. "Aerobic degradation of trichloroethylene by a new type II methane-utilizing bacterium, strain M." *Agric. Biol. chem.* 53(11): 2903-2907.

Yagi O., H. Uchiyama, and K. Iwasaki. 1994. "Bioremediation of trichloroethylene-contaminated soils by a methane-utilizing bacterium *Methylocystis* sp. M." In R. E. Hinchee, A. Leeson, L. Semprini and S. K. Ong (Eds.), *In Bioremediation of chlorinated and Polycyclic Aromatic Hydrocarbon Compounds*, PP. 28-36. Lewis Publishers, Boca Raton, FL.

EVALUATION OF METHANOTROPHIC BACTERIA DURING INJECTION OF GASEOUS NUTRIENTS FOR *IN SITU* TRICHLOROETHYLENE BIOREMEDIATION IN A SANITARY LANDFILL

Robin L. Brigmon, Denis J. Altman, Marilyn M. Franck, Carl B. Fliermans (Westinghouse Savannah River Company Aiken, SC), Terry C. Hazen, (Lawrence Berkeley National Laboratory, Berkeley, CA) and Al W. Bourquin (Camp Dresser and McKee Inc., Denver, CO)

ABSTRACT: Methanotrophic bacterial populations were quantified in an aquifer that was amended with air (oxygen), methane, triethyl-phosphate, and nitrous oxide to evaluate their effectiveness to stimulate aerobic bioremediation of vinyl chloride (VC), dichloroethylene, and trichloroethylene (TCE). Contaminants in groundwater resulted from leachate originating from a nearby landfill. Groundwater samples were taken during gas injection and analyzed for changes in bacterial populations. The methanotrophic populations were monitored in groundwater using direct fluorescent antibodies (DFA) and the most probable number (MPN) technique. Acridine orange direct counts (AODC) were used to determine the total bacterial population. Methanotrophic populations increased significantly in groundwater during the course of gaseous nutrient injections. As methanotrophic bacteria reached a maximum population in 3-4 days, contaminant levels (TCE) decreased. Cis-dichloroethylene (c-DCE) demonstrated a transient increase in concentration during the experiment but decreased rapidly over the course of the experiment. The total number of groundwater microorganisms did not change, indicating a selective stimulation of the methanotrophic bacterial population. These bacterial data were compared to physical parameters (pH, dissolved oxygen, redox) and contaminant (TCE , c-DCE, VC) concentrations within the saturated and unsaturated zone to reveal the efficiency of the system. The loss of contaminants appears to be due to cometabolic biodegradation through biostimulation since loss by volatilization was accounted for and was minimal. This work clearly demonstrates that one can effectively change the subsurface bacterial population in a relatively short period of time.

INTRODUCTION

This project was designed to evaluate the subsurface bacteria as influenced by an *in situ* bioremediation system. This system was designed for field testing of *in situ* treatment of groundwater contaminated with chlorinated solvents at the Department of Energy (DOE), Savannah River Site Sanitary Landfill (SLF) (WSRC, 1996). For over 20 years the SLF received paints, thinners, rags soaked with organic solvents including trichloroethylene (TCE) and perchloroethylene (PCE), sanitary waste, construction material, and disposable batteries. Leachates containing TCE were detected in groundwater samples during an evaluation from 1984 through 1993 (WSRC, 1996).

Subsurface gaseous nutrient injection has been found to be an effective *in situ* bioremediation treatment for TCE-contaminated groundwater (Pfiffner et al., 1997). The biodegradation of TCE was found to be sensitive to factors controlling microbial growth including injected methane and nitrogen (Travis and Rosenberg, 1997). *In situ* biodegradation is a highly attractive technology for remediation because contaminants are destroyed in place, not simply moved to another location or immobilized. Groundwater monitoring of bacteria and contaminant concentrations provides information on the efficiency of this technology. Application of these methane/air mixtures have been demonstrated to stimulate TCE-degrading methanotrophic bacteria in the subsurface (Pfiffner et al., 1997). Results from an investigation to characterize biodegradation and volatilization of TCE in a bioremediation test at the SLF are presented.

Objective. The objective of this test was to demonstrate that selective biostimulation can result in preferential changes to the status of subsurface microbial communities. We monitored how groundwater bacteria responded to multiple gaseous nutrient injections to determine the effectiveness of bioremediation of TCE and its byproducts in the SLF.

Site Description. The SRS is owned by the U.S. DOE and operated by WSRC. The SLF is located within the 310 square mile SRS facility located along the Savannah River, principally in Aiken and Barnwell Counties, South Carolina. The SLF began operating in 1974 and reached capacity in 1987 at which time expansions were added to increase its capacity to 70 acres.

MATERIALS AND METHODS

A pilot-scale gaseous injection remediation test system was designed and installed within the TCE plume to validate the use of *in situ* biodegradation of chlorinated solvents (WSRC, 1996). The test system consists of three sparge wells, a gas extraction well, and 14 nested monitoring points. Each nested monitoring location within the TCE plume has two mini-wells in the unsaturated soils (one shallow = 10 feet below ground surface [bgs]; one deep = 16 feet bgs) and two mini-wells in the saturated soils (one shallow = 30-40 feet bgs; one deep = 45-55 feet bgs). During this test, air, methane, nitrogen, phosphorous, and helium were injected for approximately 8 hours a day, followed by 16 hours of no injection (for six consecutive days). This injection contained a mixture of 15 SCFM (Standard Cubic Foot per Minute) air blended with 4% methane (CH_4), 0.07 % Nitrous Oxide (N_2O), 0.007 % to 0.01 % Phosphate (Triethyl Phosphate, TEP $(C_2H_5)_3PO$) and 1.0% Helium (He) injected as a tracer.

Once gas injection had begun water samples were collected every 24 hours from wells 2s, 9s, and 14s for microbial and groundwater contaminant data. Total bacterial counts were accomplished by the Acridine Orange Direct Count Method (AODC) methodology. Methanotrophic bacteria were quantified using the Direct Fluorescent Antibody (DFA) procedure and the Most-Probable-Number (MPN) technique (Brigmon et al., 1998b). Groundwater samples were analyzed for volatile organic compounds (VOCs) as previously described (Brigmon et al., 1998a). Vadose zone gas samples were taken to monitor possible volatilization.

RESULTS AND DISCUSSION

The DFA and MPN results on enumeration of methanotrophic bacteria from wells 2S, 9S, and 14 S are shown in Figures 1a-c. Overall, the data from each well show an increase in the methanotrophic bacteria population on or about midweek. These data are in agreement regardless of the technique employed, i.e. DFA or MPN.

FIGURE 1: Methanotrophic Cell Densities for wells 2S (1a), 9s (1b), 14s (1c) as determined by DFA and MPN during the test.

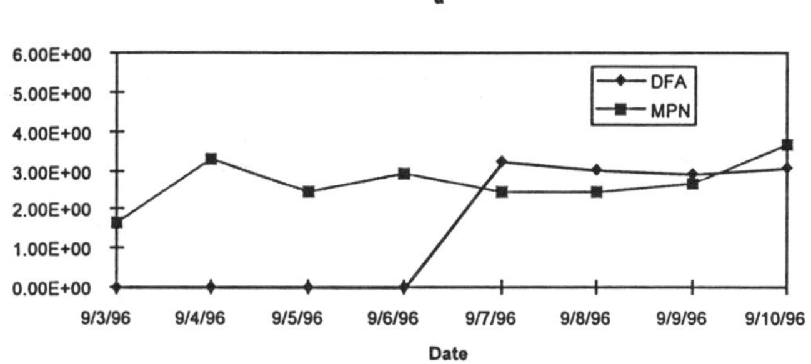

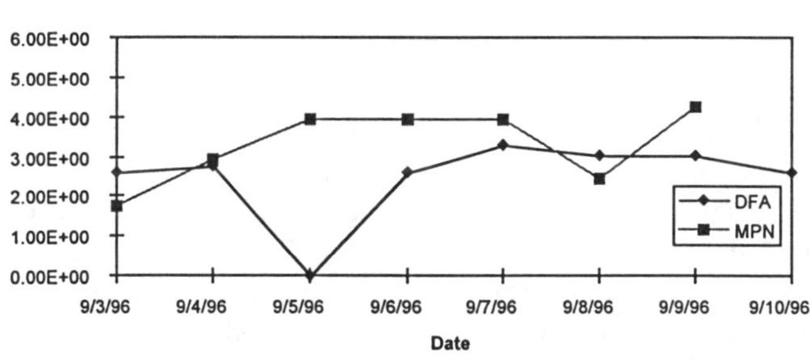

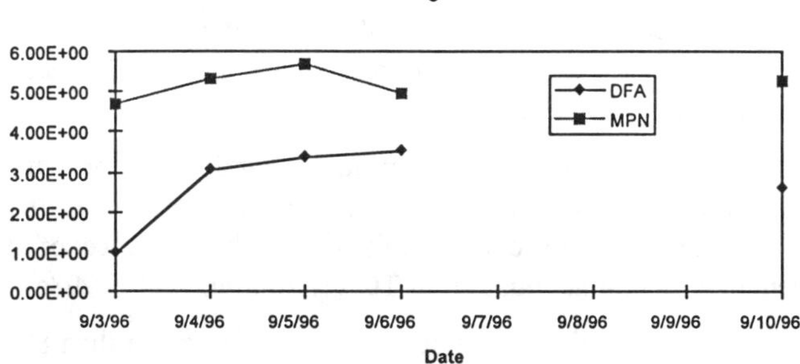

During the test, AODC densities did not significantly change (Figure 2). This lack of increase in the overall bacterial count when compared with the methanotrophic count indicates selective biostimulation during the test. Microbiological data demonstrates that the indigenous methanotrophic population requires three days to adapt to the injection of gaseous nutrients to the subsurface. In comparing the contaminant data with the bacterial densities, it is evident that as the methanotrophic bacteria population increased, the contaminant level decreased. These results show it is likely that the chlorinated hydrocarbons, TCE and c-DCE, were cometabolized by the indigenous methanotroph population.

FIGURE 2: Total Bacteria in Monitoring Wells by AODC:

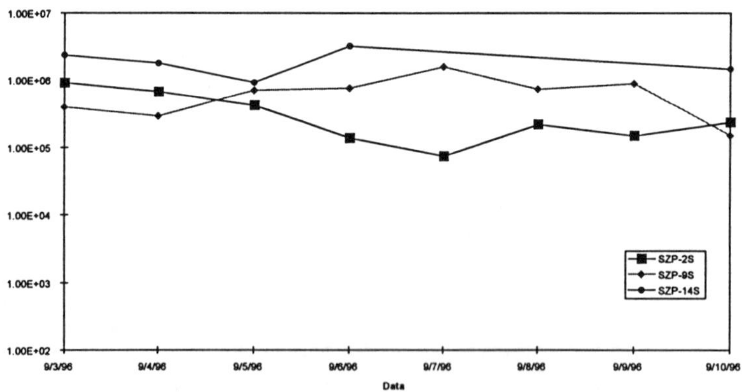

The analytical results of groundwater samples showed a decrease in the contaminant concentrations over the week of (Figures 3a, b, and c). In all three wells there was an initial increase in the quantity of c-DCE, a byproduct of the microbial biodegradation of TCE. After the middle of the week, this concentration decreased in all wells. The greatest contaminant fluctuation in all three wells was the c-DCE concentrations. Both wells 2S and 9S demonstrated an initial increase of approximately 250 ppb c-DCE that decreased below baseline levels by the end of the testing (Fig 3a and 3b). Well 14s was further from the direct influence of the injection system and results showed a lesser impact on contaminant concentrations including a 40 ppb increase in c-DCE that also decreased by the end of testing (Fig 3c).

Daughter products of TCE biodegradation including 1,2-dichloroethylene (DCE) and vinyl chloride (VC) are present in the SLF groundwater even though these compounds were never disposed of at this site. The presence of these compounds is evidence of natural attenuation of TCE (Brigmon et al., 1998a).

Bouwer (1994) has reported if the amount of c-DCE is greater than 80% of the total DCE, which it is in this case, it is a biodegradation product of TCE.

FIGURE 3: Organic Contaminants (PCE, TCE, c-DCE, VC) in groundwater for wells SZP 2S (3a), SZP 9S (3b), and SZP 14S (3C).

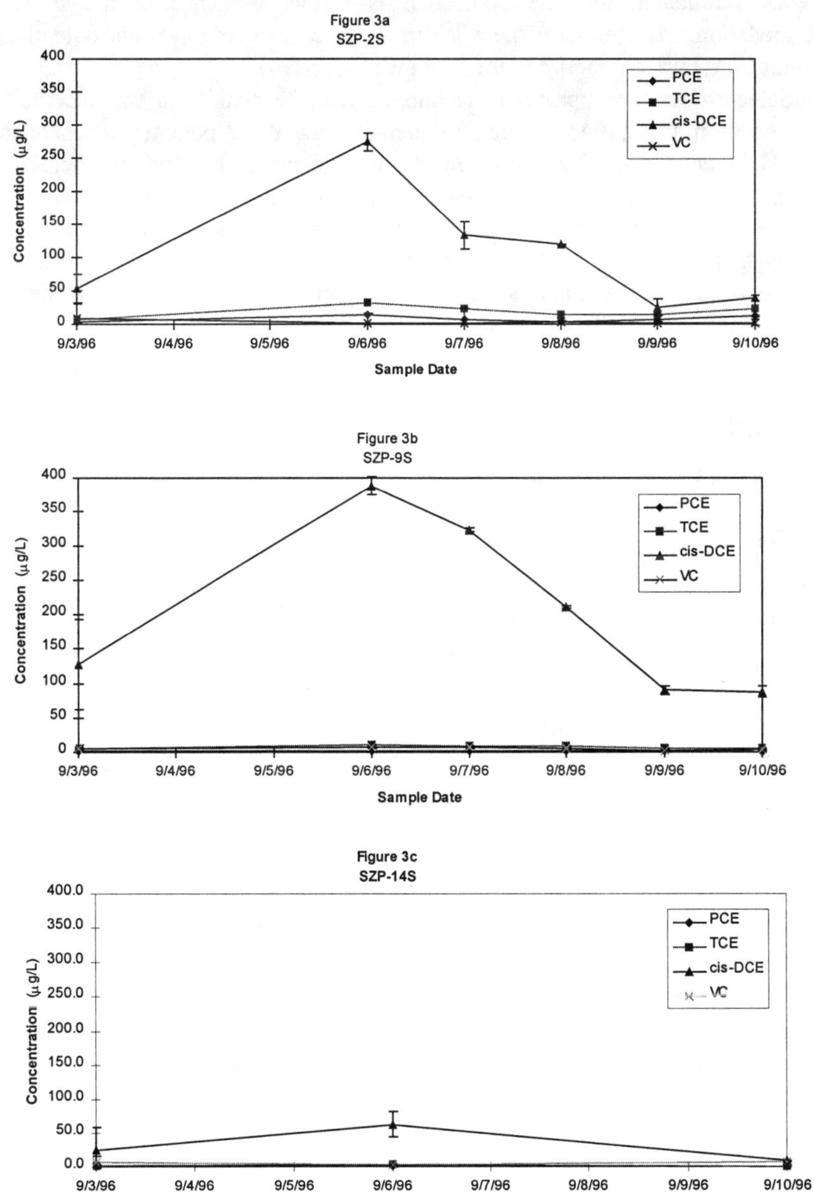

Since monitoring well information (contaminant concentrations at each depth interval for a nested well) did not indicate a contaminant mass transfer from the groundwater to the vadose zone, it is likely that c-DCE was actively biodegraded. The fact c-DCE is present in the groundwater and then after nutrient injection there is a transient increase of c-DCE is further evidence of active TCE biodegradation. In this case there is still TCE and other materials potentially leaching from buried material in the SLF (WSRC, 1996).

Selection of the appropriate technology can be critical to the success of bioremediation applications. A better understanding of the potential mechanisms involved for enhanced microbial degradation in the subsurface and the interaction between groundwater, microorganisms, and contaminants may be useful in extending the application of *in situ* bioremediation to additional contaminated sites by helping in the technology selection process. This demonstration and report represent another approach in the development of *in situ* bioremediation technologies.

REFERENCES

Altman, D.J., C.J. Berry, A. Bourquin, R.L. Brigmon, M.M. Franck, T.C. Hazen, D. Mosteller, and F.A. Washburn. 1998. Sanitary Landfill Supplemental Test Final Report, WSRC-RP-97-17, Revision 1, April 1, 1998.

Bouwer, E.J. 1994. Bioremediation of chlorinated solvents. In Norris R.D., R.E. Hinchee, R. Brown, P.L. McCarty, L. Semprini, J.T. Wilson, D.H. Kampbell, M. Rheinhard, E.J. Bouwer, R.C. Borden, T.M. Vogel, J.M. Thomas, and C.H. Ward, eds. Handbook of Bioremediation. Boca Raton, Fl. Lewis Publishers.

Brigmon, R. L., M. M. Franck, J. S. Bray, S. Lanclos, D. Scott, and C. B. Fliermans. 1998. "Direct immunofluorescence and enzyme-linked immunosorbent assays for evaluating organic contaminant degrading bacteria" *J. Microbiol. Methods*. (32): 1-10.

Brigmon, R. L., N. C. Bell, and D. L. Freedman and C.J. Berry. 1998. "Natural Attenuation of Trichloroethylene in Rhizosphere Soils at the Savannah River Site" *J. Soil Contam.* (7): 433-453.

Pfiffner, S. M., Palumbo, A. V., Phelps, T. J., Hazen, T. C. 1997. Effects of nutrient dosing on subsurface methanotrophic populations and trichloroethylene degradation. *J. Ind. Microbiol. & Biotechnol.* (18):204-212.

Travis, B.J. and N. D. Rosenberg. 1997. Modeling *in situ* Bioremediation of TCE at Savannah River: Effects of Product Toxicity and Microbial Interactions on TCE Degradation. *Environ. Sci. Technol.* (31):3093-3102.

WSRC. 1996. Sanitary Landfill *In situ* Bioremediation Optimization Test Final Report. WSRC-TR-96-0065. Westinghouse Savannah River Company, Aiken, SC.

FULL-SCALE IN SITU COMETABOLIC BIOREMEDIATION AT A PIPELINE SITE

Mark S. Nelson (Williams Gas Pipeline-Transco, Houston, Texas, USA);
Robert Legrand and Andrew J. Morecraft (Radian International, Austin, Texas; and Raleigh-Durham, North Carolina, USA);
John A. Harju (Gas Research Institute, Chicago, Illinois, USA).

Abstract: Subsurface contamination with chlorinated ethenes near a natural gas pipeline compressor station is being remediated using in situ methanotrophic treatment technology (MTT), as developed by the Gas Research Institute and the U.S. Department of Energy (Savannah River). The technology was evaluated at pilot scale at this site in 1997. Trichloroethene (TCE) levels decreased by one order of magnitude. The radius of influence of the air injection reached 30 ft (9 m). The injection system was scaled up in 1998 to remediate a larger portion of the plume. Air, methane, nitrous oxide (N_2O), and triethylphosphate (TEP) are injected via three wells. An efficient TEP diffusion system was designed and installed. Nine months into the full-scale project, most observation wells exhibit significant reductions in chlorinated volatile organic compound (CVOC) concentrations. Tetrachloroethene (perchloroethene, PCE) is not oxidized by methane monooxygenase (MMO), yet its concentration has also declined significantly from an initial level of 7600 μg/L. This suggests that MTT enhances reductive dehalogenation in nearby anaerobic microenvironments, perhaps by providing electron donor equivalents. CVOCs have been detected in some soil vapor sampling points, implying that some stripping of CVOCs is occurring. High methanotroph counts, however, strongly suggest that cometabolism is substantially responsible for the decline in concentrations. Treatment endpoints are currently being evaluated using a site-specific risk assessment.

INTRODUCTION

Methanotrophic Treatment Technology (MTT). Methanotrophic bacteria produce methane monooxygenase (MMO) to metabolize methane. This oxygenase is nonspecific and oxidizes trichloroethene (TCE) to TCE epoxide, which breaks down rapidly into oxygenated daughter products that are readily biodegradable. This is the principle behind MTT, as developed by the Gas Research Institute (GRI) in a multiyear, multidisciplinary research and development effort. MTT was applied in situ at the U.S. Department of Energy's Savannah River site in 1993 and 1994 to remediate a TCE plume. Air and methane were injected via a horizontal well to stimulate the growth of methanotrophs, produce MMO, and degrade TCE cometabolically (Hazen, in press).

The Site. Williams Gas Pipeline–Transco and GRI contracted with Radian International to evaluate in situ MTT at a natural gas pipeline compressor station, located in rural Virginia, with subsurface contamination. Depth to groundwater is typically 8–12 ft (2.4–3.6 m), although in December 98 it averaged a record low of 18.1 ft (5.5 m); average groundwater velocity is 1.2 cm/day. The formation consists of approximately 50 ft (15 m) of residual soil (hydraulic conductivity: 3×10^{-4} cm/sec) above bedrock. The groundwater initially contained up to 7600 µg/L of PCE and 2000 µg/L of TCE. The contamination is found throughout the saturated residual soil and the upper fractured bedrock. The areal extent of the plume is about 1 acre (0.4 hectare).

Pilot-Scale Evaluation. In situ MTT was evaluated at pilot scale at this site for about 4.5 months in 1997 (Legrand et al., 1998). Air and methane were injected, along with nitrous oxide (N_2O) and triethylphosphate (TEP) (biologically available nitrogen and phosphorus sources, respectively). TEP delivery was intermittent because of design issues. TCE levels decreased from 2130 to 150 µg/L in the well initially exhibiting the highest concentration. The radius of influence of the air injection was approximately 30 ft (9 m). Methanotrophic bacteria increased by six orders of magnitude and eventually dominated the subsurface microbiota. The results indicated that, as long as nitrogen and phosphorus were reliably supplied, rapid (two to four weeks) growth of methanotrophs and associated oxidation of TCE followed.

MATERIALS AND METHODS

Injection well IW-1 and the observation wells used in the pilot MTT project are described in Legrand et al. (1998). Two additional injection wells (IW-2 and IW-3) were installed to expand in situ MTT to the entire plume (Figure 1). Each injection well is constructed of 1-in. (2.5-cm) diameter, galvanized steel casing with a 1-ft (30-cm) length of 2-in. (5-cm) diameter, stainless steel screen. Wells IW-1, -2, and -3 were installed to depths of 50, 65, and 45.5 ft below land surface, respectively (15, 20, and 14 m). Well IW-2 was installed inside the 4-in. (10-cm) casing of monitoring well MW-13, which had been completed as an open-hole well into bedrock. The injection well screen was placed at the bottom of the boring, approximately 9 ft (2.7 m) below the top of bedrock.

The pilot MTT equipment is described in Legrand et al. (1998). It was upgraded to full scale in part by installing a higher capacity compressor and related equipment. Typical total airflow is 100 L/min, methane flow is approximately 4 L/min, and N_2O flow is 25 mL/min. The air pressure is set slightly above the downhole breakout pressure to ensure gas penetration into the formation, while minimizing stripping of volatiles. The system is mostly unattended, except for a brief daily inspection by a compressor station employee and during occasional sampling events.

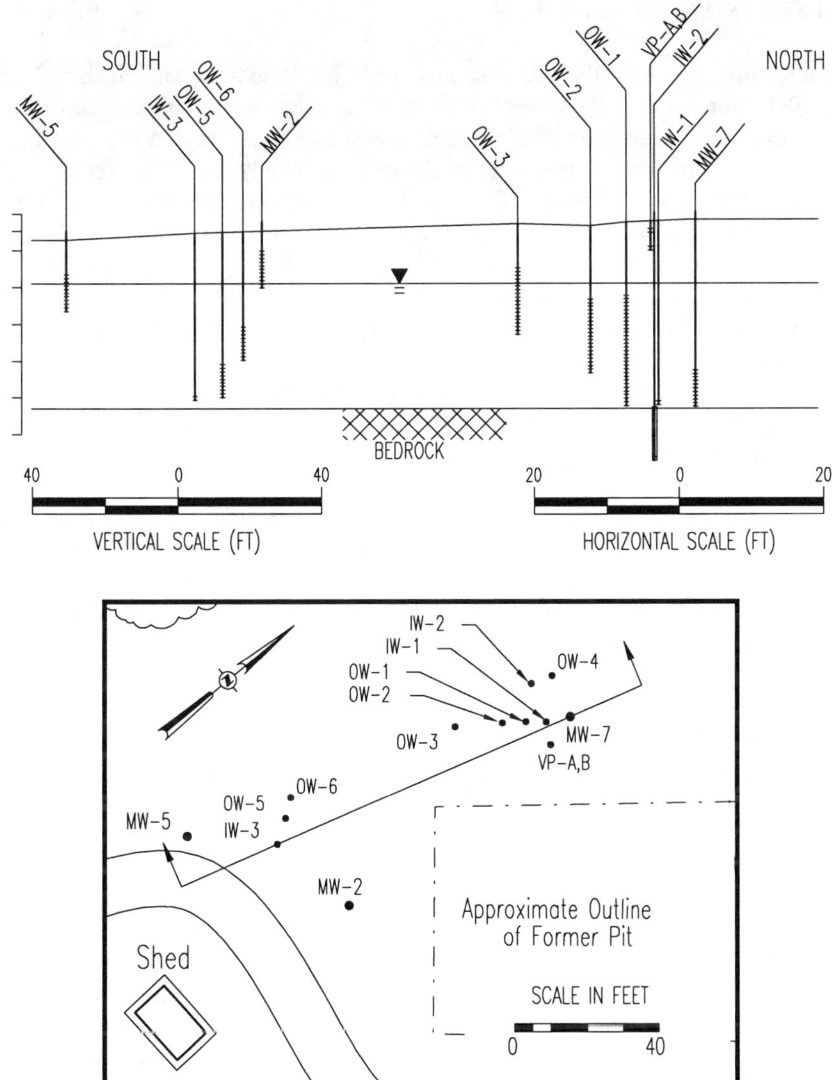

FIGURE 1: Layout and cross section of Wells; IW = Injection Well, OW = Observation Well, MW = Monitoring Well

RESULTS AND DISCUSSION

Operational History. The full-scale in situ MTT system has operated since March 5, 1998. For the first three weeks the methane flow was continuous in order to encourage maximum growth of methanotrophs and to establish a robust bacterial population. After the first three weeks, the methane supply was pulsed on a schedule of 4 hours on and 20 hours off. This was done to minimize competitive inhibition of TCE oxidation by methane, while providing an adequate carbon source for bacterial maintenance and growth. Problems were encountered feeding TEP until the TEP feed system was redesigned in late June. There were also several malfunctions of the methane feed and N_2O feed during the first two months of full-scale operation. Methanotrophs were counted in late April; low counts were found, indicating that the operational difficulties were preventing growth. Since June, however, system operation has been effective and a substantial decline of chlorinated volatile organic compound (CVOC) concentrations was observed by late August.

System Performance. Figure 2 depicts CVOC concentrations in MW-7—the well with the highest initial TCE readings. This well was targeted by the 1997 pilot-scale project and TCE declined dramatically during that period. It remained stable during the idle period between pilot- and full-scale projects, and has continued its decline since the resumption of MTT at full scale. Remarkably, *cis*-1,2-DCE rose during the idle period. *Cis*-1,2-DCE is a daughter product of the reductive dehalogenation of TCE and tetrachloroethene (PCE) under anaerobic conditions. The subsurface is largely anaerobic in the absence of MTT, so it is conceivable that reductive dehalogenation occurred here after the end of the pilot test. It may have been enhanced by the substantial pool of electron donor compounds (microbial biomass and related organics) left over from the pilot test. Other CVOCs in MW-7 have disappeared or have been reduced to trace levels.

The latest sampling event in early December 1998 showed a renewed increase in *cis*-1,2-DCE, accompanied by dissolved oxygen concentrations that declined from 7.5 to 4.5 mg/L. This suggests renewed reductive dehalogenation; the oxygen delivery may have been inadequate, possibly due to the formation of macrochannels as a result of nine months of uninterrupted sparging. Additionally, the groundwater was at a record low level during the latest sampling event, which may have reduced the radius of influence of the gas injection. Note that we have systematically observed increased VOC concentrations at many other sites under the extreme drought conditions of December 1998.

Well OW-1 shows sharply fluctuating TCE concentrations, which may reflect some interwell CVOC movement. Here too, *cis*-1,2-DCE is increasing and dissolved oxygen is down from 9.5 to 4.0 mg/L. Most other wells, however, continue to show declines in CVOCs and stable dissolved oxygen levels.

The concentration of PCE has decreased significantly during operation of

MTT—for example, in OW-6 it declined from 7600 to 320 µg/L. Some

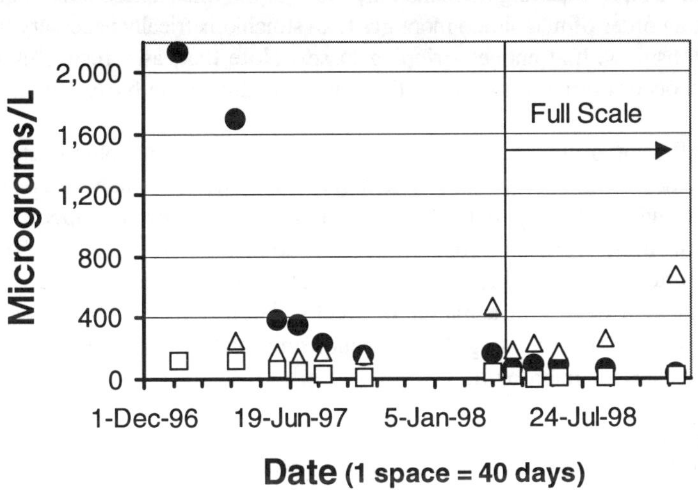

FIGURE 2: Historical CVOC concentrations in Well MW-7.

stripping of PCE is probably occurring, since it has a relatively high Henry's law constant. Additionally, although PCE cannot be oxidized by MMO, it may have been reductively dehalogenated. Although MTT should result in a nominally oxygenated aquifer, Enzien et al. (1994) conclusively documented that MTT enhances reductive dehalogenation of PCE in anaerobic microenvironments. MTT may provide otherwise limiting electron donor compounds such as bacterial biomass and organics of biological origin.

We estimate that after 4.5 months of pilot testing and 9 months of full-scale operation, 70% of the TCE and 93% of the PCE have been removed site-wide.

Stripping of VOCs from the Groundwater. The pressure in the injection wells is maintained below 140% of the breakout pressure to minimize stripping and the formation of preferential flow channels. Nevertheless, some stripping is probably occurring, as indicated by the detection of CVOCs in soil vapor points. We conducted a soil gas survey in June 1998 and found mostly very low (<1 ng/L) or nondetectable CVOC concentrations, except in the vicinity of injection well IW-3, where PCE concentrations around 25 ng/L were measured. The contribution of stripping to CVOC removal at this site was not quantified, since it was outside the scope of the project. Stripping can be a concern at certain sites because it can lead to uncontrolled release of CVOCs to the atmosphere, unless a vapor extraction system is used.

Methane is fed at a concentration of 4% in the injected air in order to remain below the lower explosive limit. Furthermore, we try to limit competitive inhibition of TCE cometabolism by injecting methane only 1/6th of the time. These constraints lead us to inject an order of magnitude more gas than stoichiometrically necessary for methane metabolism, with attendant stripping losses. Note that, as a result, more stripping should occur during in situ MTT than during hydrocarbon biosparging.

Growth of Methanotrophs. During the pilot-scale test, methanotrophic bacteria were enumerated by using the most probable number procedure. Initially, very low counts were found (1–1500), but after two months of effective operation, methanotrophs in most wells exceeded the enumeration limit of 3.28×10^6. This illustrates that after methane, air, and nutrients are reliably supplied, massive methanotroph growth occurs in a matter of weeks. Total bacterial counts were between 2×10^6 and 3×10^7, indicating that methanotrophs had become the dominant fraction of the subsurface microbiota.

CONCLUSIONS

In situ MTT using standard vertical wells is effective at rapidly reducing the concentrations of a series of CVOCs in the subsurface. Williams Gas Pipeline–Transco is discussing site closure with the Virginia Department of Environmental Quality through the use of a risk-based approach. We expect the MTT system to reduce CVOC concentrations to acceptable risk-based levels in 1999.

We are considering pulsing the entire injected air stream to minimize the formation of macrochannels and the resulting short-circuiting (see Wilson & Norris, 1997). Some stripping is likely to accompany in situ MTT. Where air emissions are a concern, soil vapor extraction should be used to control them.

REFERENCES

Enzien, M. V., F. Picardal, T. C. Hazen, R. G. Arnold, and C. B. Fliermans. 1994. "Reductive Dechlorination of Trichloroethylene and Tetrachloroethylene under Aerobic Conditions in a Sediment Column." *Appl. Environ. Microbiol. 60*(6): 2200-2204.

Hazen, T. C. (In Press). "Case Study: Full-scale In Situ Bioremediation Demonstration (Methane Biostimulation) of the Savannah River Site Integrated Demonstration Project." In D. C. Adriano and J. M. Bollag (Eds.), *Bioremediation of Contaminated Soils*, Monograph, Soil Science Society of America/American Society of Agronomy.

Legrand, R., A. J. Morecraft, J. A. Harju, T. D. Hayes, and T. C. Hazen. 1998. "Field

Application of In Situ Methanotrophic Treatment for TCE Remediation." In G. B. Wickramanayake and R. E. Hinchee (Eds.), *Bioremediation and Phytoremediation*, pp. 193-198. Battelle Press, Columbus, OH.

Wilson, D. J. & R. D. Norris. 1997. "Sparging and Biosparging: Insights through Mathematical Modeling." In B. C. Alleman and A. Leeson (Symp. Chairs), *In Situ and On-Site Bioremediation: Volume I*, pp.147-152. Battelle Press, Columbus, OH.

PASSIVELY ENHANCED IN SITU BIODEGRADATION OF CHLORINATED SOLVENTS

Maureen A. Dooley, Willard A. Murray (Harding Lawson Associates, Wakefield, Massachusetts) and Stephen Koenigsberg (Regenesis, San Juan Capistrano, California)

INTRODUCTION

Harding Lawson Associates (HLA) conducted a field demonstration to evaluate in situ enhanced anaerobic biodegradation of chlorinated solvents using Hydrogen Release Compound (HRC). The technology is designed to enhance the natural biodegradation process for a more rapid and complete degradation of the organic contaminants to non-toxic compounds (e.g., ethylene). This paper presents the results from a field demonstration that was conducted over an eleven month period.

Technical Approach. Anaerobic biodegradation of chlorinated solvents such as perchloroethene (PCE) and trichloroethene (TCE) require highly reducing conditions to stimulate anaerobic bacteria to dechlorinate the solvents. The technical approach was designed to provide a carbon or electron donor source to create the conditions necessary to enhance anaerobic biodegradation.

Hydrogen Release Compound (HRC) is a lactic acid ester that is manufactured by Regenesis. When delivered to the subsurface, lactic acid, which has been shown to be an effective electron donor, is released continuously into groundwater. Because this material can be injected directly into the subsurface it can be used to deliver nutrients passively and enhance active biodegradation of chlorinated volatile organic compounds (cVOCs).

HLA conducted a field demonstration to evaluate the performance of HRC using a system designed to recirculate groundwater and create a treatment cell. The objectives of this test were to evaluate the rate and extent of chlorinated solvent biodegradation using HRC as the electron donor source and also to gather design data such as how long HRC would last and the extent of biodegradation along the groundwater flowpath.

CASE STUDY: WATERTOWN, MA

Site Description. The site is situated in a historically industrial section of Watertown, MA. The general soil profile consists of approximately 13 feet of sand and gravel over approximately 7 feet of silty sand; then glacial till (an aquitard) is encountered. Groundwater occurs at approximately 8 feet below land surface and is contaminated with chlorinated solvents, including PCE, TCE and degradation products characteristic of natural biological reductive dechlorination.

Design, Construction and Operation. In the field demonstration, groundwater is extracted from three downgradient wells, and injected into three wells 17 feet upgradient. Figures 1 and 2 show the design of the recirculating system. Five 2-inch PVC monitoring wells are positioned between the injection and extraction wells to monitor the progress of the biodegradation process: a groundwater recirculation rate of 0.25 gallons per minute (gpm) establishes a single recirculation cell of about 30 feet in diameter (Figure 2).

Operation. The system was in full operation from February 1998 to November 1998 with a relatively constant recirculating flow rate of 0.25 gpm. HRC canisters were placed into the three injection wells beginning in February 1998 to passively deliver electron donor. VOC and other data, that measured redox conditions and the concentration of electron acceptors (dissolved oxygen, ferrous iron, sulfate, nitrate, methane) and electron donors (total organic carbon (TOC), volatile acids), were collected over the time course of the study.

The pump was shut down in November and HRC remained in the injection wells to act as a barrier. Data were collected in January 1999 , approximately 2 months after the pump was shut down, to evaluate rebound within the treatment cell and to evaluate the effectiveness of HRC as a barrier.

Results. Results from the field demonstration indicate that under anaerobic conditions significant reductions in the concentration of all cVOCs was observed (Figures 3 and 4). The average TCE concentration at the beginning of the study was 9900 ug/L and after 206 days levels were reduced to <10ug/L. Initial PCE concentrations of approximately 740 ug/L were reduced to <1 ug/L. Dichloroethene (DCE) levels, which started at an average value of 2500ug/L, remained relatively constant for the first six months then, decreased to <100ug/L. Vinyl chloride levels rose from an average 250 ug/L to 3000 mg/L after the first six months, then decreased less than 100 ug/L Ethene was detected in IN-2 and the downgradient monitoring wells indicating complete biodegradation of the cVOCs. The reduction in total mass was calculated to be 97% based on the average concentration of cVOCs in the treatment cell.

Figure 4 shows concentrations of VOCs versus time in well EPA-2. It can be seen that reductive dechlorination is not apparent until 50 days after the testing had started. After 50 days redox levels had dropped from +100mv to less than 50mv across the entire cell corresponding to the time when reductive dechlorination was first observed. Redox levels were mantained within a range of –150 to –50mv over the remaining duration of the study.

TOC and volatile acid data were collected throughout the study from the injection well. TOC levels were detected at a concentration of 1200mg/L after two months and remained constant after that. Lactic acid levels were first detected in the range

Figure 1. Biotreatment Cell
Plan View Showing Groundwater Flow Pattern and Wells

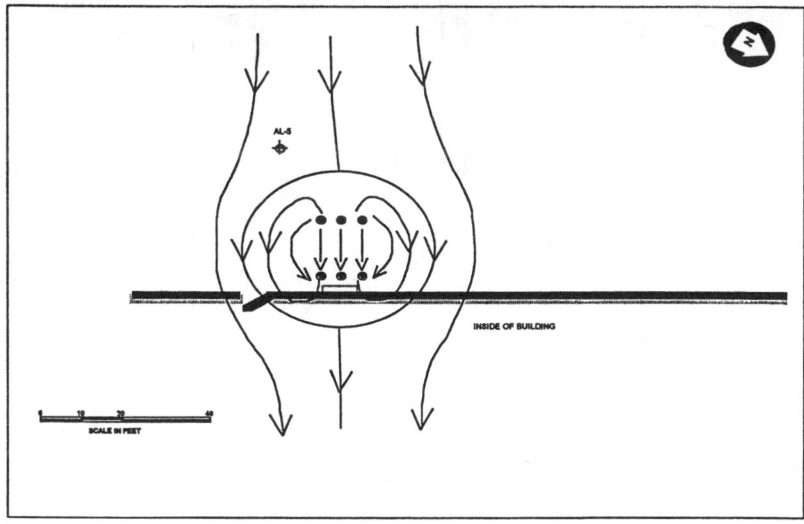

Figure 2. Vertical Section Through Biotreatment Cell

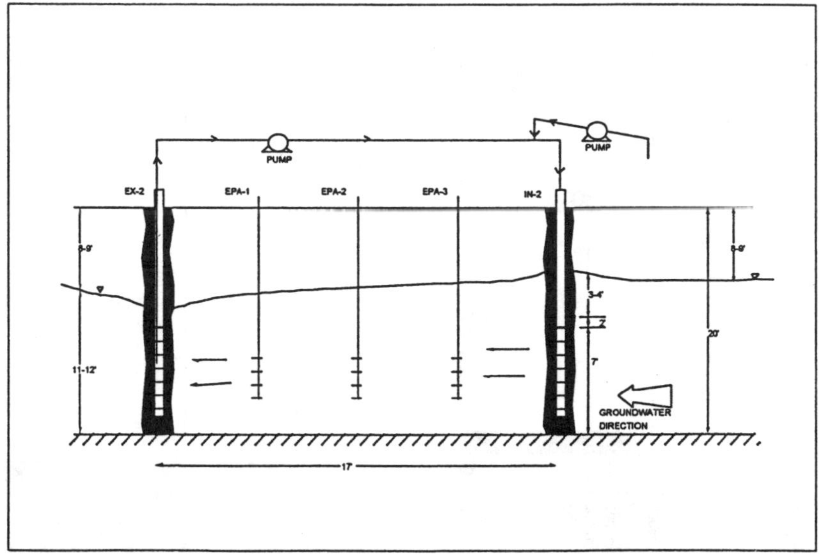

Figure 3. VOC Results From Pilot Test

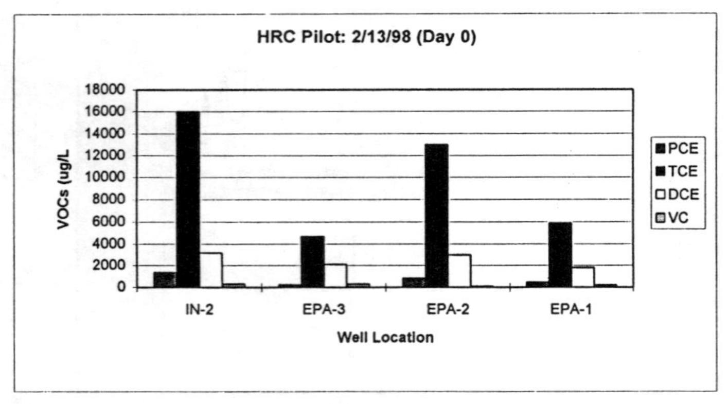

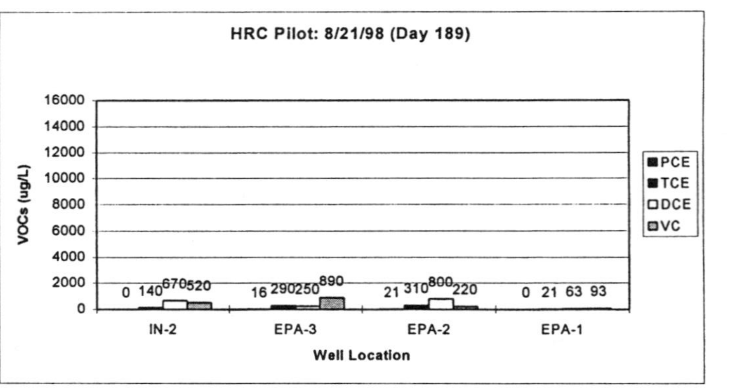

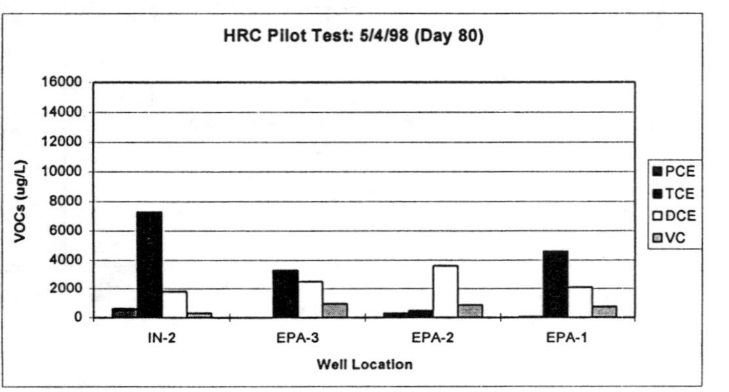

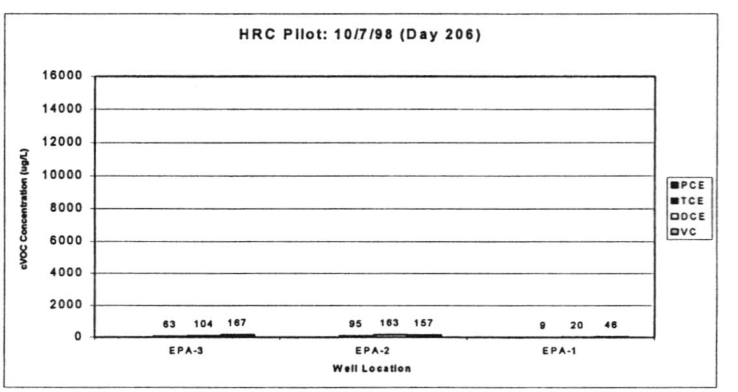

of 70 mg/L and increased to approximately 800 mg/L. The concentration of TOC and volatile acids dropped off dramatically by the time groundwater reached downgradient monitoring well EPA-3 with TOC at approximately 20 mg/L and little to no lactic acid detected. Other volatile acids such as propionic and acetic acid were sporadically detected at levels between 1-15 mg/L at downgradient monitoring wells EPA 1-3. Although TOC and volatile acid levels were significantly lower downgradient of the injection well, it was apparent that cVOC biodegradation was continuing based on the continued cVOC reduction observed along the flowpath. (Figure 3).

Elevated sulfide levels were detected in the injection well and monitoring wells EPA-3 and EPA-1. Dissolved iron levels were elevated across the cell, but there was no observed increase in methane. These results suggest that sulfate reduction and iron reduction were the predominant conditions favored across the treatment cell.

In November 1998 the recirculating pump was shut down and the HRC containing canisters remained in the injection wells. CVOC data were collected approximately 2 months after shut down. These analytical results showed that there was only a limited rebound in the concentration of cVOCs within the treatment cell. (Figure 5).

Under the conditions of this field demonstration, HRC polymer (though almost gone) was still present in the canisters after one year.

Biodegradation rates estimated for the individual cVOCs based on results from the field demonstration are: PCE 0.021 day^{-1}, TCE 0.018 day^{-1} , DCE 0.042 day^{-1} , and VC 0.044 day^{-1} .

Costs. The field demonstration was conducted for a cost of less than $30K (including analytical and monitoring). The estimated cost to expand this program to full-scale using a passive design is less than $50K. HRC injections are proposed across a source area of 100'x40' and would be delivered in holes using direct push techniques. The only other project costs beyond the $50K would include those required for long term monitoring (sampling, analytical and reporting costs). A monitoring program for this site would be estimated to cover a time period of 2-5 years and would be incorporated as part of a natural attenuation remedy. The analytical program would require collection of VOC samples as well as some natural attenuation parameters to monitor redox conditions and geochemical parameters. A second HRC injection is expected to be needed after one year, but costs may be less than the original $50K as the concentration of volatile organics will likely be lower after the first year of treatment.

Figure 4. Data From Well EPA-2

HRC-Pilot Test: EPA-2

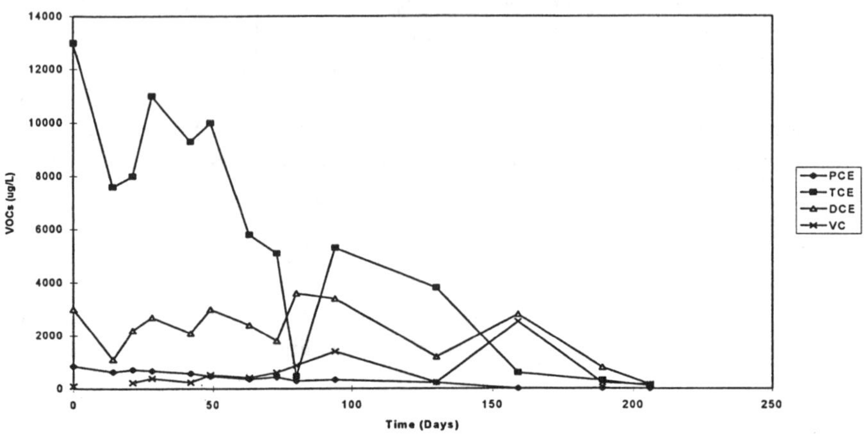

Figure 5. VOC Levels After Recirculating System Was Shut Down

Rebound 2 Months After Recirculating System Shut Down
HRC Used As Barrier (1/21/99)

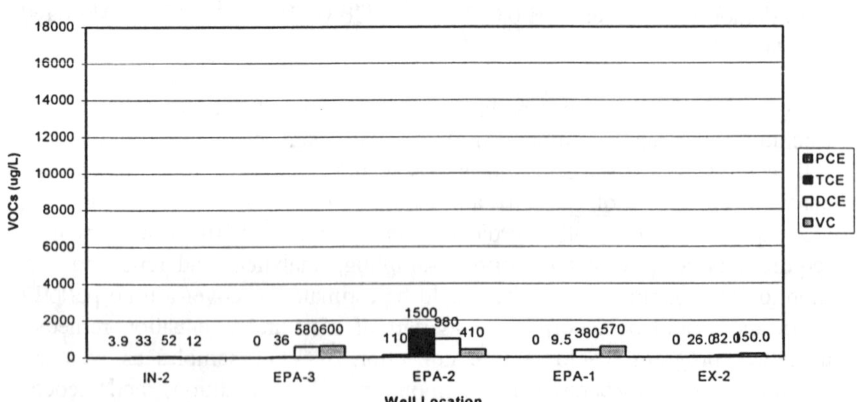

CONCLUSIONS

The specific conclusions that can be drawn from the results at the field demonstration are:

- Complete biodegradation of PCE, TCE and daughter products was demonstrated;
- DCE and VC biodegradation rates were rapid under HRC enhanced anaerobic conditions;
- Sulfate reducing and iron reducing conditions appeared to be the predominant microbiological conditions across the treatment cell;
- A lag period, which likely corresponds with the establishment of low redox conditions, appears to be required before biodegradation is enhanced;
- Enhanced biodegradation was observed at locations 17' downgradient of the injection points indicating HRC affected conditions beyond the area immediately surrounding the injection well;
- There was little rebound observed within the treatment cell after the recirculating system was shut down, suggesting significant reduction of residual cVOCs; and
- HRC can be used to enhance cVOC biodegradation passively and is a cost effective alternative to other active remediation technologies.

ACCELERATED REDUCTIVE DECHLORINATION OF PCE: A FULL-SCALE SOIL TREATMENT

TATSUO SHIMOMURA (Ebara Research Co. Ltd., Fujisawa-shi, Kanagawa, Japan)
Hiroshi Shinmura and Naoki Seki (Ebara Co., Minato-ku, Tokyo, Japan)

ABSTRACT: A newly developed bio-chemical dechlorination method was applied to clean up of soil contaminated with tetrachloroethylene (PCE). The mean initial PCE concentration was 11000 μg/kg-wet soil. A target PCE concentration was set to 100 μg/kg-wet soil in this remediation. Nutrients and trace amount of metal powder were added to 500 ton of the contaminated soil. Mixture of soil, nutrients, and metal powder produced a strong reductive environment and enhanced dechlorination. An average PCE concentration of 39 μg/kg-wet soil was achieved in the 70-day treatment without significant accumulation of intermediates such as cis-1,2-dichloroethylene (cDCE) and vinyl chloride (VC) was detected. A laboratory mass balance test demonstrated that PCE was totally converted to ethylene and ethane gas.

INTRODUCTION

Various studies have been performed to investigate chemical dechlorination. Focht et al. (1996) reported a wall consisting of granular iron. As groundwater contaminated with chlorinated organics passed through the wall, the contaminants were dechlorinated while iron was oxidized. Using this reaction, some soil clean-up methods has been proposed. However, these methods have a problem in contacting the contaminant and the metal surface, which results in low efficiency of the chemical reaction in the soil.

A recent theory for biological dechlorination emphasizes that microorganisms are capable of using chlorinated organics as electron acceptors for anaerobic respiration (Picardal et al., 1995; Maymo-Gattell et al., 1997). For biological dechlorination, it is important to keep oxidation-reduction potential (ORP) sufficiently low. Reductive environment for the conventional biological reaction yields an ORP of -100 to -350 mV. This ORP level is not low enough to enhance complete dechlorination and results in accumulation of cis-1,2-dichloroethylene (cDCE) and VC.

In this study we developed a soil remediation technology which incorporates both the chemical and biological dechlorination process and named it a terra-reduction method (TRM). In this method, we mixed a reducing agent (metal powder) and nutrients with contaminated soil. The reducing agent generates substantially low ORP (-500 to -700 mV) and leads to chemical dechlorination, while the nutrients are served as substances for microbial growth in the reductive environment. The chlorinated organics are then decomposed through anaerobic respiration.

MATERIALS AND METHODS

Reagent/Soil All reagents used as the nutrients and reducing agent satisfied industry-grade quality. Standard solutions of PCE, TCE, VC and cDCE with concentrations of 1000 mg/l were purchased from Wako Pure Chemical Ltd. as dissolved in methanol. The PCE-contaminated soil consisted of 30% brown loam and 70% pale grayish brown silt with a trace volume of fine-grain sand.

Treatment procedure This TRM was applied to the PCE-contaminated soil stored in a 1.2-meter-deep concrete pit. The initial PCE concentration was 28000 μg/kg-wet soil at maximum and 11000 μg/kg-wet soil on average. The total weight of the soil was about 500 tons. The nutrients and metal powder were added to the soil with a backhoe. The soil features extremely low gas permeability. Therefore, volatilization of PCE due to mixing was insignificant.

Sampling methods Soil samples were collected with a hand auger at depth of 50 cm and 100 cm at four different horizontal locations (A, B, C, and D), in the concrete pit. All samples were collected in sealed serum bottles, stored in an ice box, and brought back to our laboratory. They were analyzed within 24 hours.

Analytical methods ORP of the sample soil was measured by filling a serum bottle containing 25g of the sample soil with oxygen-free water and inserting an ORP probe (UK-2030, Central Kagaku Corp.) into it. The chlorinated organics were analyzed with a headspace gas chromatography equipped with a photoionized detector (GC-311/PID, HNU Systems Inc.).

Mass balance test procedure To prove that PCE was entirely converted to ethylene and ethane, we performed a test using a sealed container in our laboratory. Nutrients of 45 ml whose concentration was reduced to 1/5 of the usual treatment and uncontaminated silt soil of 5 g were put in the container. Metal powder was then added as reducing agent. After PCE of 0.25 mmol/kg-soil (equivalent to 41.5 mg/kg-soil) was injected, the container was sealed with a Teflon-lined rubber stopper, placed upside-down on a shaker, and shaken at a speed of 100 rpm at 28 °C. The concentration of chlorinated organics, ethylene, and ethane in the headspace was measured periodically, and the measurement results were compared with the initial values to check the mass balance. PCE adsorbed into the soil was not considered in this experiment. In order to suppress PCE adsorption to the soil to a negligible level, amount of the soil put in the bottle was limited to 1/9 of that of media as described above.

RESULTS AND DISCUSSION

Analytical results of chlorinated organics The target concentrations for this remediation study were set by making reference to the Japanese national guideline for soil, 100 μg/kg-wet soil for PCE, 300 μg/kg-wet soil for TCE, and 400 μg/kg-wet soil for cDCE. Figure 1 shows the temporal variations of the PCE concentration in the soil. Day 0 represents the day when the nutrients and the metal powder were mixed with the soil. The PCE concentrations for all samples decreased after Day 20, and the most of the samples achieved the target concentration of 100 μg/kg-wet soil after Day 40. Some samples, however, sporadically show high PCE concentrations even over 1000 μg/kg-wet soil. This may be attributed to incomplete mixing which produced intact soil aggregates in the concrete pit. It took 70 days that all samples showed the concentrations lower than 100 μg/kg-wet soil. The average PCE concentration was found to be 39 μg/kg-wet soil after 70 days of treatment. This implies that it took 70 days for the nutrients and reducing agent to infiltrate into the soil aggregates. Three

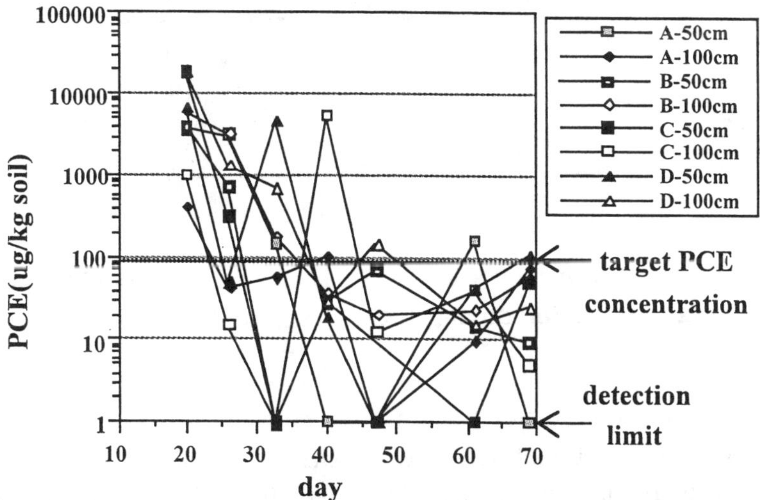

FIGURE 1. PCE Concentration in Contaminated Soil during TRM Treatment.

months after soil mixing, other samples were taken at 64 points to verify completion of this remediation. The PCE concentrations of all the samples were found to be below 100 μg/kg-wet soil.

Figure 2 shows a representative example of monitoring results for the concentrations of PCE, TCE, cDCE, VC, ethylene, and ethane measured at 50 cm deep in the concrete pit. During an initial decrease in PCE concentration, the cDCE concentration showed an increase, followed by a decrease to a value lower than the remediation target in 35 days. Insignificant accumulation was detected for other substances. Ethylene and ethane were detected to some extent although they are in gaseous phase at room temperature and hardly remain in the soil. When samples were taken for measurement at 64 sampling points three months after the operation, cDCE concentration of all the samples was found below the target level. These results indicate that the TRM is effective for decomposing PCE and satisfying the remedial objective without accumulation of intermediates.

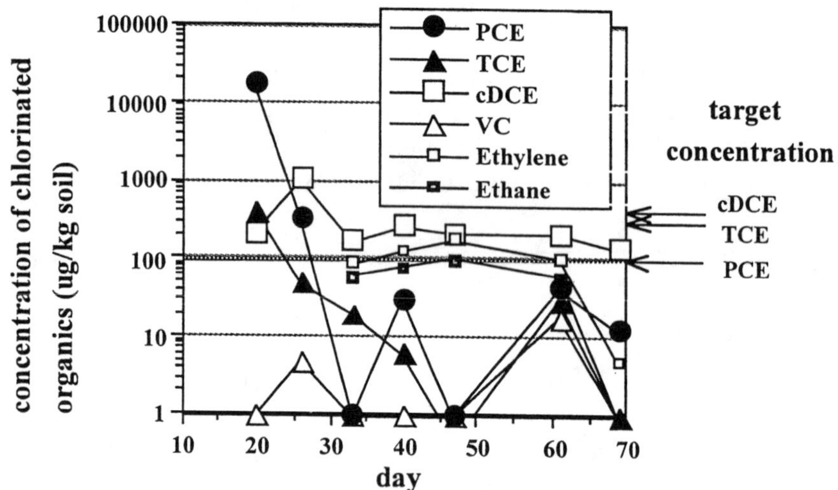

FIGURE 2. Concentrations of PCE and PCE Metabolites in Contaminated Soil during TRM Treatment.

ORP measurement results Figure 3 shows ORP results. The ORP decreased to around -600 mV immediately after the start of the treatment and gradually increased to -140 mV on Day 40. After Day 40, the ORP showed a marginal increase. Figure 3 clearly shows that the reductive environment was maintained in the soil during the treatment period, and that degree of reduction in the soil decreases gradually. No significant difference in ORP was detected between the two sampling depths.

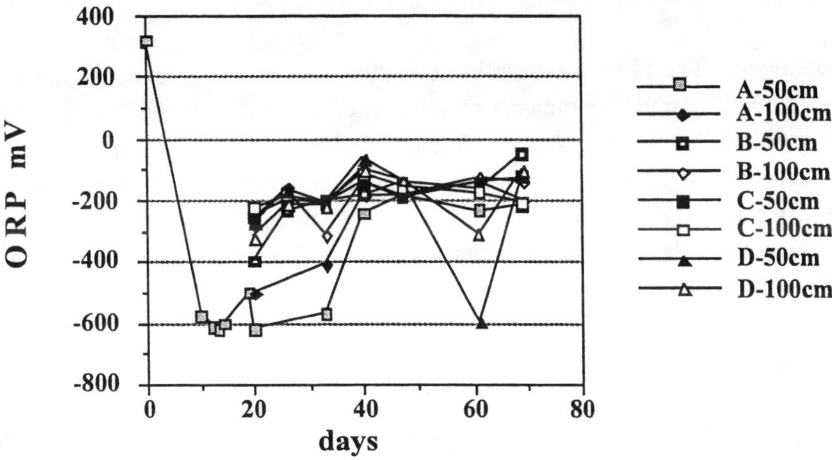

FIGURE 3. ORP in Contaminated Soil

Mass balance test result Figure 4 shows the result for the mass balance test. By Day 88, spiked PCE has completely disappeared, and 88% of the initially-spiked PCE was recovered in form of ethylene and ethane. This also

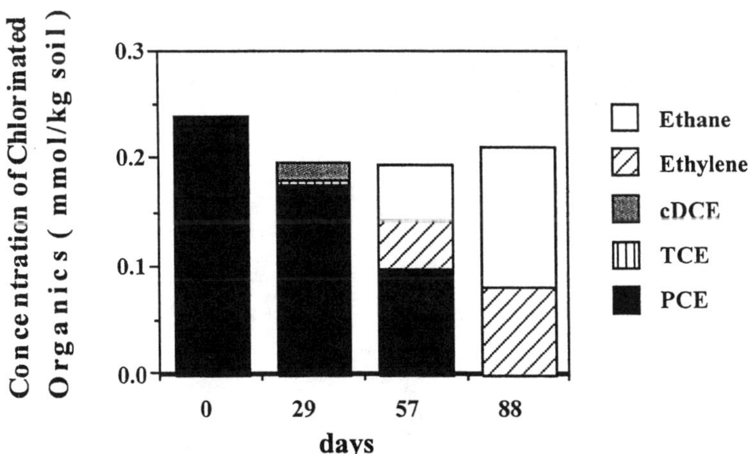

FIGURE 4. Mass Balance of PCE Metabolites Generated in Sealed Container.

demonstrates that the TRM was able to dechlorinate PCE completely. The remaining 11% of PCE metabolite which was not recovered might have leaked out of the system, or have been consumed by microorganisms.

Conclusion The TRM in which the nutrients and trace amount of metal powder are mixed with the PCE-contaminated soil created a strong reductive environment in the soil and led to successive dechlorination of PCE. Unlike the conventional methods such as the soil vapor extraction method and the pump & treat method, this new method is expected to be applied to those fields which need to be cleaned up to an extremely low contamination level and to those which contain clay and silt featuring low gas and water permeability.

REFERENCES

Focht, R., Vogan, J., and O'Hannesin, S. 1996. "Field application of reactive iron walls for In-Situ degradation of volatile organic compounds in groundwater." CCC 1051-5658/96/060381-14 *Remediation* 6:81-93

Maymo-Gatell, X., Chien, Y., Gossett, J. M., Zinder, S. H. 1997. "Isolation of a bacterium that reductively dechlorinates tetrachloroethene to ethene." *Science* 276 :1568-1571

Picardal, F., Arnold, R. G., and Huey, B. B. 1995. "Effects of electron donors and acceptor conditions on reductive dehalogenation of tetrachloromethane by Shewanellaputrefaciens 200." *Appl. Environ. Microbiol.* 61:8-12

PILOT STUDY FOR ENHANCED BIODEGRADATION OF CHLORINATED VOCS

James J. Reid, P.E., ARCADIS Geraghty & Miller, Dublin, Ohio
Denis Balcer, ARCADIS Geraghty & Miller, Dublin, Ohio

ABSTRACT: A field-scale pilot test was conducted which, through the development of more highly reducing subsurface conditions, has led to a significant decrease in the concentration of chlorinated VOCs present in Site groundwater. The objective of the pilot test was to demonstrate that VOC degradation rates could be enhanced through the development of highly reducing conditions within the aquifer. A carbon source was introduced into the aquifer to develop the desired reducing conditions. The monitoring of field parameters, biogeochemical indicator parameters and VOC concentrations proved that highly reducing aquifer conditions were developed which led to increased rates of several biodegradation processes (manganese reduction, iron reduction sulfanogenesis and methanogenesis). These conditions, in turn, resulted in significant reductions in the concentrations of chlorinated VOCs within the pilot test area. The significance of this successful pilot test is that this innovative technology, which has also proven effective at a wide range of other sites, can be cost-effectively tested at most sites and significant results can be documented in a relatively short period of time (generally 3 to 6 months).

SITE BACKGROUND

This project utilized a combination of modeling and bioattenuation assessment techniques to delay implementation of a Record of Decision (ROD) and gain approval for testing the use of an innovative in-situ enhanced reductive dechlorination process at this Superfund Site. A CERCLA RI/FS at an operating manufacturing facility determined that chlorinated volatile organic compounds (VOCs) were present in groundwater. USEPA issued a ROD requiring an air sparging/soil vapor extraction system (AS/SVE) be installed at the source, with downgradient pump and treat for plume control.

Implementation of a design investigation program and pilot testing of the remedial technologies selected in the ROD, disproved the conceptual model developed for the Site during the RI/FS. Due to the heterogeneous geologic conditions present within the source area, implementation of the AS/SVE technology was determined to be inappropriate and concerns were raised over the potential for pump and treat to achieve the remedial action objectives presented in the ROD. Using the revised conceptual model, technical impracticability screening criteria were used to evaluate the potential for this Site to be remediated using conventional treatment technologies. In support of the technical impracticability screening, a flow model and transport model were developed and used to simulate the ROD's pump and treat strategy. Modeled plume configurations for pumping and non-pumping scenarios were simulated using slow and fast degradation rates for tetrachloroethene (PCE), trichloroethene (TCE), cis-1,2-dichloroethene (cis-1,2-DCE) and vinyl chloride. These simulations indicated that for modeled conservative or aggressive degradation rates, pumping and non-pumping scenarios provided the same level of plume attenuation, indicating the need to better understand and ultimately control the rate of biodegradation occurring at the Site.

Approval was negotiated to conduct focused investigation work to characterize the intrinsic bioattenuation at the Site. Three years of supplemental groundwater sampling for VOCs and biogeochemical indicator parameters was used (7 sampling events) to document the presence of and characterize bioactivity zones. These analyses documented, through the development of multiple lines of evidence (presence of daughter products, including ethene and ethane, more highly reducing conditions coincident with the heart of the plume, and changes in the concentration of inorganic constituents which can be used as alternate electron acceptors), that intrinsic bioattenuation was occurring at the Site. However, all parties saw a benefit to the Site if the rate of intrinsic bioattenuation could be increased. Therefore, using the historic groundwater quality data, a pilot test for enhancing the in-situ reductive dechlorination process was designed and implemented.

ENHANCED BIODEGRADATION PILOT TEST

Objective. The objective of the pilot test was to demonstrate that VOC degradation rates could be enhanced through the development of highly reducing conditions within the aquifer. In the early stage of the pilot test (first 3 months), the ability to develop and maintain highly reducing conditions would be evaluated. During later stage of the pilot test (months 3 through 6), the effects of these highly reducing conditions on the chlorinated VOCs would be evaluated

Procedure. To pilot test the ability to modify conditions within the aquifer, and achieve the desired reducing conditions, mobile reactive zones were to be developed in 3 areas along the plume axis (Figure 1). These reactive zones would be developed through the introduction of a carbon source which would be mobile as the solution disperses through the aquifer with the flow of groundwater.

To facilitate implementation of the pilot test, five existing monitoring wells, a pump test well and three below grade lateral pipes present in the vadose zone above the water table were used for carbon source introduction. The carbon solution mixing and introduction system consisted of a portable mixing vessel (approximately 200 gallon skid mounted plastic tank) for preparation of the carbon/water solution and a double diaphragm air-driven pump with manual valves, tube and fittings for connection to the introduction points. This portable system could be moved throughout the Site with a forklift or pick-up truck. Batch introductions of the carbon source mixture occurred on a weekly basis. The volume and strength of the solution was either 50 gallons of a 25 to 1 (water to carbon) solution for the introduction wells or 400 gallons of a 50 to 1 solution for the lateral pipes. A total of 52,570 gallons of the carbon solution was added to the aquifer during the 6 month pilot test.

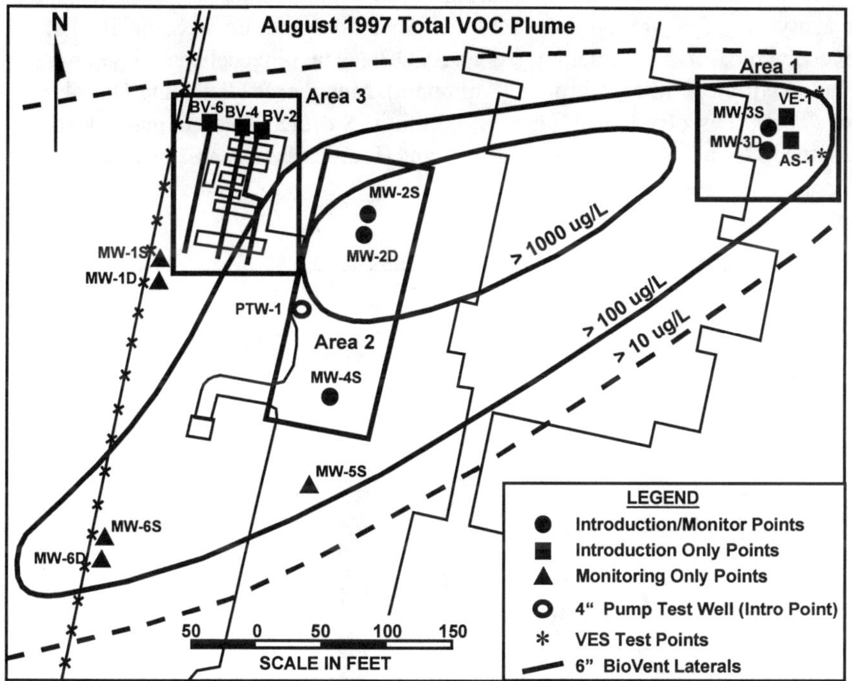

FIGURE 1. Pilot Test Introduction and Monitoring Network Showing August 1997 Pre-Pilot Test Total VOC Plume.

Field Monitoring. The pilot test was monitored frequently using field parameters to ensure reducing conditions were being developed and maintained. The field parameters included oxidation/reduction potential (ORP), dissolved oxygen, temperature, and pH. The field parameters were measured prior to each carbon solution introduction event by using a low-flow sampling procedure to remove groundwater from the monitor wells and then directing this groundwater into a flow-through cell. This technique was used to ensure the parameters measured were representative of formation conditions and to minimize the potential for the water to be exposed to atmospheric conditions which may alter the measurements. Results of the field monitoring during the initial two months indicated that the carbon solution introduction was effective in developing more highly reducing conditions within the pilot test area (Figure 2).

Groundwater Analytical Results. Groundwater sampling and analysis for biogeochemical indicator parameters and VOCs was conducted after 3 months and 6 months of carbon solution introduction. Groundwater samples were collected using a low flow sampling procedure (dedicated bladder pump installed at each monitor well). The groundwater was again directed into a flow through cell and field parameters monitored until stabilization occurred. This stabilization of field parameters was used to ensure that formational water was being sampled for analysis. The groundwater samples were analyzed for a complete list of biogeochemical indicator parameters and VOCs. The biogeochemical parameters include: Nitrate, Nitrite, Nitrogen (Ammonia), Manganese (Total and Dissolved), Iron (Total, Dissolved, and Ferrous), Sulfate, Sulfide, Total Organic Carbon, Dissolved Organic Carbon, Chlorides, Light Hydrocarbon Scan (ethane, ethene), Permanent Gases, (carbon dioxide, oxygen, nitrogen, methane and carbon monoxide).

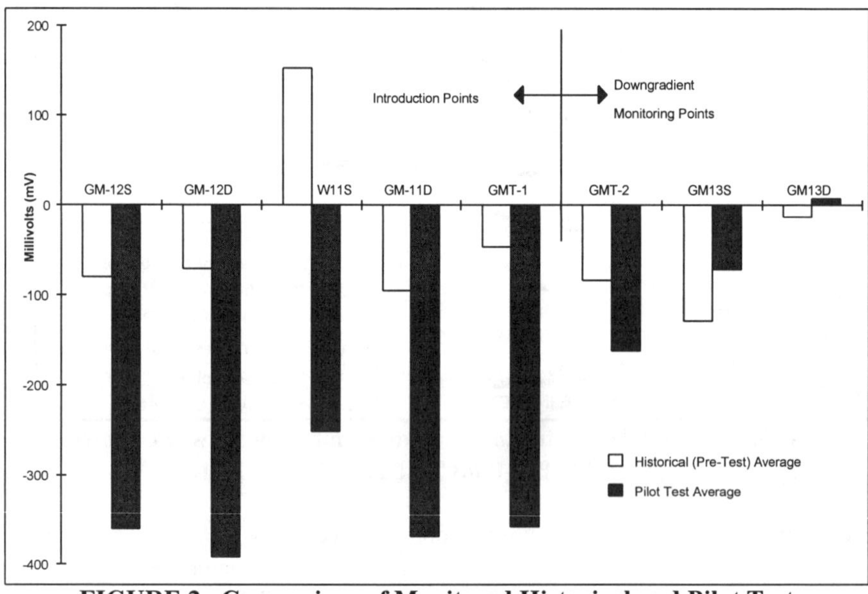

FIGURE 2. Comparison of Monitored Historical and Pilot Test Oxidation/Reduction Potential.

Significant increases in several biogeochemical indicator parameters were noted in these 3 and 6 month groundwater samples. A dramatic increase in the concentration of dissolved manganese, dissolved iron and sulfide was noted during the pilot test sampling events when compared to historical Site data (Figure 3). These parameters provide good indication that increased anaerobic and reducing conditions have been developed within the aquifer during the pilot test. Elevated concentrations of carbon dioxide and methane were also documented during the pilot test sampling events (Figure 3). Carbon dioxide represents both the ultimate degradation product of the chlorinated VOCs present at the Site and an indicator of increased biological activity. Further evidence of the strong reducing conditions developed during the pilot test is the increased presence of methane. The formation of methane is likely due to methanogenesis whereby the carbon dioxide present is being used as an alternate electron acceptor under extreme reducing conditions. It should be noted that an increase in carbon dioxide and methane has also been identified in downgradient (non-introduction) monitor wells, providing evidence that the effects of the carbon solution are migrating downgradient away from the introduction points.

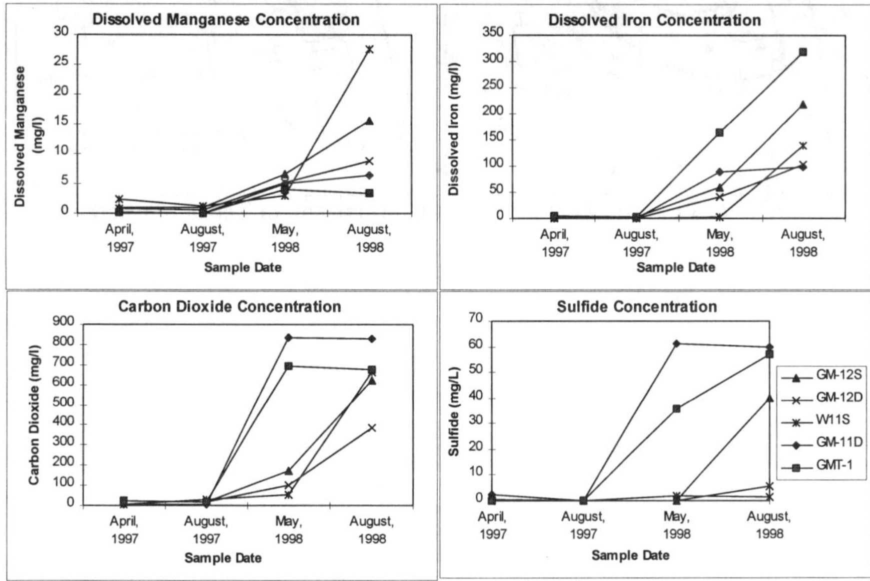

FIGURE 3. Pre- and Post-Pilot Test Biogeochemical Indicator Parameter Analytical Results

This increase in the rate of biodegradation processes led to significant VOC reductions during the pilot test. Some notable changes in VOC concentrations include:

- PCE reduced from an average concentration of 1,400 µg/L and 610 µg/L to 3 µg/L in each of the most highly impacted Site wells (Figure 4);
- Total VOC concentrations reduced from an average concentration of 2,335 µg/L to 63.1 µg/L in the most highly impacted well; and
- PCE, TCE, cis-1,2-DCE and vinyl chloride were removed from a monitor well which historically contained an average of 140 µg/L.

CONCLUSION

The rate of VOC degradation can be increased by using a carbon source introduced into the groundwater to develop more strongly reducing conditions. The significance of this successful pilot test is that this innovative in-situ technology, which has also proven effective at a wide range of other sites, can be cost-effectively tested at most sites and significant results can be documented in a relatively short period of time (generally 3 to 6 months).

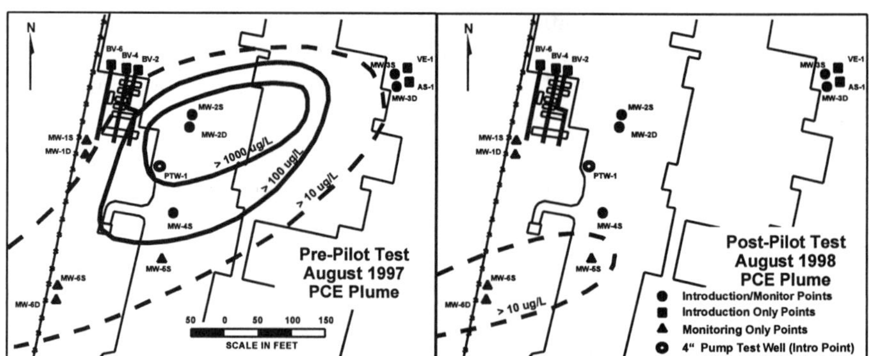

FIGURE 4. Pre- and Post-Pilot Test Comparison of Tetrachloroethene Concentrations.

BIOREMEDIATION OF CHLORINATED SOLVENTS IN PEAT AND NATURAL ATTENUATION OF PLUME

H.Slenders, T. Bosma, J. Gerritse, A. Borger and R. Baartmans (TNO Institute of Environmental Sciences, Energy Research and Process Innovation, The Netherlands), S.Hofstra (IWACO, Environmental Consultants, The Netherlands), H.de Sain. (BdS, Management Consultancy, The Netherlands), R. Hetterschijt and C. te Stroet (Netherlands Institute of Applied Geoscience TNO), and H. de Kreuk (BioSoil R&D)

ABSTRACT: The site of an industrial cleaning company in Arnhem, the Netherlands is contaminated with volatile chlorinated aliphatic hydrocarbons (CAH). The contamination can be divided into a clearly identifiable source-zone in a sandwich layer of clay-peat-clay, and a plume with mainly dichloroethylene (DCE) and vinylchloride (VC) in the underlying aquifer. The central peat layer is contaminated with perchloroethylene (PCE) upto 60.000 mg/kg. Since also the end product of reductive dechlorination ethene is found, natural degradation has high potential. In a quantitative risk-assessment and a decision analysis the in situ approach showed the most optimal cost-profile.

The plume in the aquifer was investigated for redox conditions, and compounds which fuel the anaerobic dechlorination of CAH. A groundwater model with sequential dechlorination of CAH (RT3D) showed that the plume has reached a steady state. The model indicated that with source removal the plume will attenuate in approximately 15-20 years. Without source removal the natural attenuation is estimated to take longer than 700 years.

The feasibility of an in situ approach with a dense grid of injection screens in the center of the sandwich clay-peat-clay was further investigated in an extensive project preparation program with laboratory experiments and a pilot test. About 70 anaerobic batch experiments have been executed with different electron donors and different soil types from the site. Batches with compost leachate or yeast extract showed dehalogenation and degradation until ethane within 3 months. In situ tracer-tests with bromide and compost leachate showed that the peat layer was sufficiently permeable for infiltration of substrate. This was confirmed in pumping tests (0.1 m/d). The results of the batch experiments and pilot-test are now being translated into a full-scale design.

INTRODUCTION

At the PCW-site in Arnhem the industrial cleaning company used perchloroethylene (PCE) and trichloroethylene (TCE) as solvents for the cleansing of overalls and cleaning rags. From 1979 wastewater was transported to the sewer system via a wastewater duct in the floor. Particularly the soil underneath this duct has been contaminated. Other sources, such as leakage's of the sewers in the public road, have caused additional contamination.

Within the Dutch research program NOBIS groundwater at seven chemical laundries was investigated upon its potential for natural attenuation. The site in Arnhem proved to be most promising. As a result the site owner decided to investigate the feasibility of remedial options including natural attenuation. The following preparation steps were taken:

1. Site characterization;
2. Multi component modeling of the aquifer;
3. Drafting in situ option and quantitative risk analysis;
4. Stimulation of dechlorination in batch cultures;
5. Mechanical field testing.

SITE AND SITE CHARACTERIZATION

The site is located at the industrial area "'t Broek" in Arnhem, between the rivers IJssel and Rhine, which drain the area. The regional groundwater flow originates from rainfall which infiltrates in an ice pushed ridge at the North of the area. In the forties the area was developed by applying a sand layer of 2meter. The top 5 m of the soil profile were closely examined with continuous core samples (Begeman-drillings). The soil profile can be schematized as follows:

0,0 - 2,0 m-gl	antropogenic sand layer;
2,0 - 3,0 m-gl	dense clay layer;
3,0 - 4,0 m-gl	peat layer with wood remains;
4,0 - 5,0 m-gl	clay layer with high peat content;
5,0 - 35 m-gl	aquifer, gravel and coarse sand;
35 - 45 m-gl	first aquitard, clay and fine sands.

The groundwater level is found at 1,5 m-gl. The groundwater flow in the antropogenic layer is towards surface ditches. The regional flow is in a southerly direction. The maximum concentrations of contaminants in the soil profile are given table 1 (de Vos, 1998).

Table 1: Maximum concentrations in soil profile (mg/kg dw)

	PCE	TCE	Cis-DCE	VC
Sand	74	28	18	0,3
Clay	1,400	25	130	79
Peat	55,000	6,200	9,100	350
Clay	6,200	520	150	15
Sand	300	14	17	12

Groundwater samples of the peat layer showed maximum concentrations PCE, TCE and DCE at levels exceeding the solubility of these products. Maximum concentrations in groundwater of the aquifer were found upto 10,000 μg/l (DCE) and 2,000 μg/l (VC). The plume in the 1st aquifer has a length of about 600 m. A schematic cross-section of the facility and the groundwater contamination is drafted in Figure 1.

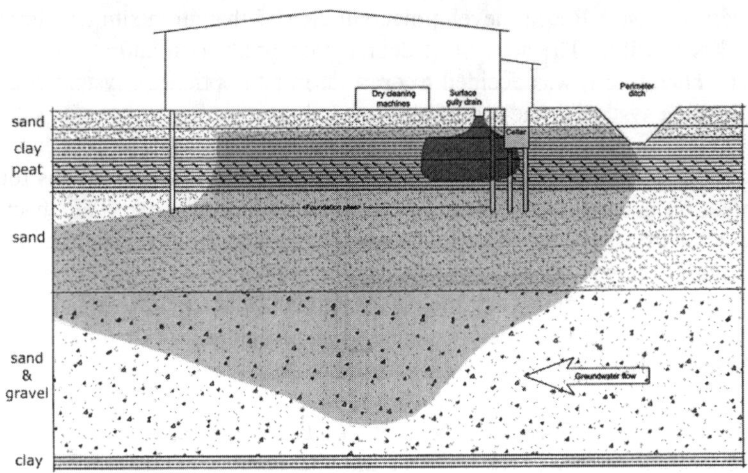

Figure 1: Schematic cross-section in north-south direction of laundry and contamination

MULTI COMPONENT MODELLING AQUIFER

The Netherlands Institute of Applied Geoscience TNO modeled the behavior of the contaminants with a multi component transport model in the code RT3D in order to predict the effectivity and duration of several remediation options; an extensive pump and treat, a hydro-geological containment system and two natural attenuation options (with and without source zone). In a previous stage a one dimensional sensitivity analysis on the degradation rates based on field data showed that degradation rates of 0.15-0.2/d for PCE, 0.021-0.003/d for TCE, 0.0004-0.0008/d for DCE and 0.004-0.008/d for VC can be expected. Calculations starting from the occurrence of the spill till now showed that the plumes have become stable.

The one dimensional approach was extended to a three dimensional modeling. Degradation rates were corrected for pore water velocities and organic carbon content. This also showed that if natural attenuation proceeds at the present rates, the plumes are in a steady state situation which means that the risk of the contaminants reaching existing groundwater extraction's are small. Remediation by natural attenuation will only be effective if the source of contamination is completely removed or contained. It will take 15 years to remediate the aquifer. The use of a pump-and-treat system will not lead to a reduction of this time period. Partial removal of the source will cause a prolonged remediation time which can be up to several decades depending on the effectiveness of the source removal. Enhanced degradation can possibly reduce remediation time to less than 3 years, but is not a suitable option due to the extreme high costs of the injection screens needed.

DRAFTING IN SITU OPTION AND QUANTITATIVE RISK ANALYSIS

With an adequate removal or isolation of the source, natural attenuation is a cost-effective option. In previous stages rough detailing and costs for isolation and excavation were assessed. Recent developments indicated that an in situ dechlorination of PCE is possible. The amount of degradation products found at this site was very high. Therefore it was decided to draft an in situ option. A system consisting of a flushing system in the top sand layer and an injection system for substrate in the center of the clay-peat-clay sandwich was suggested. The peat layer seemed permeable and contained the center of contamination. Various options for source remediation and natural attenuation were combined, and their fall back scenarios were discussed and calculated (Hofstra et al, 1998). Both excavation and isolation alternatives have high end risks. If for any reason the excavation proves not to be successful in the future, additional measures implicate a disturbance of industrial processes and are extremely expensive. The feasibility of the in situ option was further investigated.

STIMULATION OF DEGRADATION IN BATCH CULTURES

Due to microbial processes the PCE at the site has been transformed into mainly *cis*-DCE. In the in situ option complete reductive dechlorination (eventually to the non-toxic ethene) should be obtained. The dechlorination can be enhanced by the addition of an electron donor. In Figure 2 the complete dechlorination of PCE is illustrated.

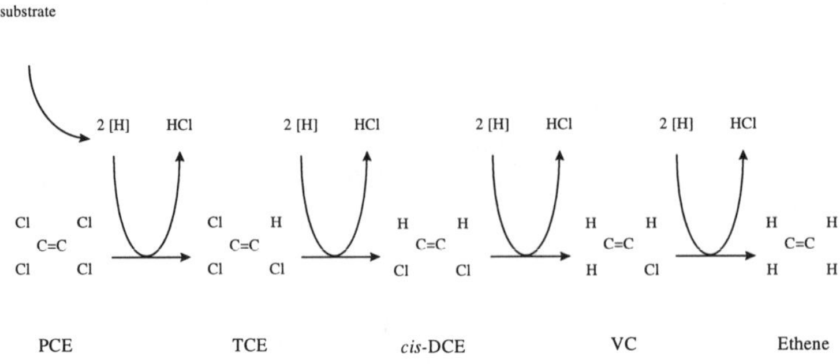

Figure 2 : Complete anaerobic dechlorination of PCE

In order to investigate whether complete dechlorination is possible with soil material from the site, batch experiments were carried out. Soil samples from the different layers were transferred in bottles, together with anaerobic ground water and different substrates (compost percolate, compost percolate with methanol, and yeast extract). Anaerobic conditions were maintained in the bottles, and the dechlorination was followed in time. In many batches a complete dechlorination of PCE to ethene and/or ethane was observed. In most of the highly contaminated batches ethene was the final product after 60-80 days, whereas in the slightly contaminated batches also ethane production was found. Stimulation with various carbon substrates contributed to shorter lag phases and half lives for PCE and its dechlorination products. Of the substrates tested, yeast extract showed the best results (i.e. Figure 3). Complete dechlorination was found in the peat and clay containing batches.

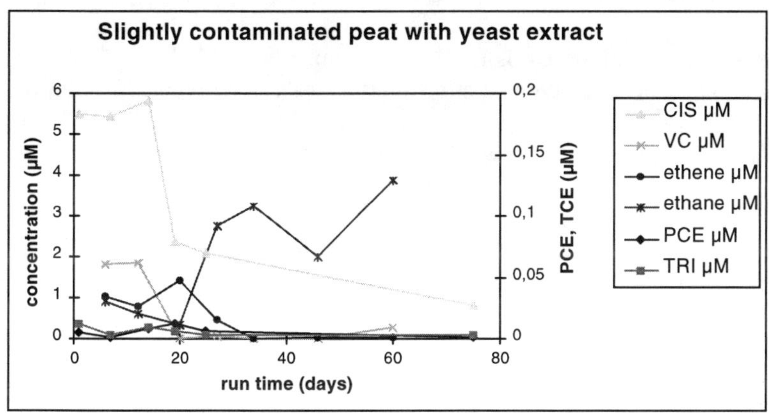

Figure 3: Concentrations of Chlorinated Alifatic Hydrocarbons in a slightly contaminated peat batch.

MECHANICAL FIELD TESTING

The possibilities for introducing the required electron donor in the peat at the site was investigated. Around two wells five monitoring wells were placed in the peat at distances of 0.3, 0.6, 0.9, 1.2 and 1.5 m. A tracer solution containing bromide as a conservative tracer and a mixture of natural organic compounds (volatile fatty acids) as the electron donor were introduced in the wells. The tracer solution could be introduced at a rate of about 2 m^3 per m^2 screen surface per day. In one well bromide showed a cylindrical pattern. In the other well some heterogeneity was observed. These experiments showed that the permeability of the peat is sufficient for performing in situ treatment of the soil. At an infiltration rate of 30 liter/hr, the fatty acids were transported over 1.5 m within 3 weeks.

CONCLUSIONS

Natural attenuation of the groundwater plume with DCE and VC at the site in Arnhem is possible within 10-20 years when the source of the contamination is removed or contained. The results of laboratory experiments and mechanical field tests show that complete dehalogenation is possible in these sediments, and that the substrate needed can be transported in situ. Although some uncertainties regarding free phase product and the realization of a uniform distribution of substrate in the subsoil remain, the results are thus promising that a full scale remediation is being prepared.

REFERENCES

Vos, I. de, “Aanvullend onderzoek P. Calandweg 2 te Arnhem”, IWACO-report 3366760, September 1998

Borger, A.R. Gerritse, J., Baartmans, R.F.W., Slenders, H., Bosma, T.N.P., 1998. "Stimulation of PCE dechlorination in NEPROMA batch cultures". TNO-MEP REPORT R98/354.

Hetterschijt, Rolf, Stroet, Chris te, Bosma, Tom. 1998. "Model study on effectivity and duration of several remediation alternatives". TNO-report NITG98-82-C.

Hofstra, S.T., Slenders, H. 1998 "Remedial action plan P.Calandweg Arnhem". IWACO-report 3368950.

Kreuk, H. de, 1998. "Field test for determining the treatability of a PCE and TCE contamination in a peat soil matrix". BIOSOIL report 6743.007.

ENHANCED REDUCTIVE DECHLORINATION OF TCE IN A BASALT AQUIFER

Kent S. Sorenson, Jr., Lockheed Martin Idaho, Idaho Falls, Idaho
Lance N. Peterson, Lockheed Martin Idaho, Idaho Falls, Idaho
Roger L. Ely, University of Idaho, Moscow, Idaho

ABSTRACT: A field evaluation of enhanced reductive dechlorination of trichloroethene (TCE) in ground water has been in progress since November 1998 to determine whether in situ biodegradation can be significantly enhanced through the addition of an electron donor (lactate). An in situ treatment cell was established in the residual source area of a large TCE plume in a fractured basalt aquifer utilizing continuous ground water extraction approximately 150 meters downgradient of the injection location. After a 1-month tracer test and baseline sampling period, the pulsed injection of lactate was begun. Ground water samples were collected from 11 sampling points on a biweekly basis and in situ water quality parameters were recorded every 4 hours at two locations. Within 2 weeks after the initial lactate injection, dissolved oxygen and redox potential were observed to decrease substantially at all sampling locations within 40 m of the injection well. Decreases in nitrate and sulfate concentrations were also observed. Both quantitative in situ rate estimation methods and qualitative measures such as changes in redox conditions, decreases in chlorine number, and changes in biomass indicator parameters are being used throughout the test to evaluate the extent to which biodegradation of TCE is enhanced.

INTRODUCTION

Test Area North (TAN) at the Idaho National Engineering and Environmental Laboratory is the site of a nearly 2-mile long trichloroethene (TCE) plume resulting from the injection of liquid waste directly into the fractured basalt of the Snake River Plain Aquifer during the 1960s. While most of the plume has TCE concentrations less than 1 mg/L, a high concentration core exists in the immediate vicinity of the former injection well (where a residual source of TCE remains) with ground water concentrations ranging from 1.5 mg/L to over 300 mg/L. Co-disposed organic materials have provided an electron source to drive limited intrinsic reductive dechlorination in the vicinity of the former injection well (Sorenson et al., 1999). The complete reductive dechlorination pathway for chloroethenes is as follows: PCE → TCE → DCE → VC → ethene (Freedman and Gossett, 1989). In each step of the process the compound is reduced through substitution of a chlorine atom by a hydrogen atom.

The evidence for intrinsic reductive dechlorination at TAN is derived largely from 9 years of ground water monitoring data. These data include significant concentrations of DCE, with the *cis*-DCE isomer predominating. Low levels of vinyl chloride, ethene, and ethane have also been observed in the former injection well. Dissolved oxygen was found to be less than 1 mg/L for a distance of about 150 m downgradient. Nitrate concentrations in the immediate vicinity of the former injection well were depressed and low levels of methane have been detected. While localized areas of strongly reducing (methanogenic) conditions seemed to exist, the bulk redox conditions in the high concentration core appeared to range from nitrate-reducing to sulfate-reducing. In addition to ground water monitoring data, the potential for complete dechlorination of TCE at TAN was demonstrated through fed-batch bioreactor studies using basalt core collected from the plume as the inoculum (Sorenson, 1998). Rapid TCE disappearance accompanied by ethene and ethane generation was observed in bioreactors fed lactate. Accumulation of ethene and ethane was not observed in the controls.

Based on the preliminary field and laboratory data, a field evaluation of enhanced reductive dechlorination was initiated in the fall of 1998. The primary objective of the evaluation is to determine whether in situ TCE degradation can be significantly enhanced through the addition of an electron donor. It is believed that the addition of a suitable electron donor will foster the growth of the indigenous microbes, driving the aquifer to a more reducing state throughout the treatment cell. This has the potential to increase the rate and extent of reductive dechlorination where it is already occurring, and to stimulate the process over a larger area.

EXPERIMENTAL DESIGN AND OPERATION

The field evaluation of enhanced reductive dechlorination of TCE at TAN entails the periodic injection of high concentrations of an electron donor solution into Well TSF-05 (the former waste injection well) with ground water monitoring throughout the treatment cell. In order to exert some control on the distribution and residence time of the electron donor in the subsurface, a hydraulic gradient was induced through pumping. The test was designed to create a 150-m long treatment cell in the high concentration core between Wells TSF-05 and TAN-29, a downgradient extraction well (Figure 1). The induced flow conditions are likely to cause a corresponding change in the distribution of contaminants and other groundwater solutes and parameters of interest independent of electron donor addition. For this reason historical groundwater sampling results cannot be used as the basis for comparison when evaluating the effect of electron donor addition.

Start-up Period. In order to separate the effect of the new flow conditions on contaminant distributions from the effect of electron donor addition, the pumping and injection system was operated and monitored for almost two months prior to lactate addition. This allowed a new baseline to be established for contaminant and other groundwater solutes and parameters of interest. Chloroethenes, competing electron acceptors, redox potential, temperature, pH, conductivity, and nutrients were all measured approximately weekly during the start-up period.

The start-up period was used not only to establish the baseline for relevant parameter distributions, but also to establish the baseline for flow and transport in the aquifer under the conditions of the field evaluation. This was accomplished by adding a conservative tracer (bromide) to the injection line at TSF-05 at the beginning of the start-up period. The objective was to determine the groundwater

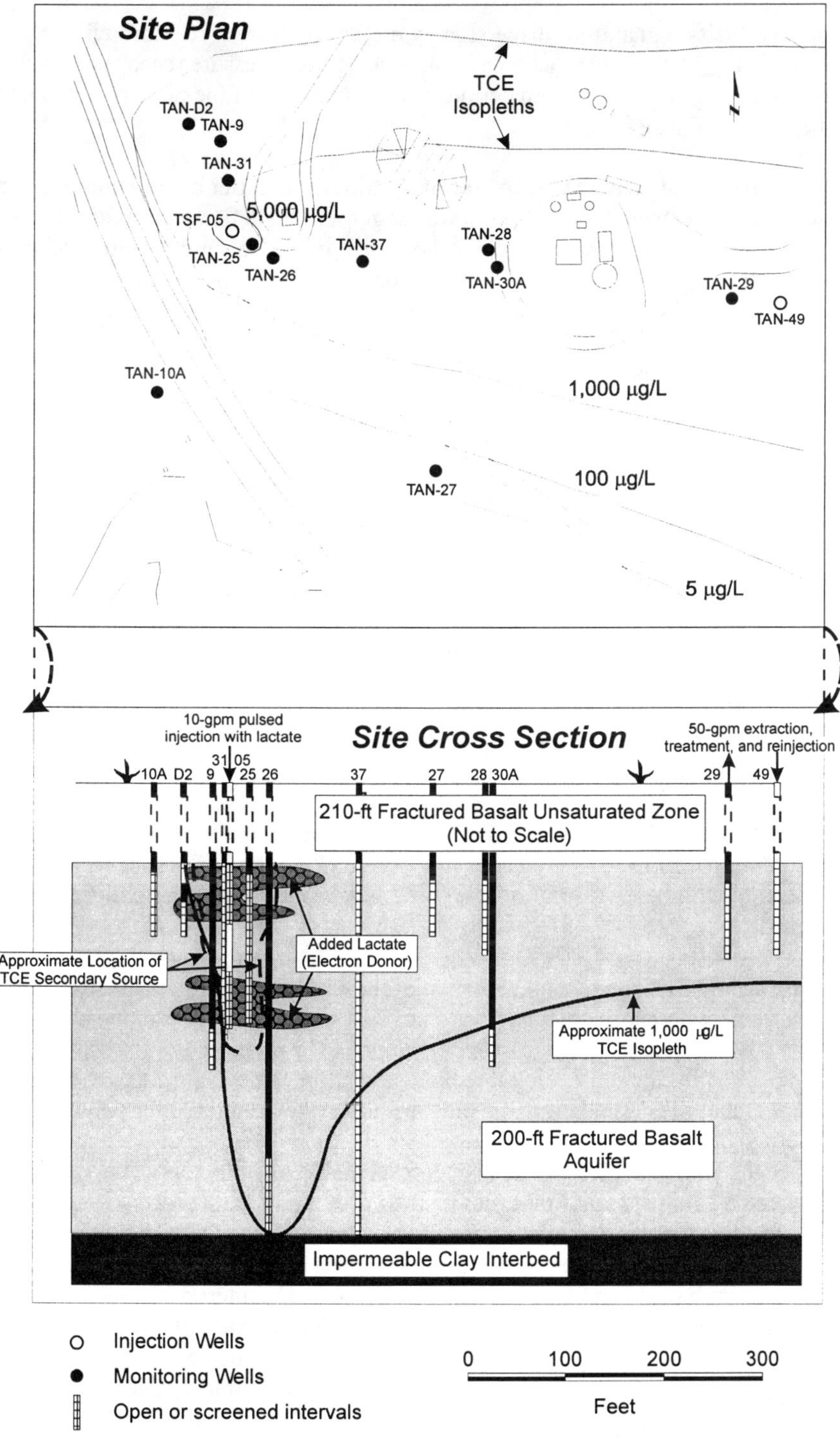

Figure 1. Layout of the TAN treatment cell.

flow velocity and aquifer dispersion by measuring tracer breakthrough at the monitoring wells in the treatment cell. The aquifer pressure response was also measured during the start-up period to aid calibration of the numerical model to be used for data evaluation.

Enhanced Reductive Dechlorination Evaluation. After completion of the start-up period, electron donor addition was begun. In the laboratory studies using enrichment cultures from TAN, lactate was found to be a much better electron donor than glucose or methanol. Based on this information and on its success in other studies (e.g., DeBruin et al., 1992; Gibson and Sewell, 1992; Fennel et al., 1997), lactate was chosen as the electron donor. The design average in situ lactate concentration was 200 mg/L based on the results of the laboratory bioreactor studies. The lactate was injected into Well TSF-05 as a 60% sodium lactate solution on a weekly basis. Pulsing should aid with in situ mixing and the high concentrations should help to prevent biofouling of the injection well because they will inhibit microbial growth. The initial pulsing frequency and concentration can be modified as necessary to improve the distribution of lactate in situ.

The monitoring network for the enhanced ISB field evaluation at TAN is shown in Figure 1. Well TAN-37 is being sampled at two depths in order to take advantage of its open-hole completion. All of the wells are being sampled using dedicated, low flow, submersible pumps. With the exception of Wells TAN-10A and TAN-27 (which are sampled monthly) all of the wells are being sampled biweekly. The parameters being monitored and their significance are given in Table 1. In addition to ground water sampling, in situ sondes are being used to collect dissolved oxygen, redox potential, specific conductivity, pH, and temperature in Wells TAN-37 and TAN-31. These were included in the monitoring design in an attempt to measure the temporal effects of lactate injection with better resolution than biweekly ground water sampling would allow.

Data Analysis. The overall objective of the enhanced ISB field evaluation is to determine whether the biodegradation of TCE through ARD can be enhanced through addition of an electron donor supply. The quantitative criterion established for demonstrating that biodegradation can be significantly enhanced at TAN is that the degradation rate must be observed to increase for two consecutive quarters of ground water monitoring with the final rate equaling or exceeding twice the baseline rate. As data become available, the biodegradation rate will be estimated using several first-order methods. A numerical model will also be utilized to estimate the rates through inverse modeling.

In addition to the quantitative evaluation criterion, many other qualitative parameters have been identified which will aid in the interpretation of the field evaluation results. In particular the time variations of the following parameters are expected to contribute to the overall analysis: chloroethene concentrations, electron donor concentrations, reaction product concentrations, electron acceptor

Table 1. Parameters monitored during the field evaluation.

Parameter	Data Need(s)	Analytical Method	Lab[1]	Precision
Bromide	Groundwater velocity, aquifer dispersivity	Ion-specific electrode	Field	±5 %
Water level changes	Aquifer hydraulic response	Pressure transducer and data logger	Field	±0.02 ft
Lactate	Time-varying electron donor concentrations	Ion chromatography	Fixed	±10 %
Acetate/propionate/ butyrate	Initial and time-varying reaction product concentrations	Gas chromatography/ flame ionization detector	Fixed	±10 %
Chloroethenes	Initial and time-varying chloroethene concentrations	Gas chromatography/ electron capture detector	Fixed	±10 %
Chloride	Initial and time-varying reaction product concentrations	Colorimetric	Field	±1%
Ethene/ethane/ methane	Initial and time-varying reaction product concentrations	Gas chromatography/ flame ionization detector	Fixed	±10 %
Temperature/pH/ conductivity	Initial and time-varying reaction product concentrations	Hydrolab probe in flow-through cell	Field	±.5% ±3 % ±1 %
Tritium	Initial and time-varying tritium concentrations	Liquid scintillation	Fixed	
Dissolved oxygen	Initial and time-varying electron acceptor concentrations	Hydrolab probe in flow-through cell	Field	±2 %
Nitrate/sulfate/iron	Initial and time-varying electron acceptor concentrations	Colorimetric	Field	±8 % ±1 % ±1 %
Redox potential	Initial and time-varying redox conditions	Hydrolab probe in flow-through cell	Field	±2 %
Carbon dioxide/ alkalinity	Initial and time-varying biomass indicator parameter values	Titrimetric	Field	±1 % ±1 %
Chemical oxygen demand	Initial and time-varying biomass indicator parameter values	Colorimetric	Field	±3 %
Phosphate/ammonia as nitrogen	Initial and time-varying biological nutrient concentrations	Colorimetric	Field	±4 % ±20 %

1. Distinguishes between measurements made in the field and those made in a fixed laboratory.

concentrations, redox potential, biomass indicators, and micronutrient concentrations. Ultimately, performance of the field evaluation will also demonstrate whether the ISB system can be adequately monitored and will provide information that can be used to estimate long-term operations cost.

PRELIMINARY RESULTS

As of January 1999, the start-up period has been completed and three weekly lactate injections have occurred. Tracer and baseline monitoring results are presented and the initial effects of lactate injection in the treatment cell are discussed.

Tracer and Baseline Monitoring. During the start-up period, potable water was continuously injected at about 76 L/min (20 gpm) in Well TSF-05 to enhance the hydraulic gradient through the treatment cell. Based on previous tests, the effective porosity in the aquifer near Well TSF-05 was known to be extremely low (presumably due to years of use as a waste injection well), so rapid tracer breakthrough was anticipated at nearby monitoring wells. This required a high sampling frequency at Wells TAN-25 and TAN-31 in particular.

Tracer was injected for a period of about 30 minutes with an average bromide concentration of 12,000 mg/L. Figure 2 shows the tracer breakthrough curves at several monitoring points. The breakthrough at Wells TAN-25 and TAN-31 was very rapid as expected, with peak arrivals in less than 1 and 2 hours, respectively. A good breakthrough was also obtained at Well TAN-D2 (about 120 ft from the injection point) where the peak arrived in just under 4.5 days. The tracer arrival at Well TAN-37 was less well defined, with a lower, broader concentration peak, although it is only slightly further from the injection well than Well TAN-D2. In any case, it was verified that the tracer reached all of these observation points and arrival times were estimated. Tracer was not observed in TAN-26, presumably because it is completed about 30 m (100 ft) deeper in the aquifer (Figure 1) than Well TSF-05.

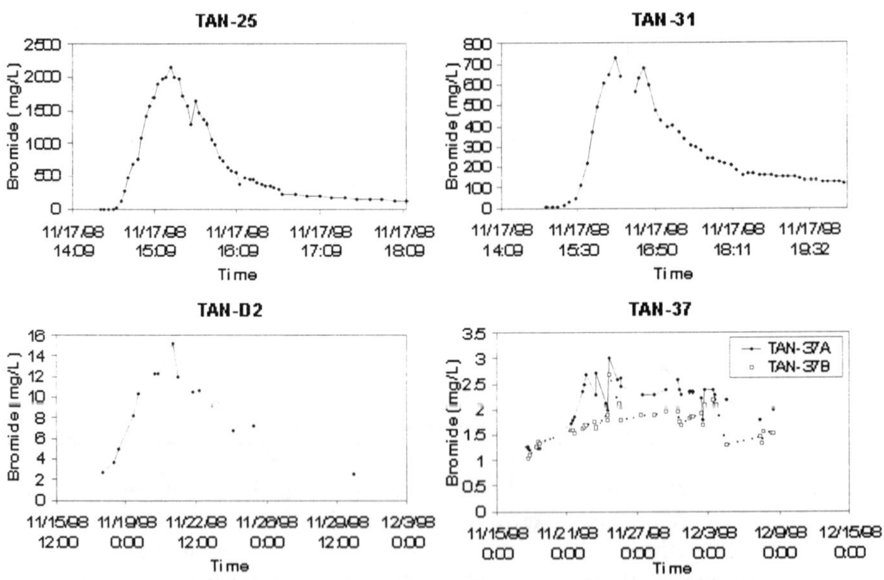

Figure 2. Bromide breakthrough curves at selected monitoring wells.

In addition to monitoring for tracer breakthrough, baseline sampling was performed at all of the wells on a weekly basis throughout the start-up period.

Table 2 provides the average values measured for selected parameters. Well TAN-29 is not included because it is affected by dilution from the water reinjected in Well TAN-49. A blank in the table indicates the parameter was not measured. If a solute was not detected, a value of 0 is entered.

TABLE 2. Average values measured for selected parameters before lactate injection.

Well	DO (mg/L)	ORP (mV)	Nitrate (mg/L)	Sulfate (mg/L)	CO_2 (mg/L)	Lactate (mg/L)	Acetate (mg/L)	Propionate (mg/L)
TAN-10A	1.4	290	1.8	40	41	0	0	0
TAN-25			0.9	34	33	0	0	0
TAN-26	2.6	310	1.5	34	43	0	0	0
TAN-27	0.5	290	1.8	37	51	0	0	0
TAN-28	0.1	250	2.4	43	37	0	0	0
TAN-30A	0.1	260	2.6	41	48	0	0	0
TAN-31	0.5	250	1.0	36	38	0	0	0
TAN-37A	0.8	120	1.4	41	38	0	0	0
TAN-37B	2.0	180	1.9	39	45	0	0	0
TAN-D2	0.2	44	1.5	40		0	0	0

Initial Lactate Injection Effects. One round of ground water sampling was completed in January 1999 following the first two weekly lactate injections. Although a lag period was expected before significant microbial utilization of lactate occurred, significant changes were observed at monitoring wells near the injection well in less than 2 weeks. Table 3 gives the values of the parameters in Table 2 measured just 11 days after the first lactate injection.

TABLE 3. Measured values for selected parameters 11 days after initial lactate injection.

Well	DO (mg/L)	ORP (mV)	Nitrate (mg/L)	Sulfate (mg/L)	CO_2 (mg/L)	Lactate (mg/L)	Acetate (mg/L)	Propionate (mg/L)
TAN-25	0.0	-230	0.2	3	54	2600	42	61
TAN-26	0.0	-150	0.2	34	56	1500	23	31
TAN-28	0.0	250	2.3	39	35	0	0	0
TAN-30A	0.0	260	2.1	41	44	0	0	0
TAN-31	0.0	40	0.1	40	31	62	0	0
TAN-37A	0.2	-42	0.2	32	37	33	0	0
TAN-37B	0.1	-42	1.2	36	40	66	0	0
TAN-D2	0.3	-143	0.1	36	51	0	0	0

Comparison of Tables 2 and 3 reveals that lactate injection had a pronounced effect on several parameters of interest at all of the wells in Table 3 except the two furthest from the injection location, TAN-28 and TAN-30A. The conditions in the ground water have become more strongly reducing as reflected by decreases in dissolved oxygen (DO), oxidation reduction potential (ORP), nitrate, and even sulfate in TAN-25. The carbon dioxide concentrations increased in the two wells with the highest lactate concentrations, probably as a result of increased microbial activity. The presence of acetate and propionate is almost certainly associated with the biodegradation of lactate.

The in situ sonde deployed in Well TAN-37 has also provided some very interesting preliminary data. The sonde is located immediately above the pump associated with sampling location TAN-37A and measures a variety of parameters on a 4-hr interval. Figure 3 shows the ORP and specific conductivity measured in situ for the period of January 18 to January 25, 1999. An inverse correlation in which local maxima for ORP correspond to local minima for specific conductivity and vice versa is readily apparent. This correlation suggests that the combination of these two parameters may serve as a surrogate for lactate. The sodium lactate has been observed to induce a dramatic increase in specific conductivity in Wells TAN-25 and TAN-26 where concentrations are highest. It seems reasonable that the ground water migrating downgradient with the lactate would have a lower ORP than the surrounding water because of microbial utilization of lactate as an electron donor. If this proves to be true, the in situ sonde will provide an excellent means for measuring transport of the electron donor in the subsurface with very fine temporal resolution.

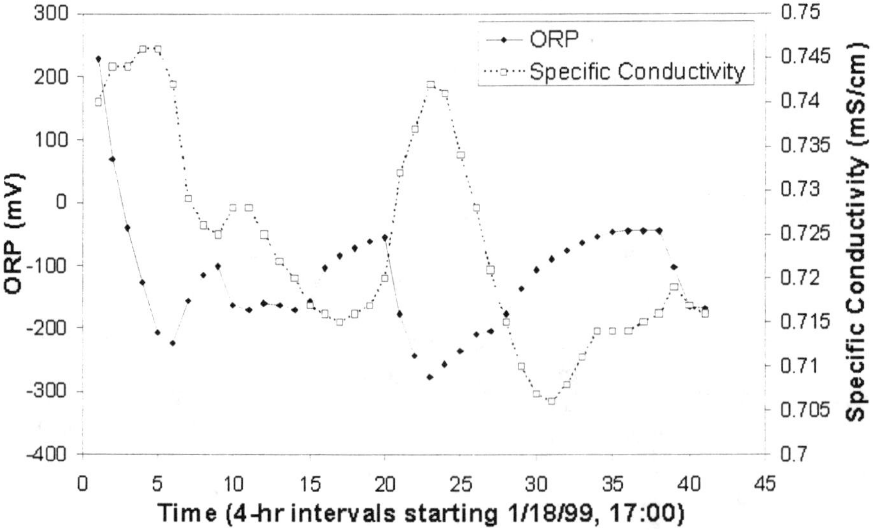

Figure 3. In situ measurements in Well TAN-37 for the period of January 18-25, 1999.

SUMMARY

A field evaluation of enhanced reductive dechlorination of TCE was begun in November 1998 at the Idaho National Engineering and Environmental Laboratory. A 150-m long in situ treatment cell was designed through which a hydraulic gradient was forced by ground water extraction at the downgradient end and intermittent injection at the upgradient end. A 1-month tracer test and baseline monitoring period was completed and revealed relatively good hydraulic communication between the injection well and the monitoring wells. Sodium lactate injection began in January 1999 to provide an electron donor to drive reductive dechlorination. Within 2 weeks after the initial injection, significant changes in several redox parameters indicated that conditions in the aquifer were becoming much more reducing. In situ data collection also revealed that it might be possible to monitor lactate transport in the subsurface through surrogate parameters with very fine temporal resolution.

The preliminary results appear to indicate that electron donor addition in the fractured basalt aquifer over 60 m below land surface at TAN can be used successfully to manipulate the redox conditions of the ground water. Continued monitoring will focus on evaluating the extent to which reductive dechlorination is enhanced in the treatment cell. If this can be accomplished, efforts will be made to refine the operation of the system in terms of the electron donor addition strategy and the hydraulic controls.

ACKNOWLEDGMENTS

This work is being performed under United States Department of Energy Idaho Operations Office Contract DE-AC07-94ID13223. Many thanks are due the project managers: Al Jantz of Lockheed Martin Idaho Technologies Company and Joe Rothermel of Parsons Infrastructure & Technology Group, Inc.

REFERENCES

DeBruin, W. P., M. J. J Kotterman, M. A. Posthumus, G. Schraa, and A. J. B. Zehnder. 1992. "Complete Biological Reductive Transformation of Tetrachloroethene to Ethane." *Applied and Environmental Microbiology. 58*(6): 1996–2000.

Fennel, D. E., J. M. Gossett, and S. H. Zinder. 1997. "Comparison of Butyric Acid, Ethanol, Lactic Acid, and Propionic Acid as Hydrogen Donors for the Reductive Dechlorination of Tetrachloroethene." *Environmental Science and Technology. 31*(3): 918-926.

Freedman, D. L. and J. M. Gosset. 1989. "Biological Reductive Dechlorination of Tetrachloroethylene and Trichloroethylene to Ethylene under Methanogenic Conditions." *Applied Environmental Microbiology. 55*(9): 2144-2151.

Gibson, S. A. and G. W. Sewell. 1992. "Stimulation of Reductive Dechlorination of Tetrachloroethene in Anaerobic Aquifer Microcosms by Addition of Short-Chain Organic Acids or Alcohols." *Applied and Environmental Microbiology. 58*(4): 1392–1393.

Sorenson, K. S. 1998. "Design of a Field-Scale Enhanced In Situ Bioremediation Evaluation for Trichloroethene in Ground Water at the Idaho National Engineering and Environmental Laboratory." Presented at Engineering Biological Processes for Environmental Enhancement, September 1998. Paper No. PNW98-113. ASAE, 2950 Niles Rd., St. Joseph, MI 49085-9659.

Sorenson, K. S., P. Martian, L. N. Peterson, R. L. Ely, R. E. Hinchee. 1999. "An Evaluation of Anaerobic and Aerobic Natural Attenuation of Trichloroethene." *Bioremediation Journal.* Submitted for publication.

FULL-SCALE BIOREMEDIATION AT A CHLORINATED SOLVENTS SITE: PROJECT UPDATE

***Susan Tighe Litherland*, P.E. (Roy F. Weston, Inc., Austin, Texas)**
David W. Anderson, P.E., P.G. (Roy F. Weston, Inc., Austin, Texas)
Blake A. Dinwiddie (Roy F. Weston, Inc., Houston, Texas)

ABSTRACT: A full-scale in situ biodegradation system was designed to increase the degradation rate of trichloroethene (TCE) at this Texas Gulf Coast property. The remediation system consists of alternating extraction and injection trenches. Through the trenches, groundwater is extracted, amended with methanol, and reinjected. Over a 40-month treatment period, the TCE concentrations have been reduced within the treatment area almost 90% (from an average of approximately 22 mg/L to approximately 2.6 mg/L). TCE concentrations in one potential "source" area have remained elevated, but this source is now being removed. When this area is excluded from the evaluation, the TCE decrease for the treatment period is in excess of 99% (from an average of approximately 12 mg/L to an average of 0.12 mg/L). Even with less than ideal conditions, concentrations of TCE have been reduced in some areas to at or near the detection limit. Methanol is currently being added at an approximate dosage of 500 mg/L. On completion of the active remediation, which is anticipated by the summer of 1999, post-remediation monitoring will be performed for approximately 2 to 4 years to verify that natural attenuation will adequately control future movement. This work is being performed under the Texas Natural Resource Conservation Commission's Voluntary Cleanup Program.

INTRODUCTION

Groundwater trichloroethene (TCE) concentrations at a facility historically used for manufacturing were observed to decrease over an extended (7+ year) monitoring period. Based on observations of the decrease of the TCE and subsequent increases and decreases of the degradation products, it appeared that the decrease was due to intrinsic dechlorination of the TCE. A program was initiated to increase the degradation rate and "stabilize" the chlorinated organic plume. The primary objective of the work performed was to actively remediate the plume to a point that natural attenuation could prevent future migration of the plume. This paper presents a description of the site and conditions prior to initiation of active remediation; and the design, implementation, and operation of a full-scale system to increase the degradation rates of the chlorinated organic compounds present by stimulating the naturally occurring biological activity. The full-scale system was installed during the summer of 1995 and operation was initiated in September 1995.

BACKGROUND INFORMATION

Site History. Manufacturing activities at this Texas Gulf Coast site occurred between approximately 1952 and 1985. One potential "source" area appeared to have been used for disposal of debris and potentially TCE. The soils in this potential "source" area were excavated in the fall of 1998.

Site Description. The area of affected groundwater is approximately 600 feet by 700 feet in an unconsolidated water-bearing zone which occurs at a depth of between approximately 12 and 20 feet below ground surface (bgs).

Monitoring Data. Due to extended litigation, periodic groundwater sampling was performed between the time of discovery of TCE in the groundwater (around 1986) and the installation of the remediation system (1995). Based on the data collected during this period, it appeared that the average concentrations of TCE had decreased from approximately 50 mg/L to approximately 22 mg/L. The data also suggested, although not conclusively, that the rate of TCE decrease was slowing. For further information regarding the site history, description, and monitoring data, see the previous proceedings paper in the 1997 *In Situ and On-Site Bioremediation: Volume 5* (Litherland and Anderson, 1997).

FULL-SCALE DESIGN AND INSTALLATION

Performance Objectives. Based on both information that suggested that the site was a good candidate for in-situ anaerobic degradation and the lack of other promising options, a full-scale system was designed. The main objectives of this technology were to accelerate the naturally occurring processes and to stabilize the site so that continued operation of a remediation system would not be needed in perpetuity. The "end" point of the remediation is assumed to be the point at which the rate of degradation slows so that continued operation of the system is not practical and in a plume that can be adequately controlled in the future by natural attenuation, so that long-term operations and/or monitoring will not be required.

Design and Installation. Other objectives of the design included circulation of one pore volume every 6 to 9 months, accommodation of property redevelopment during remediation, completion of the active remediation period within 3 to 5 years, adequate control of the plume during the treatment period, and flexibility so that other technologies could be applied. To accomplish these objectives, a series of injection and extraction trenches were designed for installation throughout the area of affected groundwater. For further information regarding the trench design and installation, refer to the 1997 Litherland and Anderson proceedings paper. Figure 1 illustrates typical injection and extraction trenches at the site, and Figure 2 illustrates the system layout, approximate extent of the plume, and groundwater flow direction.

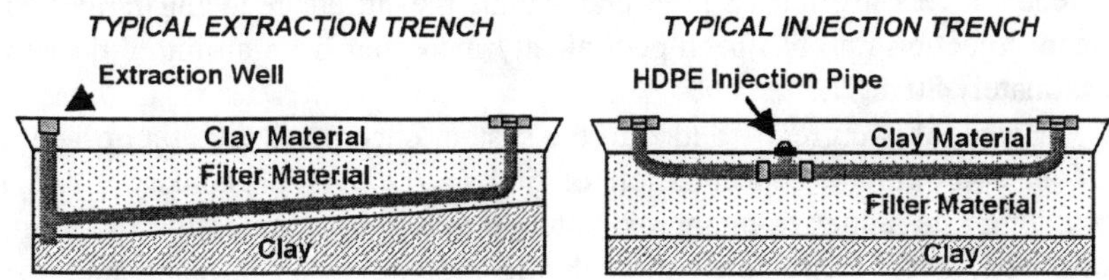

FIGURE 1 – Typical Extraction and Injection Trenches

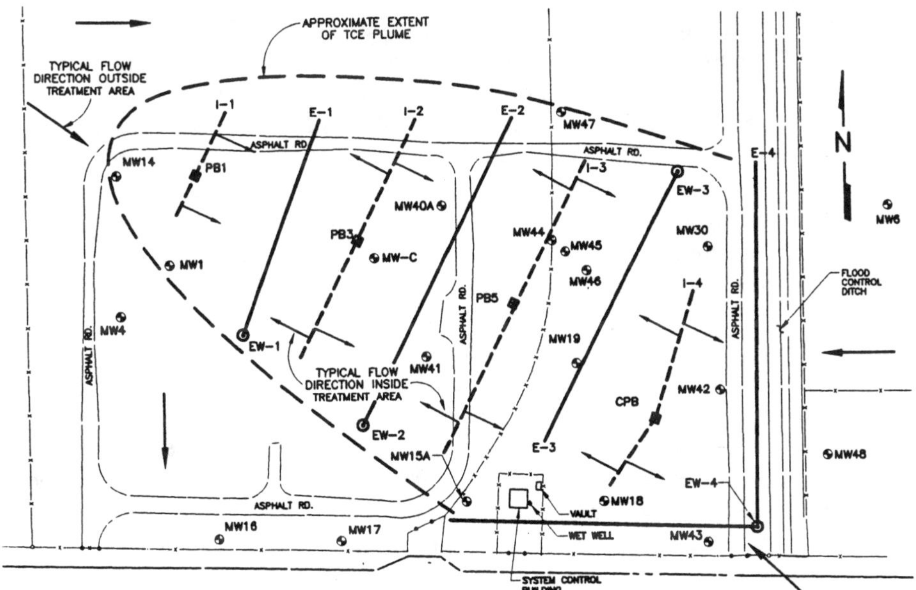

FIGURE 2 – System Layout, Approximate Extent of Plume, and Groundwater Flow Direction

SYSTEM OPERATION

System operation began in August 1995 with circulation of the groundwater to evaluate the hydraulics of the system prior to introduction of nutrients and/or substrate. A circulation rate of approximately 12 gallons per minute (gpm) was achieved. Introduction of nitrogen and phosphate began in September 1995. Although it appeared that there were initial decreases in TCE concentrations with the nutrients alone, a greater degradation rate was desired. Introduction of methanol, along with nutrient addition, began in June 1996 in an attempt to further increase the degradation rate. The purpose of the methanol was both to serve as a primary substrate and to further reduce the dissolved oxygen (DO) and oxidation-reduction (redox) potential to levels that were thought to be more favorable to anaerobic degradation of TCE. Nutrient addition was discontinued during May 1997 since injected nitrates could potentially stimulate non-dehalogenating bacteria rather than the desired dehalogenating bacteria, and keep redox levels artificially high and within the nitrate reducing range. The methanol injection rate has been periodically increased from an initial dosage of approximately 80 mg/L.

Methanol is currently added to the system once per week at a dosage that yields an approximate concentration of 500 mg/L. As of January 1999, the recirculating rate ranges from approximately 6 to 8 gpm, and a total of 12 million gallons have been recirculated through the system. This is estimated to be approximately 2.5 pore volumes of the affected groundwater.

SYSTEM PERFORMANCE

As with any full-scale in situ system, especially where funds are not available for extensive monitoring, the performance of the in situ biodegradation system at this property has been difficult to accurately determine. The primary reasons for this difficulty include: the fact that the groundwater within the treatment area was not initially well mixed, there are only a limited number of monitoring locations, the system is not "closed," and there is significant variability in the groundwater elevation. Rainwater and groundwater from beyond the treatment area potentially dilute the concentrations within the treatment area. In addition, more recent information suggests that there was a continuing source of TCE in one portion of the treatment area.

Prior to initiation of the system operation in August 1995, TCE concentrations in samples collected from eight monitoring wells within the treatment area ranged from 2.0 to 88 mg/L, with an average concentration of 22 mg/L. The latest sampling event conducted during December 1998 indicated TCE concentrations ranging from <0.005 to 20 mg/L and an average concentration of 2.6 mg/L in the same eight monitoring wells. This represents an average decrease in TCE concentrations of nearly 90% over a 40-month period. If the results for the one potential "source" area are not included, the average decrease in TCE concentrations is approximately 99% (from an average of 12 mg/L to approximately 0.12 mg/L) during the same period. In addition to the observed general decrease in TCE concentrations, TCE concentrations have now been decreased to below the detection limit of 0.005 mg/L in portions of the plume. Well MW-40, located within the treatment area, has had consistent elevated concentrations of TCE. Recent excavation within this area identified a potential source of continuing release to the groundwater that was preventing the rate of decrease observed in the remaining plume. In an effort to address the elevated TCE level in this potential "source" area, an excavation was completed during October 1998 to allow volatilization of the soils in this area. Although no evidence of dense non-aqueous phase liquids (DNAPL) was identified during excavation activities, it was apparent that various debris had been historically buried, and a 4-inch steel pipeline existed near well MW-40.

Table 1 provides a summary of the system monitoring data. Five monitoring events are included: June 1995 (prior to initiation of system operation), May 1996 (nutrient addition), June 1997 (batch introduction of methanol), March 1998 (methanol injection only), and December 1998 (most recent sampling event). The values shown represent the mathematical average of the concentrations for each group of wells. The values shown in the column "Outside Wells" are the average values for up to five wells that are assumed to be outside of the treatment zone. Values from one representative well are shown in cases where there is limited data. The values shown in the column "Inside Wells" are the average values for up to eight wells within the treatment zone for which long-term data are available. Also included in Table 1 are the average concentrations for TCE and degradation products when the results from the potential source area (MW-40) are excluded.

TABLE 1 – Summary of System Monitoring Data

	Jun-95		May-96		Jun-97		Mar-98		Dec-98	
Parameter	Inside Wells	Outside Wells	Inside Wells	Outside Wells	Inside Wells	Outside Wells	Inside Wells	Outside Wells	Inside Wells	Outside Wells
TCE	22.7	0.070	16.7	0.07	4.02	0.748	2.80	0.046	2.60	0.065
cis-1,2-DCE	1.91	<0.005	1.0	<0.005	1.31	0.049	2.90	0.006	0.682	0.009
Vinyl Chloride	0.102	<0.005	<0.005	<0.005	0.084	<0.002	0.222	<0.002	0.105	<0.002
TCE*	11.8	-	9.68	-	3.24	-	2.42	-	0.119	-
cis-1,2-DCE*	1.28	-	0.733	-	1.38	-	2.52	-	0.165	-
Vinyl Chloride*	0.078	-	0.016	-	0.095	-	0.181	-	0.054	-
Chloride	162	35	147	29	132	20	136	23	120	14
NO3-NO2	<1.0	<1.0	11	0.90	0.20	0.13	<0.10	<0.10	<0.10	<0.10
O-PO4	0.24	0.09	2.1	0.40	1.9	0.52	0.56	0.12	0.65	0.10
TKN	<1.0	<1.0	3.6	1.6	<1.0	<1.0	<1.0	<1.0	<1.0	<1.0
Ferrous Iron	-	-	-	-	0.23	<0.05	0.27	<0.02	2.1	<0.02
Total Iron	-	-	-	-	-	-	3.4	0.25	4.4	0.07
Sulfate	-	-	-	-	170	28	146	24	42	38
Sulfide	-	-	-	-	<1.0	<1.0	<1.0	<1.0	<1.0	<1.0
Methane	-	-	-	-	0.69	0.007	2.8	0.009	3.6	0.85
Ethene	-	-	-	-	-	-	-	-	0.09	0.02
DO	1.7	3.4	1.1	2.4	0.45	1.6	0.65	1.6	1.1	2.3
Redox (mV)	+292	+319	+111	+126	+85	+178	-61	+332	-116	+245
pH (SU)	6.56	6.75	6.69	6.84	6.46	6.83	6.38	6.74	6.85	7.23

Notes: (-) Not Analyzed All values in mg/L unless otherwise indicated. (*) Excludes well MW-40.

Based on a comparison of chloride values between June 1995 and December 1998, it appears that there has been a general dilution of 60% for wells outside of the treatment area, but a reduction of only 26% within the treatment area. The difference in the chloride concentrations between the inside and outside wells suggests that average concentrations of TCE could have been as high as 100 to 125 mg/L. The difference in the reduction of chloride between the two sets of wells illustrates the chlorides contributed by the degradation process.

Degradation products such as cis-1,2-dichloroethene (cis-1,2-DCE) and vinyl chloride have shown periodic increases with the decreasing TCE concentrations, but no accumulation of degradation products has been observed. The highest cis-1,2-DCE concentration reported during the most recent sampling event was 4.3 mg/L, and the highest vinyl chloride concentration was 0.37 mg/L. The lack of accumulation of the degradation products indicates that degradation beyond cis-1,2-DCE and vinyl chloride is occurring under current site conditions. This is further supported by the presence of methane and ethene at elevated concentrations (>0.1 mg/L) in wells within the treatment area.

Groundwater samples have been periodically collected from the injection trenches and from wells near or adjacent to these trenches to monitor the presence of injected methanol. Although concentrations of methanol have been detected as high as 34,000 mg/L in the injection trenches soon after dosing, samples from wells downgradient of the trenches indicate that the high concentrations of methanol remain only in the trenches and is not present throughout the treatment area. This indicates that the majority of the treatment may be occurring near the injection trenches.

DO and redox levels have remained at levels that are indicative of anaerobic conditions over the past year and appear to correlate with batch introduction of methanol into the system. DO levels are currently <1.0 mg/L in wells within the treatment area, and redox levels currently range between –100 to –200 in these same wells. DO and redox levels are generally the lowest in the injection trenches. Levels of pH have been observed to be consistently lower in wells within the treatment area.

CONCLUSIONS AND OBSERVATIONS

The most significant conclusion from the work completed to date is that anaerobic dechlorination of TCE can reduce TCE and its degradation products to non-detectable levels. A second significant conclusion is that even a small "source" can drastically affect the performance of the technology. Several significant observations have been made during operation of the system to date. It appears that the higher dosage of methanol to the system (500 mg/L) has had the greatest effect on the degradation rate of TCE in the treatment area. Small dosages of methanol appear to be quickly consumed by microbes before the methanol infiltrates from injection trenches into the shallow aquifer. Based on the low TCE concentrations and high methanol injection rates, it is apparent that not all of the methanol degradation is directly related to the degradation of TCE. It is speculated that an alternate substrate might be more efficient; however, since operation is near completion, a change in substrate is not planned. Excessive biomass was not observed when nutrients alone were added. After methanol injection began, a significant increase in the biomass (and reduced flow rate throughout the system) was observed. This was remedied, to some degree, by changing from continuous to batch introduction of methanol. Close attention should be given to system hydraulics during the design, especially systems relying significantly on gravity flow. Due to the system hydraulics, greater than anticipated efforts were needed to balance the gravity fed flow to the various infiltration trenches. It has been difficult to keep the appropriate rates of flow into each of the injection trenches, particularly during periods of high rainfall. Operation of a full-scale in situ biodegradation system requires patience; it took approximately one year to achieve something close to a well-mixed system. Data interpretation is made difficult due to the non-homogeneous nature of the initial groundwater quality and dilution due to both recharge of rainwater and clean water from beyond the planned treatment area being pulled into the system. An accumulation of degradation products, particularly vinyl chloride, has not been observed. Even under the anaerobic conditions, the degradation products are continuing to be further degraded. The upgradient groundwater does appear to have been diverted around the treatment area; however, the system may be pulling in excessive amounts of clean water from beyond the treatment area. This has not caused a significant problem.

REGULATORY INFORMATION

The work at this site is being performed under the Texas Natural Resource Conservation Commission (TNRCC) Voluntary Cleanup Program (VCP). Based on the most recent results, concentrations of chlorinated organics in groundwater meet health-based risk values to allow future residential property use. It is anticipated that system operation will be discontinued in the summer of 1999. This will be followed by 2 to 4 years of post treatment monitoring to verify that natural attenuation will adequately control migration of the chlorinated organics in the future.

REFERENCES

Litherland, Susan T. and David W. Anderson, Roy F. Weston, Inc., 1997. "Full-Scale Bioremediation at a Chlorinated Solvents Site." *In Situ and On-Site Bioremediation: Volume 5*, pp. 425-430. Battelle Press, Columbus, OH.

Weidemeier, T.H., et al, "Overview of the Technical Protocol for Natural Attenuation of Chlorinated Aliphatic Hydrocarbons in Ground Water Under Development for the U.S. Air Force Center for Environmental Excellence."

ACCELERATED *IN SITU* BIOREMEDIATION OF CHLORINATED ETHENES IN GROUNDWATER WITH HIGH SULFATE CONCENTRATIONS

Christian D. Johnson, Rodney S. Skeen, and Mark G. Butcher
(Battelle PNWD, Richland, Washington)
Daniel P. Leigh and Lisa A. Bienkowski (IT Group, Pleasanton, California)
Steve Granade (U.S. Navy, Naval Construction Battalion Center, Port Hueneme, California)
Bryan Harre (U.S. Navy, Port Hueneme, California)
Todd Margrave (U.S. Navy, San Diego, California)

ABSTRACT: A pilot-scale test of accelerated *in situ* bioremediation of chlorinated ethenes is underway at a former underground storage tank site at the Naval Air Station Point Mugu in southern California. Two phases of nutrient addition are being used in this field test. In the first phase, lactate is being added to the contaminated aquifer in regular, high concentration pulses to stimulate sulfate reduction and remove sulfate in the groundwater. This is necessary since sulfate was shown to inhibit dechlorination at the site. During the first phase, only small amounts of reductive dechlorination are expected. In the second phase, a more aggressive nutrient injection scenario is used to disperse lactate throughout the test region and initiate rapid reductive dechlorination. To date, the first injection phase has been operating continuously for 42 days with the addition of a total of 6300 lb (2860 kg) of lactic acid. Vigorous sulfate reduction is occurring at distances of up to 68 ft (21 m) from the point of nutrient injection.

INTRODUCTION

The Installation Restoration Program (IRP) Site 24 at Naval Air Station (NAS) Point Mugu in southern California has two former underground storage tank (UST) sites with contaminated groundwater. Historical records for these two UST sites (Site 23 and Site 55) show that the tanks were used for oil/water separation and for storage of waste solvents, respectively. Chlorinated ethenes and total petroleum hydrocarbons have been detected at both locations. Figure 1 shows a plan view of IRP Site 24.

In 1997 a remediation alternative analysis was conducted to determine viable remediation technologies for both Site 23 and Site 55 [OHM, 1998]. The result of this evaluation was that UST Site 55 was found to be amenable to natural attenuation because the high organic carbon content from the petroleum hydrocarbon co-contamination would support degradation of all the chloroethenes in a reasonable time frame. However, UST Site 23 had a larger mass of chlorinated ethenes and much less organic carbon. Hence a pilot test of accelerated *in situ* bioremediation was implemented for UST Site 23 to demonstrate the effectiveness of the technology and to provide operating and cost information for evaluating a full-scale system.

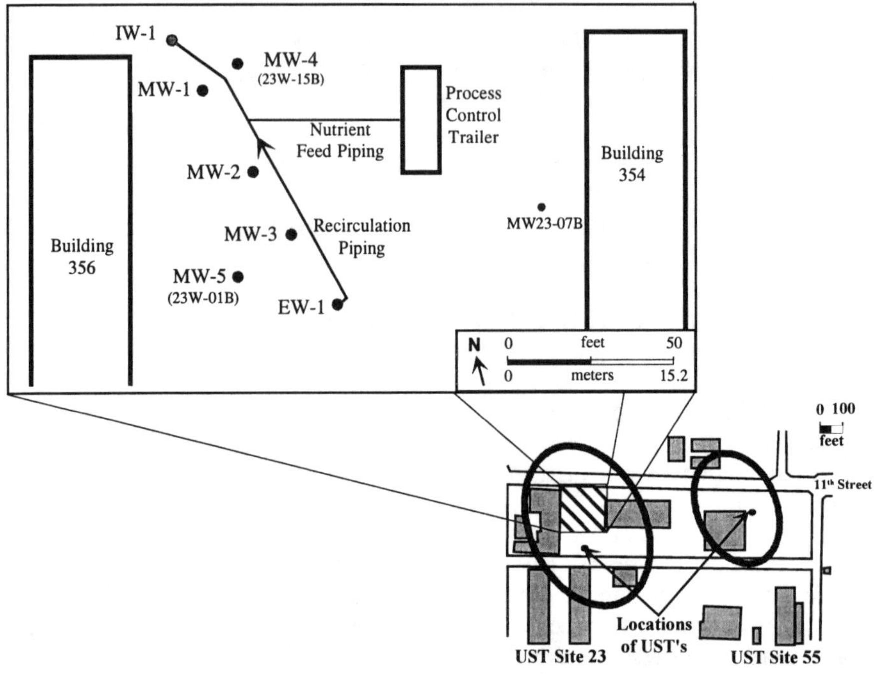

FIGURE 1. Plan view of IRP Site 24 with enlargement of the portion of UST Site 23 where the *in situ* bioremediation pilot test is being conducted. Contour lines on the view of IRP Site 24 indicate the approximate extent of chlorinated ethene plumes in the groundwater.

Site 23 Characterization. IRP Site 24 is located approximately 1 mile (1.6) inland from the Pacific Ocean. The depth to groundwater is about 5 feet (1.5 m). A 5 ft (1.5 m) thick clay layer extends from about 5 ft (1.5 m) below ground surface (bgs) to about 10 ft (3.0 m) bgs. The layer from about 10 ft (3.0 m) bgs to 30 ft (6.1 m) bgs consists of primarily sand with some silty sand and silt material and has been designated Zone B. The wells used in this pilot test are screened in this layer from 20 to 30 ft bgs. The regional groundwater gradient is approximately 0.001 m/m to the south.

The contaminants of concern at Site 23 are trichloroethene (TCE), dichloroethene (DCE; mostly cis-1,2-DCE), and vinyl chloride (VC). Sampling in August of 1997 showed that concentrations in Zone B were approximately 1700 μg/L, 750 μg/L, and 1 μg/L for TCE, cis-1,2-DCE, and VC, respectively. Because the site is so near the ocean, seawater intrusion has resulted in chloride and sulfate concentrations in Zone B of approximately 5,000 mg/L and 700 mg/L, respectively.

Pilot Test Design. The pilot test design was completed using the site model previously described [Johnson et al., 1998] and is reported in Appendix N of

OHM [1998]. Reactive flow and transport was simulated using the site model with the RT3D code [Clement, 1997]. Figure 1 shows an enlargement of the northern portion of Site 23 with a plan view of the pilot test system. Figure 2 shows the concept for an *in situ* bioremediation system (as applied at UST Site 23), where groundwater is circulated in a closed loop from an extraction well to an injection well and through the contaminated aquifer. For this pilot test, groundwater is continuously recirculated at 10 gpm (0.63 L/s) between the extraction well (EW-1) and the injection well (IW-1). Lactate is periodically added to the recirculating groundwater prior to re-injection into the aquifer. Three monitoring wells (MW-1 to 3) are placed in a line between EW-1 and IW-1. Two additional offset monitoring wells (MW-4 and MW-5) are placed to collect data on lateral flow and transport.

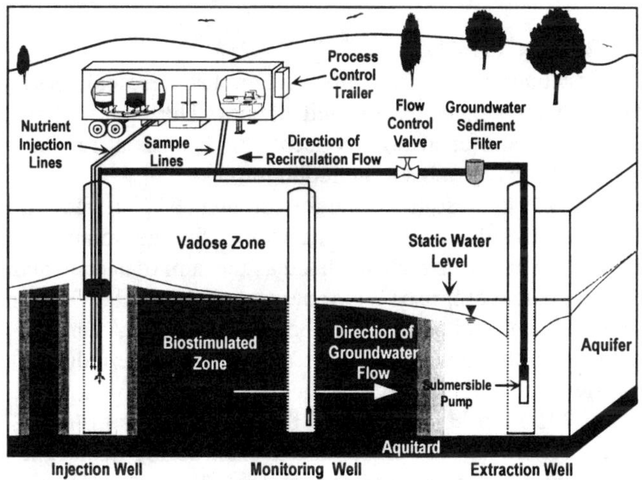

FIGURE 2. Conceptual design of an *in situ* bioremediation system.

The purpose of the pilot test system is threefold. First, it is desired to demonstrate the ability to accelerate reductive dechlorination of the chloroethenes to ethylene and other non-hazardous end products through the addition of lactate to the aquifer. The second purpose of the field test is collect sufficient field-data to design and cost a full-scale *in situ* biological treatment system for Site 23. This objective requires developing estimates of microbial transformation rates along with the stoichiometry of nutrient consumption and contaminant transformation. The third purpose is to develop a nutrient injection strategy that allows addition of an adequate amount of electron donor (i.e., lactate) without biofouling the injection well. To meet this objective the pilot test must demonstrate that the selected nutrient injection strategy delivers the required nutrients without an increase in the injection well pressure required for a specific flow rate.

The design of the accelerated *in situ* bioremediation pilot test encompasses two nutrient injection phases. The objective of the first phase is to reduce the sulfate concentration in the groundwater to the point where sulfate is no longer a major sink for added electron donor. This phase is necessary since laboratory experiments at Oregon State University have indicated that reductive dechlorination at this site will not occur under sulfate reducing conditions [Semprini, personal communication]. During the first phase of operation, approximately 17 gallons per day of 88% lactic acid is injected at regular intervals. The objective of the second nutrient injection phase of the pilot test is to provide the necessary conditions for rapid reductive dechlorination. This second phase will be conducted after removing the bulk of the sulfate from the test area of the aquifer. A high concentration pulse of lactic acid will be injected along with a non-reactive tracer and pushed out into the flow field. Once the material has reached MW-3 the recirculation will be discontinued and groundwater concentrations of the chloroethenes, ethene, lactate, propionate, and acetate will be monitored with time.

RESULTS AND DISCUSSION

In the first nutrient injection phase, samples were collected for sulfate, lactate, acetate, propionate, and chlorinated ethenes. During this phase, lactic acid was injected into the aquifer in high concentration pulses at regular intervals to stimulate biological activity and sulfate reduction.

Figure 3 shows the results to date for sulfate in MW-1, MW-2, MW-3, and EW-1. Figure 4 shows the corresponding results for the volatile organic acids (VOA). Volatile organic acids are calculated as the sum of acetate and propionate concentrations. Acetate comprises 70-90 percent of the total VOA concentration, depending on how long the bacteria have been stimulated. After only 20 days of nutrient injections, sulfate levels in MW-1 and MW-4 were reduced to less than 100 mg/L while the level of organic acids at both MW-1 and MW-4 are consistently near 1150 mg/L. By day 42, the sulfate levels in MW-2 have been reduced from approximately 700 mg/L to less than 20 mg/L. At the same time, VOA concentrations have risen to approximately 890 mg/L. Based on reactive flow-and-transport modeling with RT3D, it was anticipated that the zone of reduced sulfate concentration would move to MW-3 by day 42, which has in fact occurred. On day 42 sulfate levels in MW-3 are approximately 300 mg/L and the VOA level is approximately 225 mg/L. Taken together, these data indicate that sulfate reduction and lactate fermentation is occurring in the region between the injection well and MW-3.

One of the objectives of this test is to demonstrate a nutrient injection strategy that allows addition of adequate levels of electron donor without biofouling the injection well. Biofouling has been a problem at other accelerated in situ bioremediation test sites [Hooker et al., 1998]. To date, the system has been in operation for 35 days and over 5000 lb (2300 kg) of lactic acid have been added to the aquifer at the test site. The injection flow rate has remained constant throughout the test and pressure increases in the injection well have been less than 3 psi (0.2 bar). This result indicates that significant biofouling of the well bore has not occurred.

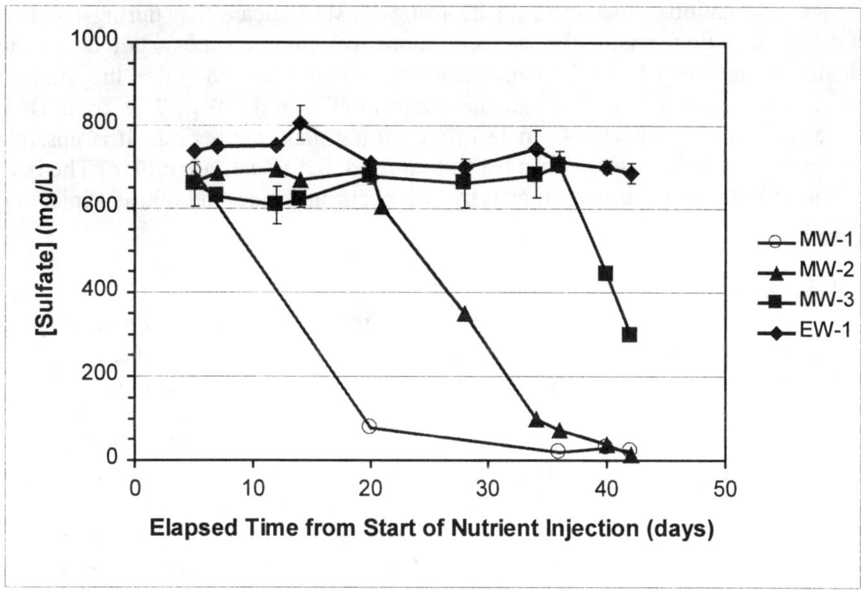

FIGURE 3. Concentrations of sulfate during the first nutrient injection phase. Where error bars are not shown, they are within the size of the symbol.

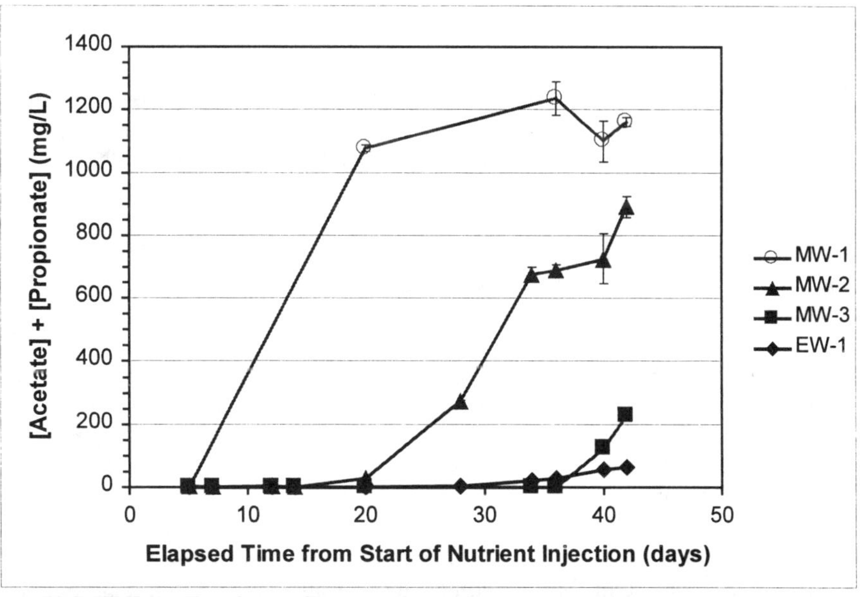

FIGURE 4. Total volatile organic acid concentrations during the first nutrient injection phase. Where error bars are not shown, they are within the size of the symbol.

Simulations modeling Site 23 using RT3D indicate that during the first nutrient injection phase the concentrations of chloroethenes should remain relatively unchanged. TCE concentrations through day 28 (after the start of nutrient injections) have been in the range of 970 to 1130 μg/L. Total DCE concentrations have been 145 to 180 μg/L during the same period. It is possible that reductive dechlorination has been stimulated, but effects are minor. The bulk of the reductive dechlorination activity will occur in the second nutrient injection phase.

SUMMARY

The pilot test of accelerated *in situ* bioremediation at NAS Point Mugu is currently underway. Injection of lactate into the aquifer has stimulated vigorous sulfate reduction activity to a distance of at least 50 feet from the injection well, as indicated by a decrease in sulfate concentration and increases in acetate and propionate concentrations. An increase in concentration of DCE at MW-3 is consistent with RT3D modeling predictions. No significant biofouling of the injection well has occurred during the first phase of nutrient injection.

The pilot test will continue with pulsed nutrient injections until the bulk of the sulfate has been removed from the test area of the aquifer. The second nutrient injection phase with a high concentration of lactate will then sustain rapid reductive dechlorination activity at this test site. Results of this pilot test will be used to show the effectiveness of the technology, to design and cost a full-scale system, and to provide an effective nutrient injection strategy.

REFERENCES

Clement, T.P. 1997. "RT3D - A Modular Computer Code for Simulating Reactive Multi-Species Transport in 3-Dimensional Groundwater Aquifers." Pacific Northwest National Laboratory, Richland, Washington. PNNL-11720. (http://bioprocess.pnl.gov/rt3d.htm)

Hooker, B.S., R.S. Skeen, M.J. Truex, C.D. Johnson, B.M. Peyton, and D.B. Anderson. 1998. "*In Situ* Bioremediation of Carbon Tetrachloride: Field Test Results." *Bioremediation Journal.* **1**(3):181-193.

Johnson, C.D., R.S. Skeen, D.P. Leigh, T. P. Clement, and Y. Sun. 1998. "Modeling Natural Attenuation of Chlorinated Ethenes Using the RT3D Code." In: *Proceedings of the Water Environment Federation 71st Annual Conference and Exposition, WEFTEC '98, Volume 3.* Water Environment Federation, Alexandria, Virginia. pp. 225-247.

OHM Remediation Services Corp. 1998. *Technical Memorandum IRP Site 24.* Naval Air Station Point Mugu, Point Mugu, California. Document Number SW4628.

ENHANCED REDUCTIVE DEHALOGENATION OF CAHs: A REMEDIAL PILOT STUDY

I. Richard Schaffner, Jr.; Christopher F. Wright; James M. Wieck; and Steven R. Lamb (GZA GeoEnvironmental, Inc., Manchester, New Hampshire)

ABSTRACT: Remedial pilot study (Study) results indicate substrate injection enhanced parent chlorinated aliphatic hydrocarbon (CAH) reductive dehalogenation in groundwater. The Study was performed to collect performance data for full-scale groundwater remediation. The remedial strategy includes yeast extract injection to stimulate parent CAH dehalogenation in the core of the plume, and natural attenuation of daughters at downgradient locations. Biostimulation was performed with groundwater monitoring to evaluate response. Biotransformation monitoring parameters included CAHs and certain indicator parameters. Biostimulation resulted in: 1) substantially reduced dissolved oxygen (DO), nitrate, sulfate and oxidation-reduction potential (ORP) values; 2) a one to two order of magnitude increase in methane concentration; 3) increased chemical oxygen demand (COD) and chloride concentrations; 4) nominal change in ethene/ethane concentration, and 5) biotransformation of 50% to 78% parent mass to daughters.

INTRODUCTION

CAHs in groundwater are present at concentrations up to 10^4 *ug*/L at a former wastewater treatment facility. CAHs include parents (tetrachloroethene, PCE; trichloroethene, TCE; and 1,1,1-trichloroethane, TCA); and daughters (1,1-dichloroethene, 1,1-DCE; *cis*/*trans*-1,2-dichloroethenes, 1,2-DCEs; 1,1/1,2-dichloroethanes, DCAs; and vinyl chloride, VC). The CAH source was an on-site pit where CAHs were co-disposed with sludges from the facility. Substantial quantities of non-aqueous phase liquid are not believed to be present based on groundwater monitoring results. Two overburden hydrologic units separated by a clayey silt aquitard are present beneath the site. An upper fine sand and silt unit extends from a depth of about 5 to 20 ft (1.5 to 6 m). A lower sandy and silty till unit extends from a depth of about 30 to 50 ft (9 to 15 m). The aquitard generally extends from a depth of 20 to 30 feet (6 to 9 m). Groundwater generally flows westward and discharges into a river along the property boundary (Figure 1).

Two distinct redox zones with differing biotransformation processes are present within the plume. Anaerobic, chemically reducing conditions are present within the core, with decreasing parent and increasing daughter concentrations with time. Aerobic, chemically oxidizing conditions are present along the periphery, with decreasing parent/daughter concentrations with time, though parents attenuate more slowly than within the core. Results of previous work suggest disposal of sludge with CAHs drives biotransformation by exerting terminal electron acceptor (TEA) demand, which stimulates certain bacteria to scavenge TEAs and drives conditions methanogenic. Methanogenic conditions support parent dehalogenation within the core and evolved methane stimulates parent/daughter co-oxidation within the aerobic-anaerobic mixing zone along the periphery. Daughter mineralization likely occurs within that zone. Historically residual sludge fueled biotransformation, however, low dissolved organic carbon (DOC) concentrations ($\leq$10 mg/L) coupled with TEA breakthrough into the core suggest transformation is becoming DOC limited since sludge disposal ceased over 10 years ago. Natural attenuation mechanisms reduce CAH concentrations below regulatory levels before discharging to the river located 400 ft (122 m) from the disposal pit source area.

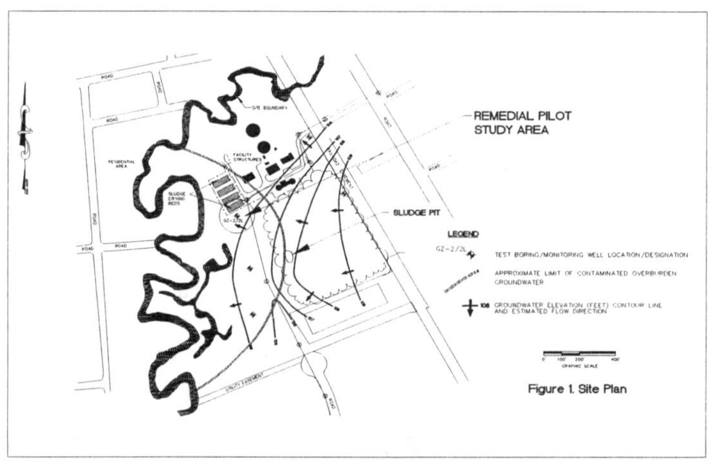

Figure 1. Site Plan

A 54-day microcosm study was performed to simulate parent CAH enhanced reductive dehalogenation in consideration of the DOC-limited groundwater condition. The proposed remedial strategy includes substrate injection to create an anaerobic parent CAH treatment zone and assumes daughters will be naturally attenuated upon reaching aerobic downgradient locations. Microcosms study results indicated yeast extract amendment stimulated parent reductive dehalogenation (Schaffner, et al., 1998) and form the basis for this Study.

The purpose of the Study was to develop design/operating criteria for full-scale groundwater remediation. Objectives included injecting substrate into an upper and lower unit test zone, and evaluating CAH and indicator parameter response. The "test zone" referenced herein refers to a well couplet location where injections/monitoring was conducted without radius of influence considerations.

METHODS

Biostimulant Injections. Well couplet GZ-2/2L screened in the upper (GZ-2) and lower (GZ-2L) units was selected for the Study in consideration of intermediate CAH concentrations and redox potentials relative to present site conditions. In July 1997, a 2,000-gallon (7.6E3 L) biostimulant solution was injected at the couplet. Biostimulant (FNI 200 Brewer's Yeast Extract supplied by American Yeast) was selected for the Study based upon literature search and microcosm study results (Schaffner et al., 1998). The solution consisted of 1,750 lbs (794 kg) of yeast extract powder dissolved in formation make-up water to minimize dilution effects upon groundwater chemistry. Two centrifugal Honda® pumps re-circulated the solution in a 2,000-gallon (7.6E3 L) capacity tank to aid dissolution. Biostimulant loading was about 500 pounds (227 kg) for the upper unit and about 1,250 pounds (567 kg) for the lower unit.

Biotransformation Monitoring. Pre-injection sampling was performed 20 days before injections. Results for that round along with those for four others performed between May 1996 and April 1997 establish baseline conditions. Up to eight post-injection sampling rounds were performed between July 1997 and November 1998. Table 1 summarizes the analytical program for the Study.

TABLE 1. Remedial pilot study analytical program.

Parameter and Method	Selection Rationale
CAHs, gas chromatographic (GC) screening	Contaminants of concern
Methane, GC screening	Methanogenesis product
Ethene and Ethane, GC screening	Daughter product
Carbon Dioxide, GC screening	Respiration product
Nitrate, U.S. Environmental Protection Agency (EPA) Method 353.2; Sulfate, EPA Method 45-OOE-SO_4	Respiratory pathway
COD, EPA Method 410.4	Substrate surrogate
Inorganic chloride, EPA Method 325.2	Reductive dehalogenation product
DO and ORP, parameter-specific electrodes	Respiratory pathway

RESULTS AND DISCUSSION

Historical CAH and certain indicator parameter data vary with time. Variability is primarily attributed to surficial hydrologic events (*e.g.*, precipitation, changing river levels) impacting groundwater quality. Short term events impact the upper unit more significantly due to greater atmospheric contact. To evaluate the data in consideration of this variability, results include pre-injection arithmetic mean (mean) concentration data to establish baseline conditions compared to post-injection mean data (Table 2). Table 2 also presents total concentrations for parents, daughters, and CAHs, and accounts for CAH variability by presenting concentration ratios of total parents to total CAHs (*i.e.*, parent ratios) to normalize the data assuming magnitudes may vary widely but ratios will not.

Results show biostimulation significantly altered groundwater chemistry. Subsequent discussion of CAH results focuses on comparisons between mean pre- and post-injection results due to the variability of the data; however, discussion of indicator parameter results focuses on temporal trends rather than mean values.

CAHs. VC, a CAH dehalogenation daughter product, was detected during two of four pre-injection upper unit sampling rounds, but was detected during seven of eight post-injection rounds at concentrations one to two orders of magnitude greater than pre-injection. The only post-injection upper unit round in which VC was not detected above the practical quantification limit (PQL) was the first, suggesting there was insufficient time for the reaction to proceed to VC. These results suggest yeast extract stimulates CAH dehalogenation within the upper unit.

TABLE 2. Summary of Remedial Pilot Study Results

Sampling Date	CAH Data (Upper Unit Well GZ-2): Parent CAHs (μg/L) TCA	PCE	TCE	Total Parents (μg/L)	Daughter CAHs (μg/L) DCE	DCEs	DCAs	CE	VC	Total Daughters (μg/L)	Total CAHs (μg/L)	Parent Ratio (%)	Indicator Parameter Data (Upper Unit Well GZ-2): Methane (μg/L)	Ethene (mg/L)	Ethane (mg/L)	Nitrate (mg/L)	Sulfate (mg/L)	COD (mg/L)	Chloride (mg/L)	DO (mg/L)	ORP (mV)	Comments
5/9/96	110	< 10	32	147	< 10	1,924	68	< 10	< 10	2,007	2,154	6.8	< 10	NT	NT	1.1	29	NT	5	NT	NT	
7/31/96	190	< 10	34	229	15	3,125	98	< 10	< 10	3,248	3,477	6.6	< 10	NT	NT	0.53	35	30	14	NT	+150	
11/25/96	< 10	< 10	10	20	< 10	395	< 30	< 10	< 10	425	445	4.5	< 10	NT	NT	3.6	17	< 5	3	NT	+150	Pre-Injection Data
4/22/97	44	< 10	23	72	< 10	1,105	39	< 10	45	1,199	1,271	5.7	56	NT	NT	1.3	23	< 5	9	6.9	+290	
7/8/97	480	< 10	42	527	32	8,978	160	20	120	9,310	9,837	5.4	370	< 0.01	< 0.01	2.0	25	25	21	7.0	+235	
				199						3,238	3,437	5.8	88	< 0.01	< 0.01	1.7	26	15	10	7.0	+206	Mean Baseline Data
7/28/97	66	<10	40	111	120	6,347	120	<10	<10	6,597	6,708	1.7	530	< 0.01	< 0.01	< 0.50	**240**	16,000	210	0.1	-50	
8/7/97	< 10	16	93	114	< 10	1,205	36	120	200	1,566	1,680	6.8	140	< 0.01	< 0.01	< 0.50	< 10	4,550	110	0.0	-45	
10/8/97	170	< 10	77	252	7,200	408	205	< 10	200	8,018	8,270	3.0	180	< 0.01	0.05	< 0.05	< 10	1,250	52	0.0	-80	
12/22/97	470	< 10	18	493	130	21,110	40	< 10	4,500	25,785	26,278	1.9	370	< 0.01	< 0.01	**0.2**	**< 50.6**	1,070	37.5	0.0	-30	Post-Injection Data
4/22/98	< 10	< 10	< 10	15	< 10	3,005	< 30	< 10	85	3,115	3,130	0.5	< 10	≤0.01	< 0.005	< 0.05	17	15	13	0.2	-10	
7/28/98	110	< 10	< 10	120	11	4,119	63	< 10	110	4,308	4,428	2.7	100	≤0.01	< 0.005	< 0.05	11	490	14	0.1	-60	
11/30/98	460	< 10	29	494	35	8,840	170	< 10	3,100	12,150	12,644	3.9	5,600	< 0.01	≤0.005	< 0.05	2.9	110	12	0.0	-55	
				228						8,791	9,019	2.9	989	0.01	0.01	0.1*	8.2*	3,355	64	0.1	-47	Mean Post-Injection Data

Sampling Date	CAH Data (Lower Unit Well GZ-2L): Parent CAHs (μg/L) TCA	PCE	TCE	Total Parents (μg/L)	Daughter CAHs (μg/L) DCE	DCEs	DCAs	CE	VC	Total Daughters (μg/L)	Total CAHs (μg/L)	Parent Ratio (%)	Indicator Parameter Data (Lower Unit Well GZ-2L): Methane (μg/L)	Ethene (mg/L)	Ethane (mg/L)	Nitrate (mg/L)	Sulfate (mg/L)	COD (mg/L)	Chloride (mg/L)	DO (mg/L)	ORP (mV)	Comments
5/9/96	85	< 10	90	180	13	1,911	50	< 10	55	2,034	2,214	8.1	34	NT	NT	0.49	30	NT	11	NT	NT	
7/31/96	NT	NT	NT	NT	NT	NT	NT	NT	NT	NT	NT	-	NT	NT	NT	NT	NT	NT	64	NT	+30	
11/25/96	190	< 10	220	415	93	20,033	100	< 10	14,000	34,231	34,646	1.2	**10,000**	NT	NT	0.58	19	5	49	NT	+130	Pre-Injection Data
4/22/97	93	< 10	140	238	20	4,432	91	< 10	130	4,678	4,916	4.8	110	NT	NT	1.1	25	10	10	6.0	+220	
7/8/97	770	14	190	974	270	39,280	1110	< 10	3,700	44,365	45,339	2.1	110	0.1	< 0.01	< 0.05	17	40	78	3.3	+105	
				452						21,327	21,779	4.1	85*	0.1	< 0.01	0.6	23	18	42	4.7	+121	Mean Baseline Data
7/28/97	68	< 10	47	120	190	13,005	340	< 10	< 10	13,545	13,665	0.9	470	< 0.01	< 0.01	< 0.50	**125**	12,500	170	0.0	-55	
8/7/97	< 10	25	16	46	< 10	4,805	190	80	< 120	5,200	5,246	0.9	780	< 0.01	< 0.01	**< 5.0**	< 10	4,750	130	0.0	-60	
10/8/97	180	< 10	< 10	190	100	9,272	225	500	460	10,557	10,747	1.8	2,100	< 0.01	0.005	< 0.05	< 10	2,950	71	0.1	-55	
12/22/97	160	< 10	14	179	13	17,039	520	< 10	13,000	30,577	30,756	0.6	2,300	< 0.01	0.01	0.057	**91.1**	2,030	84	0.0	-45	Post-Injection Data
2/25/98	82	< 10	14	101	28	6,237	260	120	2,700	9,345	9,446	1.1	1,600	< 0.001	≤0.01	**6.4**	14	500	19	**3.9**	**+80**	
4/22/98	60	< 10	24	89	36	9,744	30	< 10	2,300	12,115	12,204	0.7	2,000	≤0.01	< 0.005	0.32	19	134	15	0.2	+10	
7/28/98	130	< 10	< 10	140	31	7,142	80	< 10	3,000	10,258	10,398	1.3	9,100	≤0.01	< 0.005	< 0.05	10	800	20	0.0	0	
11/30/98	68	< 10	< 10	78	21	95	340	< 10	10,000	10,461	10,539	0.7	50,000	< 0.01	≤0.005	< 0.05	<10	114	100	0.0	-50	
				118						12,757	12,875	0.9	8,544	0.01	0.01	0.1*	9.7*	2,972	76	0.0*	- 36 *	Mean Post-Injection Data

Notes:

1. "*" indicates the mean does not consider values in **boldface**, as those values are considered outliers.
2. "NT" indicated respective parameter was not tested.
3. Total parents, total daughters and total CAHs as well as mean values were estimated using a value of one half the practical quantification limit (PQL) for cases in which parameters were not detected above PQLs, and the actual PQL value for cases in which parameters were detected below PQLs.

Post-injection lower unit total parent, daughter, and CAH mean values were less than pre-injection values, suggesting substrate injection stimulates reductive dehalogenation within that unit. Concentration reductions are attributed to biotransformation mechanisms because there was greater percent reduction of mean total parents ($74\%_{Red.}$) than daughters ($40\%_{Red.}$). Significantly, TCE was detected during all pre-injection lower unit sampling rounds, but was detected in five of eight post-injection rounds. Post-injection TCE concentrations for the lower unit were significantly lower than pre-injection concentrations (*i.e.,* two-tailed unpaired P value = 0.0162 in an unpaired t test with Welsh correction, calculated using GraphPad, 1998, conservatively assuming non-detected TCE concentrations were present at the PQL).

Parent ratios for both units indicate significant parent CAH mass reduction due to substrate injection. For example, mean upper and lower unit pre-injection parent ratios were 5.8% and 4.1%, respectively, whereas post-injection ratios were 2.9% and 0.9%, respectively. Parent ratios reflect a $50\%_{Red.}$ for the upper unit and a $78\%_{Red.}$ for the lower unit. Greater lower unit parent CAH removal likely reflects oxygen loading of the upper unit from precipitation events that inhibits anaerobic processes until oxygen becomes depleted.

Methane. Methane was detected in two of five pre-injection upper unit sampling rounds, and in all pre-injection lower unit rounds. Post-injection methane concentrations exceed baseline by an order of magnitude for the upper unit and two orders for the lower unit, reflecting establishment of methanogenic conditions resulting from substrate injection. Elevated methane demonstrates yeast extract provided sufficient electron donor to exert TEA demand and drive methanogenesis.

Ethene/Ethane. Ethene/ethane data were stable, suggesting dehalogenation is not proceeding to completion, which agrees with CAH results. While lower unit data indicate limited daughter CAH transformation, parent dehalogenation is more significant. Given the CAH data suggesting dehalogenation to be more important for parents and the ethene/ethane data indicating incomplete transformation, parents may be preferentially selected for dehalorespiration over less oxidized daughters.

Nitrate. Nitrate was detected in the upper unit during each pre-injection sampling round and during all but one lower unit round, indicating residual sludge-derived DOC does not exert substantial nitrate demand. Following injection, nitrate was not detected in groundwater samples above the PQL for six of seven upper unit rounds and for five of eight lower unit rounds, suggesting yeast extract stimulated denitrification, which scavenged nitrate. Upper/lower unit post-injection nitrate concentrations were generally below the 1.0 mg/L inhibitory threshold for reductive dehalogenation reported by Wiedemeier et al., (1996).

Sulfate. Sulfate was detected in groundwater samples collected during each pre-injection sampling round. Following substrate injection, sulfate was detected during the first post-injection round at an order of magnitude greater concentration than baseline. Elevated sulfate immediately following injections suggests sulfate is a yeast extract component, which is consistent with vendor-supplied information. Subsequently, sulfate concentrations decreased below baseline suggesting injections stimulated sulfate reduction, which scavenged sulfate. Post-injection sulfate levels for both units were generally below the 20 mg/L inhibitory threshold for reductive dehalogenation (Wiedemeier et al., 1996).

COD. Baseline COD concentrations in groundwater samples were relatively low for both units, and increased three orders of magnitude after injections. Elevated COD following injections reflects the presence of yeast extract, which is organic carbon enriched and exerts TEA demand. Subsequently, COD concentrations decreased as a function of microbial utilization and hydrodynamic dispersion, but remain elevated 14 months after injections (*i.e.,* >100 mg/L for both units during November 1998) reflecting the continued presence of yeast extract. According to Wiedemeier et al., 1996, DOC concentrations >20 mg/L drive dehalogenation.

Chloride. Mean baseline upper and lower unit chloride levels were 10 mg/L and 42 mg/L, respectively. Concentrations increased an order of magnitude following injections, and then generally decreased with time. Elevated chloride reflects the presence of yeast extract that contains $3\%_{weight}$ chloride. The subsequent decrease reflects hydrodynamic dispersion of yeast extract-supplied chloride.

DO/ORP. Baseline groundwater was aerobic, chemically oxidizing, with higher upper unit values reflecting greater atmospheric contact. There was a highly significant DO/ORP decrease in response to injections, with the resulting shift yielding anaerobic, chemically reducing conditions within each unit. DO/ORP data are consistent with other TEA data, suggesting yeast extract exerted TEA demand.

CONCLUSIONS

Pre-injection parent CAH intrinsic bioremediation was DOC limited. Yeast extract amendment provided DOC that stimulated certain bacteria to scavenge TEAs, re-established anaerobic, chemically reducing conditions, and provided electron donor for parent CAH reductive dehalogenation.

REFERENCES

GraphPad, 1998. InStat Version 3.01. GraphPad Software, Inc.

Schaffner, I.R., Wieck, J.M., Lamb, S.R. 1998. *Enhanced reductive dehalogenation of CAHs at a former wastewater treatment facility: A microcosm study*, in proceedings, Northeast Focus Ground Water Conference, p. 115-125 (NGWA)

Wiedemeier, T.H., Swanson, M.A., Moutoux, D.E., Wilson, J.T., Kampbell, D.H., Hansen, J.E., and Haas, P. 1996. *Overview of the technical protocol for natural attenuation of chlorinated aliphatic hydrocarbons in ground water under development for the U.S. Air Force Center for Environmental Excellence*, in proceedings, Symposium on Natural Attenuation of Chlorinated Organics in Ground Water, EPA, EPA/540/R-96/509, p. 35-63 (EPA)

A PILOT STUDY USING HRC™ TO ENHANCE BIOREMEDIATION OF CAHs

Madeline Wu (Water Restoration Inc., Fort Lauderdale, Florida)

ABSTRACT: The effect of Hydrogen Release Compound (HRC™) in enhancing the anaerobic bioremediation of chlorinated aliphatic hydrocarbons (CAHs) was observed in a single well pilot study. HRC is a proprietary, environmentally safe, food quality, polylactate ester formulated for the slow release of lactic acid upon hydration. Microbes metabolize the lactic acid releases by HRC and produce hydrogen, which can be used by reductive dehalogenators to dechlorinate CAHs. The objective of the pilot study was to enhance anaerobic reductive dechlorination by reducing the redox potential while providing increased levels of dissolved hydrogen.

For the pilot study, eleven four foot long canisters containing HRC were placed into a five inch monitoring well containing trichloroethylene (TCE), cis-1,2-dichlorethylene (cis-1,2-DCE), and vinyl chloride (VC). Initial TCE, cis-1,2-DCE, and VC concentrations in the well were 148 ppb, 2,030 ppb, and 471 ppb, respectively. After one month of treatment, dissolved oxygen was depleted from the well and highly negative redox levels were achieved. Data from the final monitoring event, taking place approximately three months after HRC installation, indicated reductions of TCE, cis-1,2-DCE, and VC concentrations of 96%, 98%, and 96%, respectively. Based on these results, a full scale application of HRC is being planned.

INTRODUCTION

A 3-month pilot study testing the efficacy of HRC (manufactured by Regenesis, San Jan Capistrano, California) in enhancing anaerobic bioremediation of chlorinated aliphatic hydrocarbons (CAHs) was conducted at a site in Florida. HRC is a proprietary, environmentally safe, food quality, polylactate ester formulated for the slow release of lactic acid upon hydration. Microbes metabolize the lactic acid releases by HRC and produce hydrogen, which can be used by reductive dehalogenators to dechlorinate CAHs. Contaminants of concern include TCE, cis-1,2-DCE and vinyl chloride. The contaminant plume, contained in a fine to medium grained sand aquifer, measured 120 feet in length by 60 feet in width (see Figure 1). Due to a flat gradient, groundwater velocity is estimated to be less than 0.1 foot per day.

MATERIALS AND METHODS

Application. For the pilot study HRC was packed in four-foot long, three and one half diameter PVC canisters designed to be inserted into existing monitoring wells. Each canister contained several hundred small holes to allow for the contact of HRC with groundwater passing through the well. On March 30, 1998, eleven canisters were placed vertically into a five inch recovery well RW100 (see Figure 1) covering a vertical distance of approximately 45 feet.

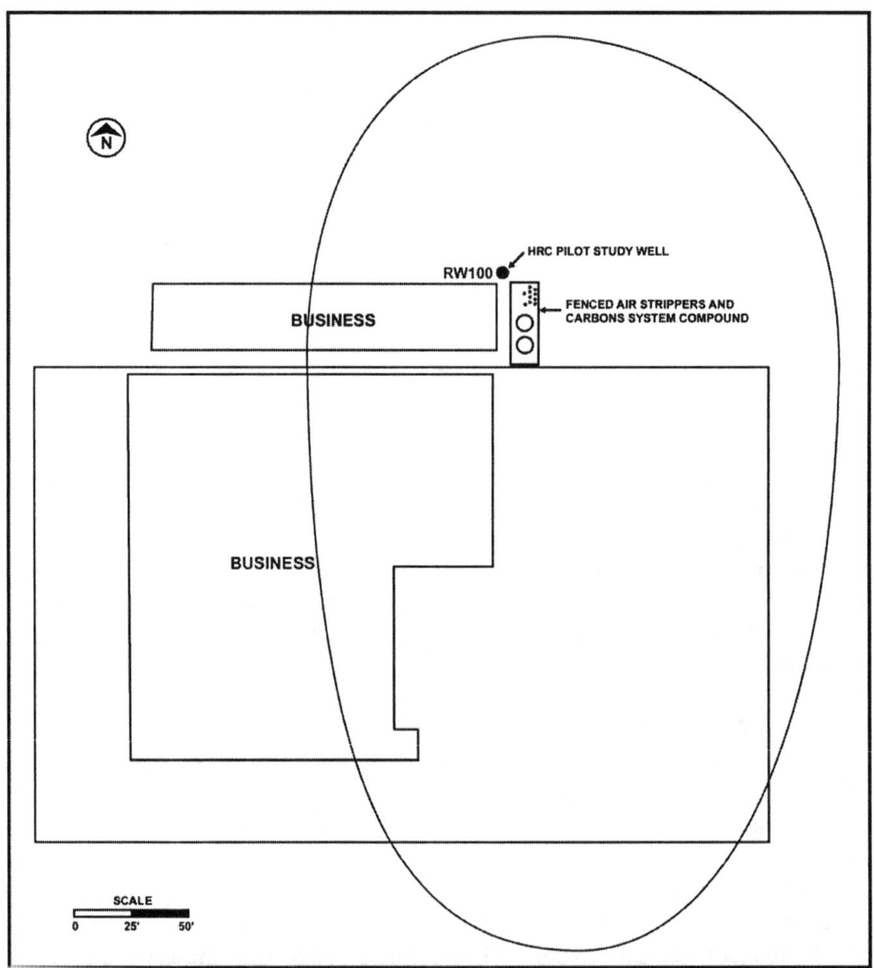

FIGURE 1. Tetrachloroethene plume map identifying HRC pilot study well (RW100) location.

Monitoring Program. For sampling purposes a polyester tube was attached to the canister assembly at a vertical distance of 24 feet. The monitoring program included sampling of the following: 1) EPA method 8010 (Chlorinated VOCs), 2) Gases (methane, ethene, ethane), 3) Nutrients, electron acceptors, inorganics, 4) Alkalinity, TOC, and conductivity, 5) Volatile acids, 6) pH, DO, ORP and temperature, 7) Microbial (total anaerobes & sulfate reducing bacteria). Baseline sampling was performed just prior to the installation of the canisters and at two, four, eight, and twelve weeks following installation. The final sampling event took place on June 22, 1998.

RESULTS AND DISCUSSION

Following approximately 82 days of treatment with HRC, reductions in TCE, cis-1,2-DCE, and VC were 96%, 98%, and 99%, respectively. These reductions are graphically presented in Figure 2. TCE, cis-1,2-DCE, vinyl chloride, acetic acid, lactic acid, propionic acid, sulfate, and total anaerobic plate count data are presented in Table 1. The absence of DO and highly negative redox levels confirmed the existance of a highly reduced environment required for HRC to be effective. Sampling data strongly support the effectiveness of HRC in this single well application. Due to the production of lactic acid breakdown products acetic acid and propionic acid, it is evident that the lactic acid released from HRC was successfully metabolized leading to the production of hydrogen and the subsequent reductive dechlorination of the chorinated compounds. All contaminant concentrations in the well were decreased substantially over the twelve week period of the test, supporting the effectiveness of HRC in this pilot scale application. Based on these results a full-scale application of HRC on the site is being planned.

TABLE 1. 12 week pilot study sampling data.

	Day 0	Day 14	Day 28	Day 55	Day 82
TCE (ug/L)	59.2	3	16.6	4.5	4.9
cis-1,2-DCE (ug/L)	634	60.8	43.3	22.7	34.2
VC (ug/L)	156	26.9	18.3	11	<1
Acetic Acid (mg/L)	2.2	-	382	-	261
Lactic Acid (mg/L)	<1	-	788	-	63
Propionic Acid (mg/L)	<1	-	281	-	302
Sulfate (mg/L)	21.6	-	5.5	-	7.1
DO (mg/L)	1.6	0	0	0	0
Redox (eV)	-50	-91	-136	-156	-113
Anaerobic TPC	42	-	-	-	12,708

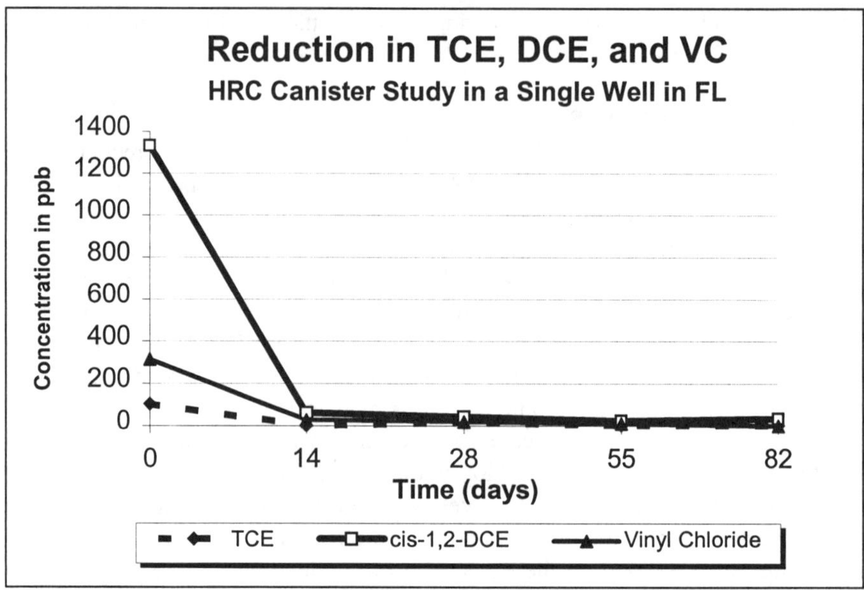

FIGURE 2. Reduction of TCE, cis-1,2-DCE, and vinyl chloride.

ENHANCED BIOREMEDIATION OF CHLORINATED SOLVENTS - A SINGLE WELL PILOT STUDY

S. Kallur (Environmental Strategies & Applications, Inc. Somerset, New Jersey)
S. Koenigsberg (Regenesis, San Juan Capistrano, California)

ABSTRACT: A slow-release lactic acid product, Hydrogen Release Compound (HRC^{TM}), was used in a pilot study to enhance anaerobic reductive dechlorination of chlorinated aliphatic hydrocarbons (CAHs) in a single monitoring well. The study was performed to evaluate the performance of HRC on a small scale, which would then serve as the basis for a larger scale application. The objective of adding lactic acid is to generate hydrogen in order to drive reductive dechlorination. The objective of using a slow-release form of lactic acid is to minimize operation and maintenance activities on the site as well as to potentially facilitate reductive dechlorinators in their competition with methanogens.

Comparison of baseline and final monitoring event data, indicated reductions of PCE, TCE, and cis-1,2-DCE of 61%, 91%, and 57%, respectively. The results of the degradation of these compounds are promising even though there was a transient spike in the concentrations of PCE, TCE and cis-1,2-DCE which was observed during the eighth week of sampling; possible reasons being desorption of residual concentrations and re-oxygenation during sampling (by methods which were discontinued). Overall the data suggests that the HRC had a positive impact on the aquifer in the vicinity of the well and has justified project scale-up.

INTRODUCTION

A single well pilot study utilizing the product Hydrogen Release Compound (HRC^{TM}) (manufactured by Regenesis, San Juan Capistrano, California) was conducted at a New Jersey site contaminated with moderate levels of tetrachloroethene (PCE), trichloroethene (TCE) and cis-1,2-dichloroethene (cis-1,2-DCE). HRC is a proprietary, environmentally safe, food quality, polylactate ester formulated for the slow release of lactic acid upon hydration. Microbes metabolize the lactic acid released by HRC and produce hydrogen, which can be used by reductive dehalogenators to dechlorinate CAHs.

MATERIALS AND METHODS

Application. In order to effectively apply the HRC in the monitoring well, the HRC was contained in four foot long, three and one half inch diameter PVC canisters. Each canister contained several hundred small holes to allow for the contact of HRC with groundwater passing through the well.

The specific form of HRC used was sorbitol polylactate ester (SPL) which has the consistency of a hard gel; consequently it simply beaded out through the holes where it made contact. In the future, direct injections of a less viscous form of HRC (glycerol polylactate ester, GPL) is proposed which will eliminate the need for wells in favor of push-point application. On April 14, 1998, three HRC-containing canisters were linked together and placed into a single four inch monitoring well, covering the vertical distance of the water column.

Monitoring Program. The monitoring program included sampling of the following: 1) EPA method 8010 (Chlorinated VOCs), 2) Gases (methane, ethene, ethane), 3) Nutrients including common electron acceptors, 4) Alkalinity, TOC, and conductivity, 5) Volatile acids, 6) pH, DO, ORP and temperature and 7) microbial counts (total anaerobes & sulfate reducing bacteria). Baseline sampling was performed just prior to the installation of the canisters and at two, four, eight, and twelve weeks following installation. The final sampling event took place on June 30, 1998.

RESULTS AND DISCUSSION

Sampling data support the effectiveness of HRC in this single well application. Due to the production of lactic acid breakdown products, pyruvic acid and acetic acid, it is evident that the lactic acid released from HRC was successfully metabolized leading to the production of hydrogen and the subsequent reductive dechlorination of the chlorinated compounds under discussion. Over the first twelve weeks of the pilot test, concentrations of PCE, TCE, and cis-1,2-DCE have decreased by 61%, 91%, and 57%, respectively. A spike in the concentrations of PCE, TCE and cis-1,2-DCE was observed during the eighth week. This phenomenon may be due to desorption of unknown residual concentration. The degradation profile occurred in spite of the fact that the system was disturbed (re-oxygenated) during each measurement - when canisters were removed and returned during sampling. The re-oxygenation phenomenon may be partially responsible for the "sawtoothed" pattern of the data because 1) in five other studies conducted by Regenesis the same pattern (2 cases) or no reduction (3 cases) was observed 2) in two other studies when the system was left undisturbed due to installation of permanent tubing - there was continuous and steady decline of the CAHs.

Organic acid data indicate that the lactic acid was released and moved through various intermediates on the way to hydrogen formation. Lactic acid concentrations do not increase readily in the initial phase of the experiment, presumably because of rapid consumption. This is supported by the concomitant rise in acetic acid and the rapid decline in CAHs. As degradation of CAHs is temporarily stalled, presumably by re-oxygenation, lactic acid levels rise while residual acetic acid is putatively metabolized to carbon dioxide - in preference to lactic acid. The presence of the surplus pool of accumulating lactic acid could then account for a recovered decline in CAH levels.

Other data on sulfate levels reflect the expected pattern of decrease as a function of a depressed redox condition. Even though further monitoring is required, the initial data suggests that the HRC has had a positive impact on the aquifer in the vicinity of the well and will be the basis for a scaled-up effort using this passive approach to enhanced natural attenuation.

The full data is presented in Table 1. Changes in PCE, TCE and cis-1,2-DCE are presented in Figure 1. Changes in the organic acids are presented in Figure 2. Changes in sulfate, iron, and chloride are presented in Figure 3, in which the chloride and iron data are atypical, probably due to re-oxygenation.

CONCLUSION

This simple demonstration was one of the earliest proof-of-concept field experiments conducted on the use of HRC. Even though there were some anomalies in the data, possibly due to the disturbance of the system, the total mass of the CAHs in the system was significantly reduced. The initial data show that HRC in the form of sorbitol polylactate ester, can be a safe and effective means of delivering lactic acid, to a subsurface environment contaminated with chlorinated aliphatic hydrocarbons, to facilitate their remediation by reductive dechlorination.

TABLE 1. Single Well Pilot Study Monitoring Data.

	Day 0	Day 14	Day 30	Day 50	Day 80
PCE (ug/L)	11	5.4	1.6	8.9	4.2
TCE (ug/L)	103	74	9.3	96.9	8.9
cis-1,2-DCE (ug/L)	50	71.5	29.1	134.5	21.7
Lactic Acid (mg/L)	<1	-	39	-	1,359
Acetic Acid (mg/L)	<1	-	10.5	-	<1
Sulfate (mg/L)	54.5	-	58.5	-	36.8
Iron (mg/L)	35.9	-	33.8	-	24.5
Chloride (mg/L)	31.5	-	7.5	-	8.7

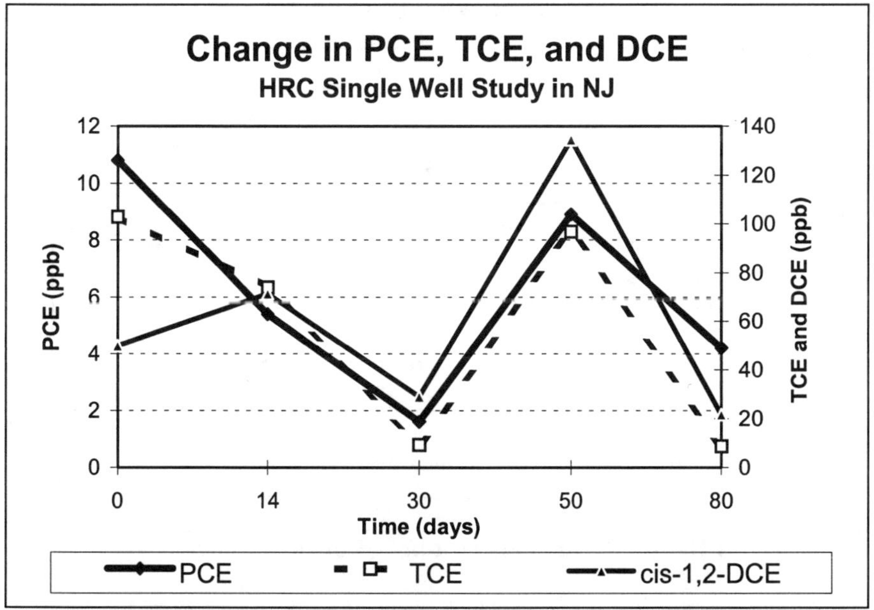

FIGURE 1. Changes in PCE, TCE, and cis-1,2-DCE.

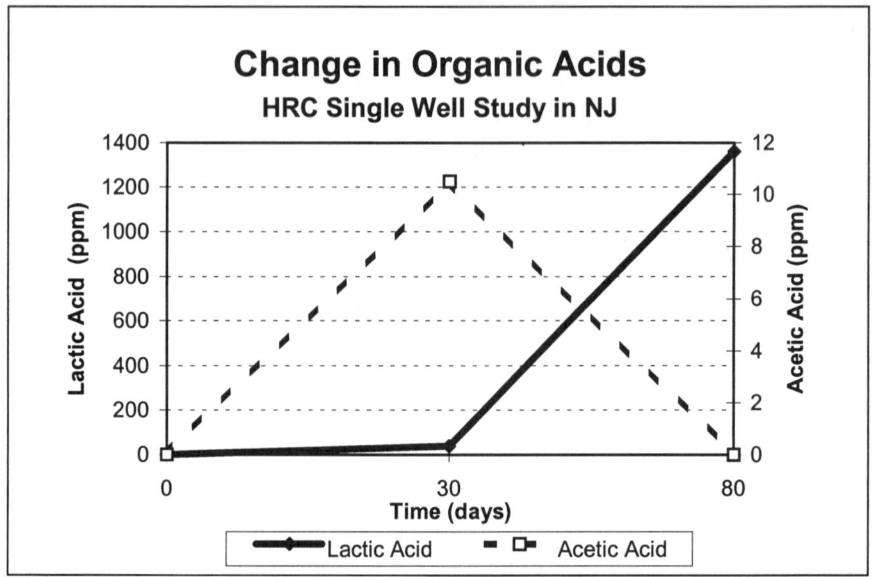

FIGURE 2. Changes in lactic acid and acetic acid.

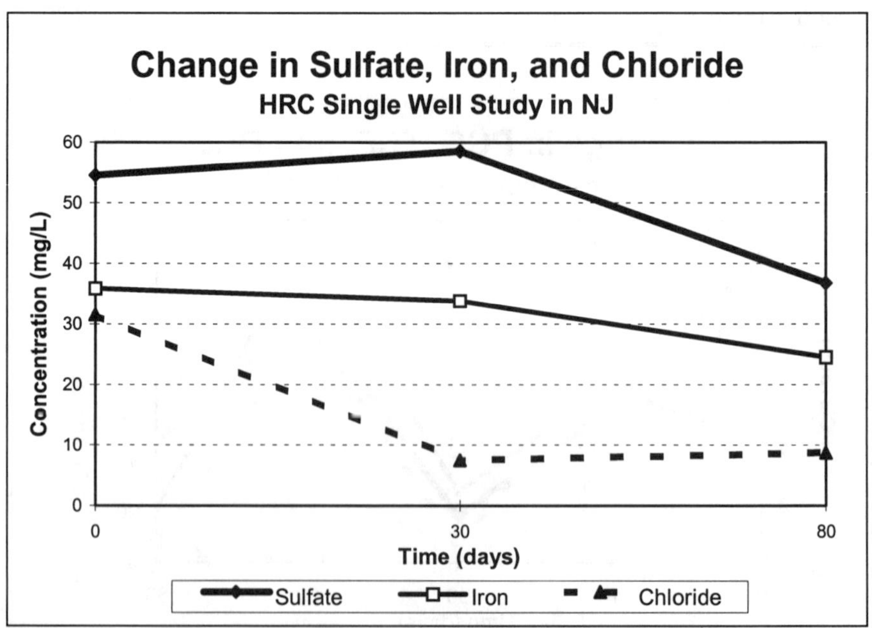

FIGURE 3. Changes in sulfate, iron, and chloride.

TREATABILITY STUDIES OF HYDROGEN-ENHANCED BIOREMEDIATION OF CHLORINATED SOLVENT-IMPACTED MEDIA

R. Todd Fisher, Groundwater Services, Inc., Houston, Texas, USA
Charles J. Newell, Groundwater Services, Inc., Houston, Texas, USA
Patrick E. Haas, Air Force Ctr. for Environ. Excellence, Brooks AFB, Texas, USA
Joseph B. Hughes, Rice University, Houston, Texas, USA

ABSTRACT: Direct hydrogen addition, wherein hydrogen is delivered without the use of fermentation substrates or carbon sources, is a new in-situ bioremediation technology for chlorinated solvent plumes (PCE, TCE, etc.) that is currently under development. A project to field test the applicability and feasibility of direct hydrogen addition has been initiated by the Technology Transfer Division, Air Force Center for Environmental Excellence (AFCEE/ERT), Brooks AFB, Texas. The test program consists of both short-term (1 week) treatability tests in the saturated zone, short-term vadose zone treatability tests, and long-term (1 year) pilot tests at select Air Force installations. The initial treatability tests are designed as site screening tests that evaluate hydrogen utilization by indigenous microorganisms via a "pull-push-pull" field test method developed by one of the authors (Haas).

Results from a pull-push-pull treatability test conducted at Offutt AFB in Nebraska in November 1998 showed > 99% reduction of the amended hydrogen and > 99% reduction of the cis-DCE observed in the test volume over the 41 hour test period. Analysis of test quality (e.g., tracer recovery, concentrations of competing electron acceptors) indicate the presence of biological reductive dechlorination of the cis-DCE during the test. A vadose zone treatability test also showed apparent utilization of hydrogen delivered to the vadose zone over a 68-hr test period in both clean and affected soils.

Future tests of direct hydrogen delivery include evaluation of low-volume pulsed biosparging of hydrogen gas at two locations in Florida in February 1999, and additional dissolved-hydrogen treatability and pilot tests.

INTRODUCTION: At sites where natural dechlorination is occurring, organic substrates such as aromatic hydrocarbons (BTEX), landfill leachate, or other non-chlorinated organics provide a source of dissolved hydrogen, produced through slow fermentation of these organics (Wiedemeier et al., 1999). The hydrogen is then rapidly utilized as an electron donor by naturally-occurring bacteria to achieve reductive dechlorination of chlorinated compounds in the subsurface (Carr and Hughes, 1998). In-situ bioremediation via direct hydrogen addition represents an extension of these naturally-occurring processes. Direct hydrogen addition simply eliminates the rate-limiting step (i.e., fermentation), by providing naturally-occurring dechlorinating bacteria with substantive quantities of hydrogen, the key growth substrate (Hughes et al., 1997; Newell et al. 1997; Newell et al., 1998).

Direct hydrogen addition is a relatively flexible process that may be implemented in a variety of process configurations for either dissolved plume management or reduction of NAPL source zones, depending upon the goals of the intended remediation effort. Potential delivery methods include:

- Dissolved hydrogen injection via groundwater recirculation system
- Low-volume pulsed biosparging (Fisher et al., 1998)
- In-situ controlled release reaction (hydrogen releasing agent)
- Injection of hydrogen gas aphrons (surfactant/hydrogen gas microbubbles)
- In-situ electrolysis of water (as described by Hughes et al., 1997)

TREATABILITY TEST: As part of the AFCEE project, a field treatability test for direct hydrogen development was conceived by one of the authors (Haas) using a "pull-push-pull" field test method:

1) *Initial Groundwater Extraction ("pull"):* Extraction of a known quantity of groundwater (e.g., 750 L) from within the test area through an existing monitoring well.

2) *Amendment Addition:* Addition of known quantities of hydrogen and various volatile and non-volatile tracers (e.g., bromide, helium, sulfur-hexafluoride (SF_6)) to the extracted groundwater, followed by thorough mixing to create a homogeneous test solution.

3) *Initial Sampling:* Collection of a representative test solution sample which is analyzed for chlorinated organic compounds, hydrogen, tracers, and other constituents of interest (e.g., oxygen, nitrate, sulfate, etc.).

4) *Re-Injection of Groundwater Test Solution:* Pulse injection ("push") of amended groundwater into the saturated zone through the same monitoring well used for groundwater extraction.

5) *Final Groundwater Extraction:* Extraction ("pull") of the test solution/groundwater mixture from the test well following a contact/reaction period (typically 24 to 48 hr). Sampling is conducted during the extraction.

6) *Final Sampling:* Collection of a final representative test solution sample which is again analyzed for chlorinated organic compounds, hydrogen, tracers, and other constituents of interest.

During the injection phase, the test solution enters the test zone through the screened area of the monitoring well. Within the test zone, biologically reactive components of the test solution (e.g., hydrogen and chlorinated organics) are utilized by the indigenous microorganisms. During the final extraction phase, the test solution is recovered and solute concentrations are measured to determine the quantities of reactants used (e.g., hydrogen, PCE, TCE) and/or products formed (e.g., DCE, vinyl chloride, ethene, ethane). The tracers are used to evaluate abiotic losses of reactants during the test process. For example, if the test results indicate that hydrogen is consumed, chlorinated solvents are consumed or transformed, and the tracers are recovered, then biological reductive dechlorination during the treatability test is indicated.

OFFUTT AFB TREATABILITY TEST: A pull-push-pull treatability test was performed at the Fire Protection Training Area 3 (FPTA 3) site at Offutt AFB, Nebraska. At this site, fire protection training exercises using both waste fuels and solvents were conducted at the site from 1960 until the spring of 1990. Across the majority of the site, groundwater occurs at 8 to 10 ft below ground surface (BGS) within a sand unit. Starting concentrations of key contaminants at the site were less than 1 mg/L of cis-DCE and less than 1 mg/L of total BTEX.

TREATABILITY TEST RESULTS: The test results are summarized below:

1) Cis-1,2-DCE concentrations dropped by more than 99% (0.43 to < 0.005 mg/L) over the course of the 41-hour test, while much lower reductions in Total BTEX was observed (34% reduction) (see Table 1). Due to the expected similar volatilization and adsorption behavior of DCE and BTEX (based on similar Henry's Law and organic carbon partitioning coefficients), the observed substantial loss of DCE does not appear to be fully accounted for by these physical mechanisms, and a biological loss is presumed.

Constituent	Initial Conc. (mg/L)	Final Conc. (mg/L)	% Reduction
cis-1,2-DCE	0.43	<0.005	>99%
Benzene	0.088	0.081	8%
Ethylbenzene	0.023	0.014	39%
Toluene	0.037	0.029	22%
Xylenes	0.145	0.068	53%
BTEX	0.293	0.192	34%

TABLE 1. Reduction of organic constituent concentrations during treatability test

2) Tracer recovery results indicate a minimum 39% mass loss of hydrogen to biological consumption over the course of the 41-hour treatability test at FPTA 3 (0.012 mg/L hydrogen consumed) (see Table 2).

Constituent	Initial Conc. (mg/L)	Final Conc. (mg/L)	% Recovery	Sample Type
Hydrogen	0.031	0.00018	0.6%	in-line grab
Bromide	298	276	92.6%	composite/grab
Helium	0.017	0.0068	40.2%	in-line grab
SF6	0.015	0.010	66.6%	in-line grab

TABLE 2. Recovery of hydrogen and tracers during treatability test

3) Based on the average observed BTEX recovery of 66%, the observed >99% reduction in cis-1,2-DCE over the course of the test cannot be fully accounted for by physical loss mechanisms (e.g., volatilization and adsorption). Although not confirmed by observed changes in vinyl chloride or ethene concentrations, a biological reduction of cis-1,2-DCE is indicated.

4) A significant reduction in the concentration of oxygen and a significant increase in the concentration of methane were observed over the course of the test. While some consumption of oxygen, and some production of methane, may have been due to the competitive utilization of hydrogen, the magnitude of the concentration changes observed indicate that other physical factors were the likely source of the observed changes (e.g., the non-biological reaction of oxygen with ferrous iron and the volatile loss of methane from the composite test samples). Competitive effects did not appear to stop reductive dechlorination processes during the tests.

5) Based on observed declines in the ratio of hydrogen to helium in soil gas samples collected over a 68-hour period following gas injection, field data indicate an apparent utilization of hydrogen to have occurred in the soil vadose zone in both affected and non-affected soil areas of the FPTA 3 site. Hydrogen half-lives in the vadose zone were 9-11 hrs in affected soils, and 19 hrs in clean soils.

6) Additional work is required to minimize volatilization of gaseous tracers in the test equipment and test volume.

FUTURE TESTING: Current plans by AFCEE for developing the direct hydrogen delivery technology include additional pull-push-pull treatability tests, year long pilot tests of hydrogen delivery via closed-loop recirculation, and two low-volume pulsed biosparging tests in Florida.

REFERENCES

Carr, C.S., and J.B. Hughes. 1998. "Enrichment of High Rate PCE Dechlorination and Comparative Study of Lactate, Methanol, and Hydrogen as Electron Donors to Sustain Activity." Environmental Science and Technology. v. 32, no. 12, pp. 1817-1824.

Fisher, R.T., C.J. Newell, J. B. Hughes, P.E. Haas, and P.C. Johnson, 1998. "Direct Hydrogen Addition and Pulse Biosparging for the In-Situ Biodegradation of Chlorinated Solvents," Proceedings of the Petroleum Hydrocarbon and Organic Chemicals in Ground Water Conference,, NWWA, Houston, Texas, November 1998 (in press).

Hughes, J.B., C.J. Newell, and R.T. Fisher. 1997. Process for In-Situ Biodegradation of Chlorinated Aliphatic Hydrocarbons by Subsurface Hydrogen Injection. U.S. Patent No. 5,602,296, issued February 11, 1997.

Newell, C.J., R.T. Fisher, and J.B. Hughes. 1997. "Direct Hydrogen Addition for the In-Situ Biodegradation of Chlorinated Solvents." *Proceedings of the Petroleum Hydrocarbons and Organic Chemicals in Ground Water: Prevention, Detection, and Remediation Conference.* November 12-14, 1997, Houston, TX. pp. 791 - 800.

Newell, C.J., J.B. Hughes, R.T. Fisher, and P.E. Haas, 1998, "Subsurface Hydrogen Addition for the In-Situ Bioremediation of Chlorinated Solvents," Designing and Applying Treatment Technologies, Remediation of Chlorinated and Recalcitrant Compounds, G.B. Wickramanayake and R.E. Hinchee, eds, Battelle Press, Columbus, Ohio.

Wiedemeier, T.H., Rifai, H.S., Newell, C.J., and Wilson, J.W. in press. *Natural Attenuation of Fuels and Chlorinated Solvents,* John Wiley & Sons, New York. in press.

EFFECT OF FENTON'S REAGENT ON SUBSURFACE MICROBIOLOGY AND BIODEGRADATION CAPACITY

James R. Kastner (University of Georgia, Athens GA), Jorge Santo Domingo (USEPA, Cincinnati Ohio), Miles Denham (Westinghouse Savannah River Co., Aiken SC), Marirosa Molina (USEPA, Athens, GA), Robin Brigmon (Westinghouse Savannah River Co., Aiken SC)

ABSTRACT: Microcosm studies were conducted to determine the effect of Fenton's reagent on subsurface microbiology and biodegradation capacity in a DNAPL (PCE/TCE) contaminated aquifer previously treated with the reagent. Groundwater pH declined from 5 to 2.4 immediately after the treatment, and subsequently rose to a range of 3.4 - 4.0 after 17 months. Groundwater microbial direct counts were approximately two orders of magnitude lower in the treated zone compared to the control after one year. Limited bacterial growth and TCE degradation were detected in the treated zone (pH 3.37 and TCE 5 mg/L) with CH_4 and phenol amendments. In contrast, methane addition to groundwater from the up-gradient control well stimulated considerable bacterial growth (pH 4.9 and TCE 0.7 mg/L), indicated by methane consumption, fluorescent antibody analysis, phospholipid based markers, and rDNA probes. TCE degradation was measured in the control microcosms, but only when phenol was added. These results suggest that high TCE concentrations, as well as the lower groundwater pH in the treated zone (i.e., pH 3.37 vs. 4.9) could have inhibited methanotrophic TCE co-metabolism. Bioremediation at the leading edge of plume may be possible, as long the local soil is able to buffer the groundwater pH (e.g., > 4.0-4.5). Alternatively, the Fenton's reagent process could be designed to operate at a higher pH (e.g., ≥ 4.0-4.5) to minimize detrimental effects, providing an optimal environment to couple advanced oxidation processes with bioremediation technologies.

INTRODUCTION

Dense non-aqueous phase liquids (DNAPL), such as TCE and PCE, are a major source of groundwater pollution throughout the United States. In-situ remediation of DNAPLs requires elimination of source zones. Advanced in-situ oxidation technologies offer a potential solution to the problem of source zone removal for DNAPL. One such process involves the use of catalyzed hydrogen peroxide for the in-situ oxidation of DNAPL. Fenton's reagent, (reactant, H_2O_2) and ferrous iron (catalyst, Fe^{2+}), react to produce hydroxyl radicals. The hydroxyl radical is a strong oxidizing agent that can rapidly degrade organic compounds to CO_2 and H_2O. Fenton's reagent has been used to treat a wide range of organic contaminants, both at the bench and field scale level (Brown et al., 1996; Ravikumar et al., 1994).

Aggressive conditions are required because the overall degradation process is mass transfer limited (Ravikumar et al., 1994). However, aggressive conditions (e.g., oxidative and acidic) can reduce microbial levels and viability within the aquifer. Low pH can significantly reduce microbial growth rates, limit microbial populations from reaching levels similar to those before the treatment, and alter or reduce the diversity of the microorganisms that repopulate the local groundwater and soil. This would have implications on any attempts to induce biostimulation of the aquifer for treatment of residual contaminants.

Objectives. Limited data is available on the effect of oxidizing agents on subsurface microbiology (Brown et al., 1996, Wolf et al., 1997). Thus, the purpose of this research project was to quantitatively determine the effect of Fenton's reagent on the microbiology and biodegradation capacity of a treated aquifer. Specific objectives were to determine the effect of pH and inducer addition (i.e., CH_4 and phenol) on microbial growth, TCE co-metabolism, and microbial diversity, and determine if a phosphate source, triethylphosphate (TEP), would stimulate microbial growth and TCE degradation.

METHODS

Site Description. The field site was the location of a previous demonstration using Fenton's reagent to oxidize a TCE/PCE DNAPL (Jerome et al., 1997). A DNAPL zone (144 ft below surface) of 64,000 ft^3 was treated via batch injections of Fenton's reagent (100 ppm $FeSO_4$, 50% solution of H_2O_2, and H_2SO_4 for pH adjustment; Jerome et al., 1997). Initially, PCE and TCE groundwater concentrations ranged between 100-160 mg/L and 15-25 mg/L respectively. After treatment, PCE and TCE groundwater levels were close to non-detect, but subsequently rebounded to near pre-test levels. The groundwater control well was located up-gradient from the test site (PCE and TCE concentrations of 19.85 mg/L and 5.65 mg/L respectively).

Sample Collection. Groundwater samples were aseptically obtained from three monitoring wells within the treated zone (MOX-5, 7, and 8) and one well outside the zone of influence of the catalyzed hydrogen peroxide (MSB-59D). Groundwater parameters were measured using pre-calibrated meters (YSI Inc., Models 6820 and 380, Yellow Springs, Ohio) and water samples were stored at 4°C until use. Subsamples of the groundwater were used to prepare acridine orange direct counts and plate counts (Lorch et al., 1995).

Enrichments and Microcosms. Methanotrophic enrichments were cultivated by adding approximately 28% CH_4 to the headspace of groundwater in pre-sterilized 120-ml serum bottles (50-ml liquid). No additional nutrients were added to the groundwater and the bottles were incubated at 23-25°C and 150 rpm. Periodically, every 2-3 weeks, transfers were made to fresh groundwater. Microcosm studies were performed in 60-160 ml serum bottles in triplicate (10-50 ml liquid; Fan and Scow, 1993). In one set of microcosms, phenol or toluene was added (50 mg/L) and in another a phosphate source was added. Inorganic phosphate and triethylphosphate (TEP) was added separately to achieve final concentrations of 0.01% (m/v), 0.07% v/v and 0.007% v/v, respectively. Controls consisted of systems that contained sterile water plus the metabolic inducer (CH_4, toluene or phenol) and groundwater minus the metabolic inducer. In a separate experiment, subcultures from previous methane enrichments were transferred into 60-ml amber serum bottles containing 25 ml of fresh groundwater (no other nutrients were added). Methane levels and TCE levels were measured via headspace analysis and liquid concentrations estimated using partition coefficients.

Microbial Analysis. Total bacterial counts were performed by the acridine orange direct count (AODC) method (Lorch et al., 1995). Cells from ten microscopic fields were then counted using epifluorescence microscopy and the average used to calculate total cell numbers. Culturable heterotrophic bacteria were enumerated by colony counts on PTYG medium (Balkwill, 1989). Methane enrichments and microcosms were also analyzed via fluorescent antibody staining, phospholipid fatty acid (PLFA) analysis (Vestal and White 1989) and rRNA targeting probes to determine the presence of methanotrophic bacteria (Santo Domingo et al, 1997). A direct immunofluorescent stain was used to tentatively identify and enumerate methanotrophic bacteria in the enrichments (Brigmon et. al., 1997).

Analytical Methods. Chloride, nitrate, nitrite, phosphate, and sulfate were quantified using a Dionex Model QIC2 ion chromatograph equipped with an IonPac Fast Anion column (Alef and Nannipieri 1995). VOC content of the microcosm headspace was analyzed using a Hewlett-Packard 5890 GC. The partition coefficient was used to estimate the TCE concentration in the liquid phase of the microcosms (Gosset 1987). A van't Hoff relationship taken from the literature was used to calculate Henry's constant at different temperatures (Gosset 1987). The TCE partition coefficient did not significantly deviate in actual groundwater when compared to deionized water (0.37 ± 0.048 from five different wells at 26°C).

RESULTS AND DISCUSSION

After catalyst injection, average groundwater pH declined from 5.71 to 2.44 and recovered slowly (Fig. 1). Additionally, dissolved oxygen was 2-3 times higher within the treated wells, than in control wells (24 ± 6.8 mg/L vs. 8.8 ± 1 mg/L). Average conductivity was also higher in the groundwater treated with Fenton's reagent, coinciding with elevated levels of Cu (e.g., 0.002 in MSB-59D vs. 5 mg/L in MOX-5), Mn, Al, and Si, relative to the control well.

Direct counts from the treated zone (groundwater) ranged from 1.2×10^4 to 1.8×10^5 cells/ml and plate counts ranged from 220 to 3665 CFU's/ml. However, direct counts were much higher in the control groundwater (MSB-59D) and ranged from 4.6×10^5 to 6.2×10^6 cells/ml. Based on direct counts, methane and TEP amendments stimulated microbial growth in all groundwater except MOX-5, indicating that although the pH of the groundwater was low, microbial growth was still possible if a carbon and/or phosphate source was added (pH from 3.6 to 4.9).

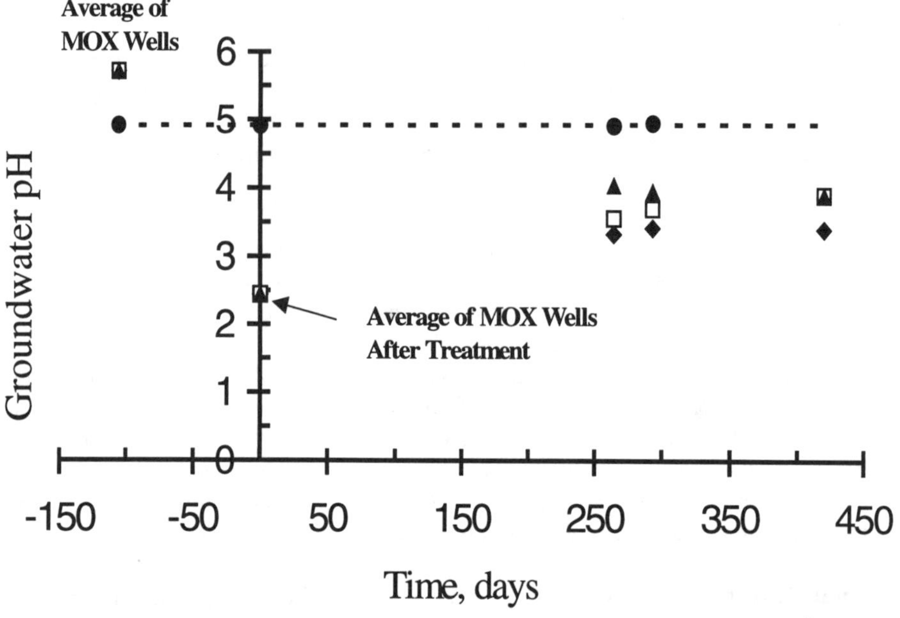

FIGURE 1: Effect of Fenton's Reagent on Groundwater pH (MOX-5 [◆], 7[□], and 8 [▲] are wells within the treated area; MSB-59D [●] was outside the zone of influence).

Bacterial cells from the MSB-59D enrichments gave strong positive signals to fluorescent antibodies specific for methanotrophs and to rDNA probes specific for type II methanotrophs. Moreover, PLFA analyses confirmed the presence of 18:1ω8, a biomarker for Type II methanotrophs, in the control enrichments (i.e., the untreated zone), but was not found in the Fenton treatments.

The type of microorganism and morphology was significantly different at pH 3.37 (MOX-5). Microscopy revealed hyphae and yeast-like microorganisms in the methane enrichments from MOX-5. Thus, the low pH of MOX-5 groundwater inhibited the growth of methanotrophic bacteria and may have selected for yeast or fungi. Yeast strains capable of oxidizing methane have been isolated from soils at pH 3.5 (Hanson and Hanson, 1996). The distinct differences in the type of microorganisms that developed in MOX-5 (pH 3.37) and MSB-59D (pH 4.9) suggest that pH will have a significant impact on the diversity of microorganisms capable of re-growth in the aquifer, as well as the rate of re-growth. Methane consumption was measured in the subcultures from the control at pH 4.9 (Fig. 2), when grown in minimal salts medium, but not in the treated system at pH 3.37. Microbial growth rates were inhibited by low pH in the MSB-59D enrichments;

the specific growth at pH 6.7 and 4.9 was 0.0696 day^{-1} (t_d = 239 hr; R^2 = 0.92) and 0.03864 day^{-1} (t_d = 430 hr; R^2 = 0.64) respectively. The lag time was significantly longer when the pH declined below 6.7 (Fig. 2).

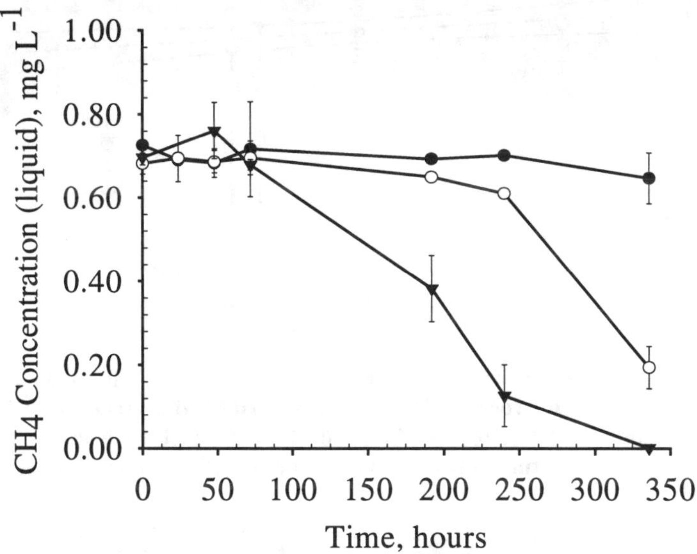

Figure 2: Effect of pH on growth and methane consumption (•, control; ○, pH 4.9; ▼, pH 6.7)

Initially methane consumption was not observed in microcosms designed to measure TCE degradation in groundwater, regardless of whether phosphate was added to the systems. After 70 days, phenol was spiked into the microcosms (~10 ppm). Subsequently, TCE loss was noted in the reactors with phenol addition, but only in the control well system and in those microcosms that contained a phosphate source (Fig. 3). Methane consumption and TCE loss was not observed in the MOX-5 microcosm (data not shown). In contrast, batch groundwater microcosms utilizing subcultures from methane enrichments did result in measurable CH_4 consumption, while significant TCE degradation was not observed during this period (Fig. 4).

Taken together these data suggest that the high TCE concentrations, combined with low pH, inhibited TCE degradation in the MOX-5 enrichments. Increased susceptibility to acute TCE toxicity by methanotrophs may also explain why phenol stimulated TCE loss in the MSB-59D enrichments, but methane did not. Although Fenton's reaction clearly reduced the number and viability of microorganism within the treated zone, it did not completely eliminate their presence. Thus, the long-term microbiological recovery of the treated aquifer is

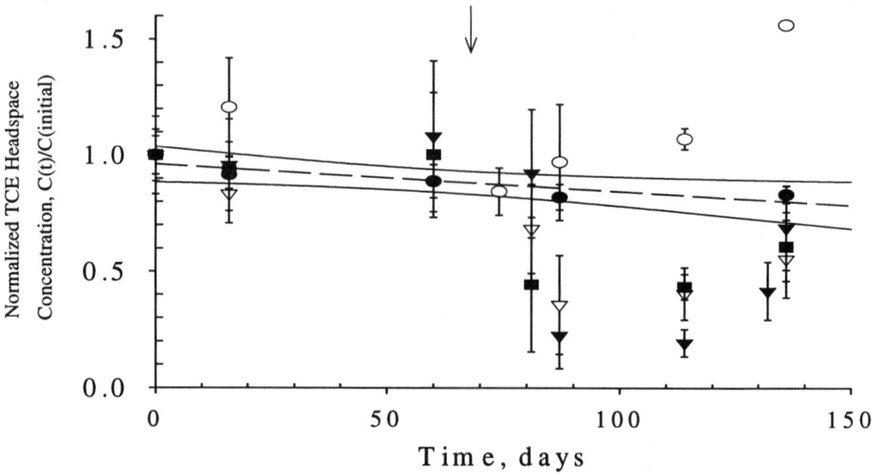

Figure 3: Effect of nutrient and metabolic inducer addition on TCE degradation in batch microcosms for the control well MSB-59D; control without CH_4, ●; CH_4 only, ○; CH_4+PO_4, ▼; CH_4+0.07% TEP, ∇; CH4 + 0.007% TEP, ■. Methane (28% at time zero) and phenol (arrow, 10 ppm) were added.

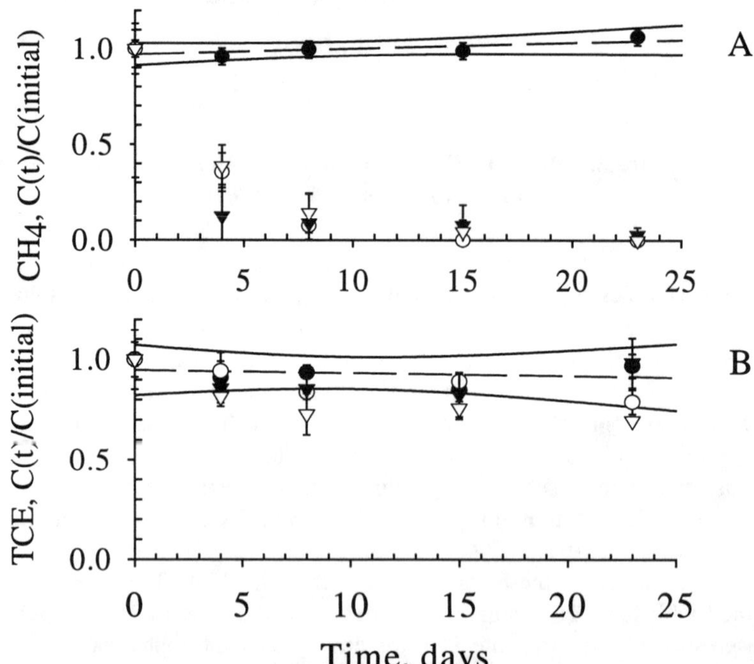

Figure 4: Effect of nutrient addition on methane consumption (A) and TCE loss (B) in methanotrophic microsoms of MSB-59D groundwater; control without CH_4 (B, ●) and sterile water (A, ●); CH_4 only, ○; CH_4+PO_4, ▼; CH_4+0.07% TEP, ∇.

possible. Infiltration of groundwater and the slow increase in groundwater pH should allow the growth and re-establishment of microbial populations.

However, based on this study, bioremediation of TCE via methane addition may not be practical after treatment with Fenton's reagent under aggressive conditions, especially if methantrophic concentrations are initially low. Even if cell densities are relatively high (e.g., a landfill) the effective coupling of bioremediation and chemical oxidation will depend on cell decay rates as a function of the oxidation conditions, TCE concentrations, and pH. Also, methanotrohic biomediation would be limited if the oxidation process releases high copper concentrations that inhibit the synthesis of sMMO.

Methanotrophic biobarriers down-gradient from the treatment zone may be possible at the plume edge. As the plume migrates outward from the treatment area, pH should increase and TCE concentrations decrease, enabling microbial growth and TCE degradation. A pH $\geq$ 4.0 seems to be a reasonable value that would allow stimulation of methanotrophic populations. Alternatively, stimulation of aerobic heterotrophs down-gradient (e.g., phenol) from the treatment zone may enhance TCE degradation rates. This is due to the fact the initial number of aerobic heterotrophic microorganisms is much higher than methanotrophs, thus minimizing the effect of Fenton's reagent. Additionally, aromatic degrading microorganism may have a wider pH tolerance compared to methanotrophs and may be less susceptible to TCE toxicity. While Fenton's reagent has recently been used to oxidize a wide range of organics in situ, a better understanding of its effect on the subsurface microbiota would improve the potential coupling of chemical oxidation with bioremediation processes. Such hybrid technology might prove effective in DNAPL remediation since oxidation processes can enhance the solubility of organic compounds and thus increase bioavailability.

Acknowledgements
We would like to thank G Victoria, M Frank, F Washburn, and H Findley for their technical assistance. J.S.D. was supported in part by an appointment to the U.S. DOE Postgraduate Research Program at the SRS administered by the ORISE. This project was funded by DOE (OTD, Subsurface Contaminants Focus Area) under the TTP, SR1-8-SS-31.

References

Alef K and Nannipieri P. 1995. "*Methods in Applied Soil Microbiology and Biochemistry*," Harcourt Brace & Company, New York, NY.

Balkwill, D. L., 1989. "Numbers, diversity, and morphological characteristics of aerobic, chemoheterotrophic bacteria in deep subsurface sediments from a site in South Carolina." *Geomicrobiol. J.* 7:33-52.

Brigmon RL, MM Frank, JS Bray, DG Scott, KD Lanclos, 1997. "Direct immunofluoresence and enzyme-linked immunosorbent assays for evaluating organic contaminant degrading bacteria." *J Microbiol Methods* 32:1-10.

Brown RA, C Nelson, M Leahy, 1996. "Combining Oxidation and Bioremediation for the Treatment of Recalcitrant Organics." In-Situ and On-Site Bioremediation: Vol 4: 457-462.

Fan S and KM Scow. 1993. "Biodegradation of Trichloroethylene and Toluene by Indigenous Microbial Populations in Soil." *Appl. Environ. Microbiol.* 59(6): 1911-1918.

Greenberg RS, T Andrews, Kakarla KC, Watts RJ, 1998. "In-situ Fenton-Like Oxidation of Volatile Organics: Laboratory, Pilot, and Full-Scale Demonstrations." Remediation Spring, p.29-42.

Gosset JM. 1987. "Measurment of Henry's Law Constants for C_1 and C_2 Chlorinated Hydrocarbons." *Environ. Sci. Technol.* 21:202-208.

Hanson RS and TE Hanson. 1996. "Methanotrophic Bacteria." *Microbiol Reviews.* 60(2):439-471.

Jerome KM, B Riha, B Looney. 1997. "*Final Report for Demonstration of In Situ Oxidation of DNAPL Using the Geo-Cleanse Technology.*" Westinghouse Savannah River Company. Aiken, SC. WSRC-TR-97-00283.

Lorch HJ, G Benckieser, JCG Ottow. 1995. "Basic methods for counting microorganisms in soil and water: *Methods in Applied Soil Microbiology and Biochemistry,*" Chapter 4: Enrichment, isolation and counting of soil microorganism, Alef K and Nannipieri P (eds.), p. 146. Harcourt Brace & Company, New York, NY.

Ravikumar JX and Gurol MD. 1994. "Chemical Oxidation of Chlorinated Organics by Hydrogen Peroxide in the Presence of Sand." *Environ. Sci. Technol.*, 28: 394-400.

Santo Domingo JW, CJ Berry, TC Hazen. 1997. "Use of conventional methods and whole cell hybridization to monitor the microbial response to triethylphosphate." *J. Microbiol Methods.* 29:145-151.

Vestal, J. R. and D.C. White. 1989. "Lipid analysis in microbial ecology." *BioScience* 39:535-541

Wolf JL, S Jones, Lawrence DM, Mutch SD, 1997."In-situ Remediation of MTBE Using OxyVac Technology." *Remediation Management*, Third Quarter, p.40-44.

TREATMENT OF TCE-IMPACTED GROUNDWATER USING IN SITU CHEMICAL OXIDATION

Christopher H. Nelson, IT Corporation, Englewood, Colorado
Craig S. Barker, IT Environmental (Australia) Pty Ltd, Adelaide, South Australia

ABSTRACT: Manufacturing operations over the past 50 years at an industrial site in Adelaide, South Australia resulted in the release of chlorinated hydrocarbons (i.e., TCE) to groundwater beneath the facility. Groundwater flow has caused off-site impacts to the surrounding suburban communities, prompting the decision to address the groundwater contamination. Groundwater remediation was further complicated by the clayey nature of the aquifer matrix. Traditional remediation methods such as pump and treat and dual phase extraction were evaluated as potential remedial technologies. This analysis indicated that traditional remedial approaches would be ineffective and/or expensive. For example, pump and treat cost estimates were in the $15 MM range. Given these constraints in-situ chemical oxidation was proposed as an innovative remedial approach.

In-situ chemical oxidation technologies include the subsurface injection of ozone, Fenton's reagent (hydrogen peroxide and iron), or potassium permanganate. Potassium permanganate injection was selected for this project due to technical and economic considerations.

Laboratory treatability studies were conducted using soil and groundwater samples collected from the site. Slurry systems were used to model the groundwater aquifer system. These tests showed that moderate levels of permanganate (50 ppm) could degrade high levels of TCE (5,000 ppb) to non-detectable levels within six hours with residual levels of permanganate remaining. Based on these results, field pilot tests were conducted to further explore the oxidation chemistry and mass transport challenges of uniformly distributing permanganate in a clayey aquifer. Several injection methods were tried all of which included soil fracturing.

The field test results confirmed the importance of permanganate mass transport and the need for controlled delivery in the subsurface. By taking these factors into account, a successful pilot test was conducted which showed dramatic TCE reductions, with TCE levels as high as 8,600 ppb reduced to below detection limits in six days. Based on the results of the laboratory and field tests, a full-scale potassium permanganate remediation system is under implementation at the site.

INTRODUCTION

Common chemical oxidants include ozone, Fenton's reagent (hydrogen peroxide and iron), and potassium permangante. Chemical oxidation is widely used for the treatment of wastewaters and drinking water. Chemical oxidation of wastewater is typically employed for the treatment of recalcitrant contaminants that are resistant to biological degradation or are not amenable to separation processes. Chemical oxidation is also used as a pretreatment step to enhance biological treatment of recalcitrant contaminants.

Potassium permanganate has been used since the early 1900's for the treatment of drinking water. It's oxidation potential (-1.695 V) is less than the oxidation potentials for ozone and hydrogen peroxide, however, potassium permanganate is an effective oxidizer of chlorinated ethenes such as trichloroethylene (TCE) and aromatics such as polyaromatic hydrocarbons (PAH) (Ladbury and Cullis, 1958). Recently, chemical oxidation has been investigated for in-situ remediation applications using ozone (Nelson, et al. 1997), Fenton's reagent (Greenberg et al. 1997), and potassium permanganate (Gates, Siegrist, and Cline, 1995 and Schnarr et al., 1998). This paper describes the laboratory and field demonstration of in-situ oxidation using potassium permangante for TCE impacted groundwater at an industrial site in Adelaide, South Australia.

The industrial facility where the testing was conducted has been in operation since the 1940's. During this time a variety of chlorinated and non-chlorinated hydrocarbons were used in manufacturing processes; the primary contaminant released into soils and groundwater at the site was TCE.

Site remediation was complicated by the clayey nature of the shallow aquifer beneath the facility. Site geology consists of an upper filling to a depth of up to one metre, over Alluvial (Quaternary) deposits, consisting of silts, silty clays and silty sands. These extend to a depth of four metres, and are underlain by clay and clayey silt, to a depth of six metres. Beyond this depth, soils grade to a sandy gravelly clay, extending to at least 10 metres, which in turn overlies sandy clays and clay extending to at least 25 metres. A shallow aquifer occurs uniformly across the site ranging in depth from 2.4 to 4.9 metres. The aquifer is typically unconfined and of low yield, with a uniform flow gradient of between 0.0065 to 0.007. For the saturated sandy/gravelly clays and silts, a low hydraulic conductivity (K) of 0.1 m/day has been estimated. Where more permeable sands, silty sands or silts are present, a K value of up to 10 m/day has been estimated. Total dissolved solids (TDS) in groundwater at the site average 1,593 mg/l. Surrounding groundwater usage within a 1.5 kilometre radius from the site indicates that at least 35 groundwater extraction wells are present. These wells were of primary importance in evaluating the need for remediation according to a risk assessment performed for the site.

METHODS

Laboratory Test. Before field pilot studies commenced, a laboratory study was initiated to determine the feasibility of oxidation of TCE by potassium permanganate. The focus of the laboratory study was to obtain useful information for a field pilot study. Groundwater and soil samples were obtained from the site for use in the study. The aims of the laboratory study were to investigate TCE oxidation by a potassium permanganate solution and to investigate the change in potassium permanganate concentration during the oxidation of TCE.

To simulate field conditions, a 10% weight to volume (w/v) slurry of site soil in groundwater was prepared. The TCE concentration in the groundwater was adjusted to 4.95 mg/l to simulate site conditions. Potassium permanganate

($KMnO_4$) solution was added to the slurry to give a starting concentration of 50 mg/l. The slurry was stirred periodically. Samples of supernatant were taken periodically by syringe.

The potassium permanganate concentration declined rapidly then stabilized at an equilibrium concentration of 28 mg/l after 300 minutes. The residual permanganate remaining after 24 hours was 28 mg/l. The pH of the supernatant after 24 hours of treatment with $KMnO_4$ solution was unchanged at pH = 7.5. There was no visual evidence of the formation of manganese dioxide or any associated precipitates. TCE was rapidly oxidized after potassium permanganate addition. The initial TCE concentration of 4.95 mg/l was reduced to below detection limits within 400 minutes. The rate constant for this process was estimated to be 0.014 min^{-1} and the half-life was estimated to be 49.5 minutes.

Field Tests. Information from the laboratory study was used to size and focus the field pilot tests. Two pilot tests for deployment of potassium permanganate were designed to evaluate the available chemical injection alternatives and to provide full-scale transport design data. From past site assessment work, it was observed that the top two metres of the shallow unconfined aquifer at the site generally consists of a clayey silt or sandy silt material (from 4.0 metres to 6.0 metres typically below gsl). Below this depth, aquifer materials consist of a clay and sandy clay, extending to at least 25 metres. As such, the pilot tests were designed to assess the performance of oxidant delivery through each of these geologic units on a separate basis.

Injection Well Testing. The first test (Test #1) for delivering oxidant involved dosage application to a nested injection well. The injection wells were nested and screened at six and ten metres with 1.5 metres of screen. Each dosage test was designed to treat a subsurface zone approximately 9 metres in diameter by 1.5 metres thick. Data from aquifer pumping tests was used to design the injection rate and volumetric additions necessary to impact the test area. Oxidant injection occurred over a period of two days. Prior to injection of any oxidant, groundwater was extracted from the pilot test well to establish steady state conditions. Three water samples were collected over time for laboratory analysis of TCE and related VOCs. Samples for TDS, total organic carbon (TOC), alkalinity, and cations/anions were also taken. Field samples were taken for the measurement of water quality parameters (temperature, pH, dissolved oxygen (DO), redox potential (Eh), and specific conductance). Eh was used as a routine field measurement to monitor for both the lateral and vertical spreading of the oxidant front over time.

A concentrated oxidant solution was mixed using a batch mixing system with make-up water. The oxidant solution was mixed and injected continuously over the test time period. The permanganate injection rate and volume were metered and monitored. Monitoring of standing water levels in monitoring wells was also used as an indicator of permanganate injection influence. At 24 hours before and six weeks after the test, and during injection periods, five groups of surrounding nested PVC monitoring wells adjacent to the injection well and within the immediate zone of influence, together with three single shallow wells, were monitored for water head level and sampled periodically for water quality and VOC analysis. The multi-level monitoring wells allowed both lateral and vertical injection influence to be monitored. Before and after the test commencement, water samples from the injection well were also collected and analysed.

Pressure Injection Testing The second test (Test #2) evaluated oxidant injection under pressure into separate depth intervals of the aquifer (4.5 to 6.0 metres and 7.5 to 9.0 metres) as a concentrated solution via direct push boreholes. The layout consisted of two lines of injection boreholes (eight boreholes in total) orientated perpendicular to the interpreted flow direction for the shallow aquifer. The boreholes had a typical spacing of 2 metres. Oxidant injection was conducted over two consecutive days.

Prior to borehole injection, a total of seven nested PVC groundwater monitoring wells were drilled and placed at selected locations within and surrounding the injection borehole gallery. Following development, wells were level surveyed and gauged for standing water levels to confirm local flow direction. All wells were sampled and tested for VOCs and associated water quality parameters, major cations/anions, TDS, and alkalinity. Following the collection of baseline water samples, the injection boreholes were driven using a direct push drilling rig. Hydraulic fracturing (fracing) of the aquifer unit was initiated with low volume-high pressure liquid addition into the formation.

During and following pressure injection work, groundwater levels in adjacent monitoring wells were monitored with time. This monitoring program was then extended to cover a six week period. Water quality parameters were also monitored at the various monitoring wells with time. Frequent measurement of Eh were made within all monitoring wells to assess the migration of the oxidant front and three separate rounds of water sampling were undertaken at all monitoring wells (following initial baseline sampling).

Because of the clayey nature of the soils at the site, soil hydraulic fracturing was necessary to allow flow of permanganate solution from the injection points out into the saturated aquifer sediments. Based on the depth of the soil column overlying the groundwater table and assuming the soils were highly consolidated (the over-consolidation ratio was estimated to be at least 3), the minimum break-out pressure was estimated to be 340 kPa. Aquifer pore space water in Test Areas 1 and 2 was estimated to be 122,522 litres. This value was used to calculate average permanganate concentrations in the test areas after injection.

RESULTS

TCE reduction results for Test #1 and Test #2 are presented in Tables 1 and 2. The results from the injection of potassium permanganate through injection wells (Test #1) were variable for TCE removal. Table 1 shows the positive influence of injecting permanganate into well MW2BD on TCE removal; reducing initial TCE concentrations of 513 ppb to less than 5 ppb in 96 hours. TCE concentrations remained below detection levels in this well for the duration of the test. This was in contrast to the results from well MW3BS where TCE levels remained virtually unchanged for the duration of the test. Results from MW5BD indicate increasing then decreasing TCE trends. Overall, results from Test #1 indicate that permanganate is an effective oxidizer of TCE if sufficient concentrations are available for the oxidation reaction to occur in a given area. Therefore, permanganate mass transport considerations are particularly important especially for a less permeable site such as this one.

The results from the injection of potassium permanganate through direct push injection (Test #2) were much more positive for the removal of TCE. Of the twelve monitoring wells in Test Area 2, ten wells showed substantial reductions of one to two orders of magnitude over the six week testing period. Representative TCE reduction data for wells MW6CS and MW4CS is shown in Table 2. Starting TCE concentrations of 8,600 ppb and 6,500 ppb respectively, were reduced to below detection limits in less than 144 hours (6 days) and these non-detectable results were maintained for at least 960 hours (40 days). These results are very encouraging as compared to other treatment options for in-situ TCE treatment. Some TCE rebound was observed in three of the twelve monitoring wells as is shown in Table 2 for well MW7CS. This is a result of the mass transport challenges of injecting permanganate into a clayey aquifer as was discussed previously and the reduced permanganate injection volumes used in some injection locations.

DISCUSSION AND CONCLUSIONS

Laboratory and field pilot tests were conducted to evaluate potassium permanganate injection for the in-situ treatment of TCE at an industrial site in Adelaide, South Australia. The following conclusions were developed as a result of these tests:

- Potassium permanganate is a very effective oxidizer of TCE. In the laboratory study, 50 ppm of permanganate oxidized 5,000 ppb of TCE to below detection limits in a groundwater slurry from the site within six hours. During field pilot tests, TCE concentrations as high as 8,600 ppb were reduced to below detection limits within 144 hours;
- There is a direct correlation between permanganate concentrations and TCE reduction. During the field tests, wells with low to non-detectable levels of permanganate showed minimal TCE reduction, wells with permanganate levels above 50 mg/l to 100 mg/l showed significant TCE reductions. Therefore, optimising the transport and distribution of permanganate in the subsurface is essential to ensure successful in-situ treatment;
- Pressure injection of permanganate through injection wells (Test #1) was less successful in uniformly distributing permanganate in the subsurface as compared to direct push pressure injection (Test #2) which was much more effective in achieving a uniform distribution;
- Based on the overall positive results from Test #2, full-scale application of permanganate is on-going at the site.

In summary, the laboratory and field tests conducted to demonstrate permanganate treatment of TCE at this site were successful. They provided critical information on the feasibility of permanganate treatment and full-scale application of the technology. Based on these results, in-situ chemical oxidation using potassium permanganate offers many technical and cost advantages over other remedial approaches for the in-situ treatment of TCE.

Table 1. TCE Concentrations in Groundwater (Test #1) (All units in $\mu g L^{-1}$ except where otherwise noted)

Well Location	Time From Injection (Injection Start at T = 0 h)			
	T = 0 h	T = 96 h	T = 552 h	T = 1056 h
MW2BD (Deep Well)	513	< 5	< 5	< 5
MW3BS (Shallow Well)	355	352	359	299
MW5BD (Deep Well)	987	1,170	1,250	646

Table 2. TCE Concentrations in Groundwater (Test #2) (All units in $\mu g L^{-1}$ except where otherwise noted)

Well Location	Time From Injection (Injection Start at T = 0 h)			
	T = 0 h	T = 144 h	T = 432 h	T = 960 h
MW6CS (Shallow Well)	8,600	< 5	< 5	< 5
MW4CS (Shallow Well)	6,500	< 5	< 5	< 5
MW7CS (Shallow Well)	5,230	< 5	< 5	422

ACKNOWLEDGEMENTS

Many individuals contributed to the success of this project in the lab and the field, including: Tony Briggs, Shaun Wybrow, Joe Pedicini, Marc Andrews, Susan Richards, Greg Brickle, David Burns, Dacre Bush, and Robert Siegrist.

REFERENCES

Gates, D.D., R.L. Siegrist, and S.R. Cline, 1995. "Chemical Oxidation of Volatile and Semi-Volatile Organic Compounds in Soil." Air and Waste Management Association Conference, June.

Greenberg, R.S., et al., 1997. "In Situ Fenton-Like Oxidation of Volatile Organics: Laboratory, Pilot and Full-Scale Demonstrations." Water Environment Federation 70th Annual Conference, Workshop on In Situ Chemical Oxidation.

Ladbury, J.W. and C.F. Cullis, 1958. "Kinetics and Mechanism of Oxidation by Permanganate." *Chemical Reviews 58*:403-437.

Nelson, C.H., et al., 1997. "Ozone Sparging for the Remediation of MGP Contaminants." Fourth International Symposium on In Situ and On-site Bioremediation, Vol. 3, pp. 457-462.

Schnarr, M. et al., 1998. "Laboratory and Controlled Field Experiments Using Potassium Permanganate to Remediate Trichloroethylene and Perchloroethylene DNAPLs in Porous Media." *Journal of Contaminant Hydrology, 29*:205-224.

INTEGRATED ANAEROBIC DEGRADATION OF TETRACHLOROTHYLENE BY VITAMIN-B_{12} AND ZERO-VALENT IRON

Ying-Chih Chiu (National I-Lan Inst. of Tech., I-Lan, Taiwan 260, ROC)
Chih-Jen Lu, Mon-Fan Yeh, and Chung-Chih Yang
(National Chung Hsing University, Taichung, Taiwan 402, ROC)

ABSTRACT: In this laboratory study, vitamin B_{12} and zero-valent iron was added to enhance the process of anaerobic dechlorination of PCE by a mixed culture of *Methanosaeta concilii* and *Methanosaeta sp.*. Experimental results showed that the initial PCE removal rate increased with the addition of vitamin B_{12}. When the concentration of vitamin B_{12} increased from 0 nM to 240 nM, the initial PCE removal rate increased from 4.8 μg/L-d to 20.1 μg/L-d. However, the rate increase insignificantly when the concentration of vitamin B_{12} was higher than 240 nM. The addition of iron powder also resulted in decreasing the formation of chlorinated intermediates, ethene was the major end-product, vinyl chloride was under the detection limit. The presence of microbial cells and iron powder (5.0 g-Fe^0/L) presented a PCE pseudo-first-order degradation reaction constant of 0.43 day^{-1}. The PCE removal rate was further enhanced to 0.49 day^{-1} when integrated with 240 nM of vitamin B_{12}. The effluents were examined by anaerobic toxicity assay (ATA) and found the maximum rate ratio (MRR) were greater than 0.95. Therefore, using anaerobic microbes with addition of zero-valent iron and vitamin B_{12} is a feasible method to treat PCE-contaminated groundwater.

INTRODUCTION

Halogenated aliphatic compounds are prevalent groundwater contaminants and significant components of hazardous wastes and landfill leachates. Chlorinated compounds are the best known and most studied VOCs because of the highly publicized problems associated with DDT, other pesticides, and numerous industrial solvent, such as cleaning solvents in dry-cleaning operations and semiconductor manufactures. In Taiwan, groundwater contributes about 22% of drinking water for more than 20% of population. Based on monitoring data, some groundwaters in the north of Taiwan are contaminated by chlorinated aliphatics, such as tetrachloroethylene (PCE) and carbon tetrachloride.

Halogenation is often implicated as a reason for persistence of these compounds in the environment. However, microbial transformation of the polychlorinated hydrocarbons, like carbon tetrachloride, tetrachloroethylene, and 1,1,1-trichloroethane, are catalyzed only under anaerobic conditions by dehalogenation mechanisms (Vogel and McCarty, 1987; Chaudhry and Chapalamadugu, 1991). The addition of catalysts, such as B_{12} and zero-valent iron, enhances the process of anaerobic dechlorination of tetrachloroethylene (Gillham and O'Hannesin, 1994; Appleton, 1996; Ballapragada *et al.*, 1997; Fennelly and Roberts, 1998; Novak *et al.*, 1998). The zero-valent iron can serve as a subsurface reactive barrier to capture the pollution plume in groundwater and degrade the contaminants (Bradley and Chapelle, 1997). However, a process of contamination remediation can only be applied if the transformation products are environmentally acceptable. The goal of this project is to comprehend the biotransformation of PCE enhanced by addition of vitamin B_{12} and zero-valent iron, the intermediates were also under monitoring.

MATERIALS AND METHODS

Seed. The microbial consortium collected from a winery anaerobic digester was acclimated by 20 µg/L PCE in a chemostat reactor at a hydraulic detention time (HRT) of 20 days. Sodium acetate of 7,000 mg/L was injected into the reactor and used as the primary substrate. The basal inorganic nutrients listed in Table1 provided the trace elements and alkalinity for the system. After identification, this mixed culture was predominated by *Methanosaeta concilii* and *Methanosaeta sp.*.

TABLE 1. The composition of basal inorganic nutrients.

Inorganic	Conc. (mg/L)	Inorganic	Conc. (mg/L)
NH_4Cl	1200	NH_4VO_3	0.5
$MgSO_4 \cdot 7H_2O$	400	$CuCl_2 \cdot 2H_2O$	0.5
KCl	400	$ZnCl_2$	0.5
$Na_2S \cdot 9H_2O$	300	$AlCl_3 \cdot 6H_2O$	0.5
$CaCl_2 \cdot 2H_2O$	50	$NaMoO_4 \cdot 2H_2O$	0.5
$(NH_4HPO_4)_2$	80	H_3BO_3	0.5
$FeCl_2 \cdot 4H_2O$	40	$NiCl_2 \cdot 6H_2O$	0.5
$CoCl_2 \cdot 6H_2O$	10	$NaWO_4 \cdot 2H_2O$	0.5
KI	10	Na_2SeO_3	0.5
$(NaPO_3)_6$	10	Cysteine	10
$MnCl_2 \cdot 4H_2O$	0.5	$NaHCO_3$	1000

Batch Reaction. Anaerobic dechlorination of tetrachloroethylene by acclimated microbial consortia was conducted in batch reactors to evaluate bioremediation of contaminated groundwater. An 80 mL aliquot of mixed liquor of acclimated microbial consortia and basal inorganic nutrients was injected into each of a set of six 125 mL serum bottles. Based on previous study (Chiu *et al.*, 1997), sodium acetate at 60 mg/L was the optimum concentration as the primary substrate for this system. Blank and sterilized sets were pre-examined to estimate the loss resulted from volatilization and adsorption.

Analytical Methods. A Hewlett Packard gas chromatography (HP GC-5890II) with a electron capture detector (ECD) and a 30 m glass capillary column (J & W

Scientific, DB-1) was used to detect PCE, trichloroethylene (TCE), and dichloroethylene (DCE). Bromoform acted as the internal standard during the extraction of PCE, TCE, and DCE by n-pentane. Nitrogen gas of 10 psi served as the carrier and makeup gas of the 1 μL injected sample. Temperature of detector and injection port were 280 °C and 150 °C, respectively. A temperature program was applied to get better result; initial column temperature of 60 °C was kept for 13 min then the temperature was raised at 25 °C/min to 180 °C. Vinyl chloride (VC) and ethene (ETH) were measured by Hewlett Packard gas chromatography (HP GC-5890A) with a flame ionization detector (FID). A 2.4 m stainless packed column (Supelco, Carbopack B/1% sp-1000) was used for VC detecting. Nitrogen gas of 40 mL/min served as the carrier and makeup gas of the 0.5 mL injected sample. Temperature of detector and injection port were 300 °C and 200 °C, respectively. Initial column temperature of 40 °C was kept for 6 min then the temperature was raised at 25 °C/min to 150 °C. Ethene was measured by a 1.83 m packed column (Supelco, porapack R) with nitrogen gas of 30 mL/min served as the carrier and makeup gas. The other operating parameters were the same as VC detection.

RESULTS AND DISCUSSION

Vitamin B_{12} Effect. Experimental results showed that the initial PCE removal rate increased with the addition of vitamin B_{12}. Based on Figure 1a, when the concentration of vitamin B_{12} increased from 0 nM to 240 nM, the initial PCE biotic removal rate increased from 4.8 μg/L-d to 20.1 μg/L-d. However, there were no significant difference between the final amounts of PCE transformations. A concentration of vitamin B_{12} higher than 240nM did not contribute significant benefit, based on the reaults shown on Figure 1b.

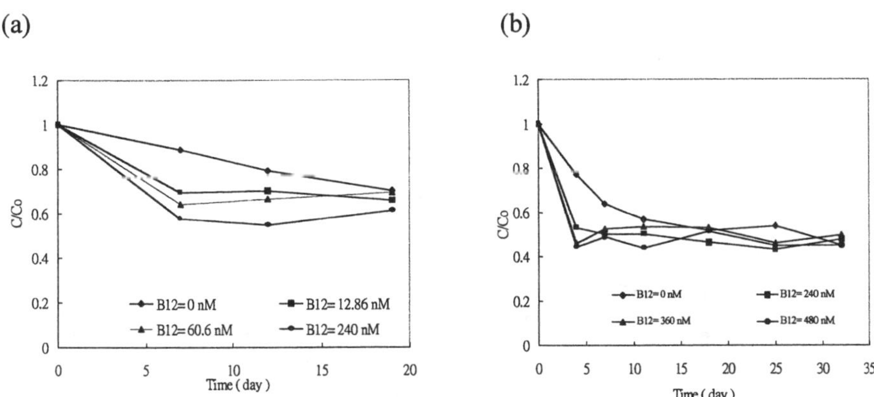

FIGURE 1. The biotransformation of tetrachloroethylene (170 nmol/bottle) under different concentrations of vitamin B_{12}.

Zero-Valent Iron Effect. Figure 2 shows the integration of acclimated microbial consortia with substrate sodium acetate (NaAc) at 60 mg/L and acid-washed (1 M HCl for 2 min.) Fe^0 (5g/L) powder. The combination resulted in higher PCE removal rate than the performations of Fe^0 or microbes along. However, the addition of Fe^0 significantly enhanced the biotransformation of PCE to trichloroethylene (TCE) then to ethene as shown on Figure 3 and Figure 4. Without the addition of Fe^0, the concentration of biotransformated TCE was still getting higher after 52 days. The accumulation of VC was not observed in this system.

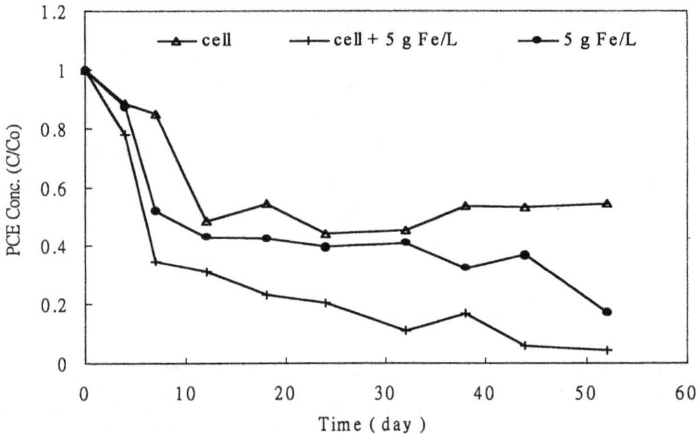

FIGURE 2. Effect of Fe^0 addition on tetrachloroethylene transformation with initial concentation of 150 nmole/bottle.

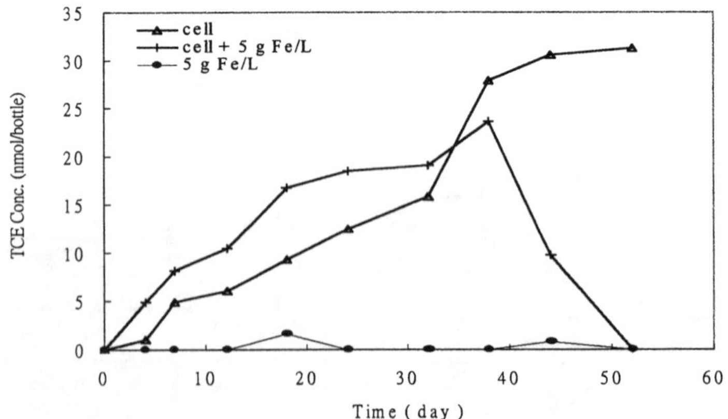

FIGURE 3. Effect of Fe^0 addition on trichloroethylene accumulation from transformations of PCE (150 nmole/bottle).

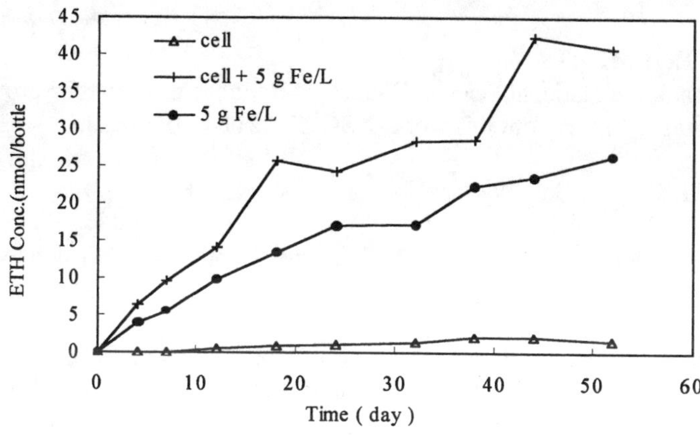

FIGURE 4. Effect of Fe^0 addition on ethene accumulation from transformations of PCE (150 nmole/bottle).

Comparing to the performance of Fe^0, the additions of vitamin B_{12} contributed limited improvements, as shown on Table 2. However, integrated degradation by Fe^0 and vitamin B_{12} can get highest removal of PCE and prdouction of ethene.

TABLE 2. Transformation of Tetrachloroethylene under different conditions.

Addition	Pseudo-first-order reaction constant (day^{-1})	R^2
Cell	0.105	0.92
5.0 g-Fe^0/L	0.430	0.99
5.0 g-Fe^0/L+ Cell	0.444	0.96
5.0 g-Fe^0/L+ Cell +240nM vitamin B_{12}	0.490	0.95

Anaerobic Toxicity Assay. To check the environmental acceptability of this process, the effluents were examined by anaerobic toxicity assay (ATA). Maximum rate ratio (MRR) is defined as the gas production ratio of a sample and a controlled blank. The MRRs of all the effluents in this study were greater than 0.95, and meant no inhibition (Owen *et al.*, 1979). Therefore, using anaerobic microbes with addition of zero-valent iron and vitamin B_{12} is a feasible method to treat PCE-contaminated groundwater.

CONCLUSIONS

Tetrachloroethylene was biotransformated to ethene in this system. Vitamin B_{12} resulted in increasing the initial PCE biotic removal rate, but did not contribute significant benefit in this system. To remedy PCE contamination in groundwater, zero-valent iron not only can be used as barrier system material, but can be effectively integrated with biotic system.

ACKNOWLEGEMENTS

We thank the National Science Council of Republic of China for providing financial assistance to carry out this work (NSC-87-2211-E-197-002).

REFERENCES

Appleton, E. L. 1996. "A Nickel-iron Wall Against Contaminated Groundwater." *Environ. Sci. Technol.* 30: 536A-39A.

Ballapragada, B. S., H. D. Stensel, J. A. Puhakka, and J. F. Ferguson. 1997. " Effect of Hydrogen on Reductive Dechlorination of Chlorinated Ethenes." *Environ. Sci. Technol.* 31: 1728-1734.

Bradley, P. M. and F. H. Chapelle. 1997. "Kinetics of DCE and VC Mineralization under Methanogenic and Fe(III)-Reducing Conditions.", *Environ. Sci. Technol.*. 31: 2692-2696.

Chaudhry, G. R. and S. Chapalamadugu. 1991. "Biodegradation of Halogenated Organic Compounds." *Microbio. Rev.*. 55 (1): 59-79.

Chiu, Ying-Chih, Chih-Jen Lu, and Su-Ying Huang. 1997. "Anaerobic Dechlorination of Tetrachloroethylene to Ethene Using a Winery Microbial Consortium." *In Situ and On-Site Bioremediation-The 4th International Symposium*. 3: 51-56, New Orleans, Louisiana, USA, Apr. 28-May 1.

Fennelly, J. P. and A. L. Roberts. 1998. "Reaction of 1,1,1-Trichloroethene with Zero-Valent Metals and Bimetallic Reductants." *Envrion. Sci. Technol.* 32 (13): 1980-1988.

Gillham, R. W. and S. F. O'Hannesin. 1994. "Enhanced Degradation of Halogenated Aliphatics by Zero-Valent Iron." *Ground Water.* 32: 958-967.

Novak, P. J., L. Daniels, , and G. F. Parkin. 1998. "Enhanced Dechlorination of Carbon Tetrachloride and Chloroform in the Presence of Elemental Iron and *Methanosarcina barkeri, Methanosacina thermophila*, or *Methanosaeta concillii.*" *Envrion. Sci. Technol.* 32(10): 1438-1443.

Owen, W. F., D. C. Stuckey, J. B. Healy, Jr., L. Y. Young, and P. L. McCarty. 1979. "Bioassay for Monitoring Biochemical Methane Potential and Anaerobic Toxicity", *Water Res.* 13: 485-492.

Vogel, T. M. and P. L. McCarty. 1987. "Abiotic and Biotic Transformation of 1,1,1-Trichloroethane under Methanogenic Conditions." *Environ. Sci. Technol.* 21(12): 1208-1213.

BACTERIAL TRANSPORT ISSUES RELATED TO SUBSURFACE BIOBARRIERS

Robert R. Sharp (Manhattan College, New York)
Robin Gerlach and Al Cunningham (Center for Biofilm Engineering, Bozeman, Montana)

ABSTRACT: Numerous studies aimed at developing static and reactive biobarriers in porous media using starved bacterial cultures have been carried out. These studies focused on the transport and resuscitation of starved bacterial cultures in porous media. The studies used 1 to 50 ft long, continuous flow, packed porous media columns. The studies evaluated the transport properties of starved bacterial cultures for use in the development of biobarriers. Biobarriers are developed by transporting starved cultures into the subsurface and resuscitating them so that they may carry out their desired reactions *in situ* (i.e. biofilm production, BTEX degradation, TCE degradation, redox reactions and others).

Results from column transport studies demonstrated that starved bacterial cultures have enhanced transport properties, as compared to vegetative cultures. Vegetative cells tend to attach to porous media close to the injection point while starved cells distribute themselves relatively evenly along the porous media flow paths. In addition, a significant number of the starved cells were detected at distances as great as 50 ft away from the injection point. Further column studies using biofilm producing cells *(Klebsiella oxytoca)* indicated that the starved cultures that are distributed along the column length can effectively be resuscitated back to their vegetative state *in situ*. The resuscitation and subsequent growth led to the, in this case, desired plugging of the columns.

The exact reasons for the enhanced transport properties of starved cells are not fully understood, however, the characteristics of starved cells that may lead to their enhanced transport are discussed.

INTRODUCTION

Bacterial transport plays a vital role in a number of bioaugmentation based remediation technologies including the design and development of static and reactive biobarriers. The ability to effectively transport a bacterial population through porous media can greatly enhance the effectiveness and development of bioaugmentation technologies.

The transport of bacteria through porous media is influenced by many parameters, including the properties of the bacterial cells, porous media characteristics, solution chemistry, and physical parameters, such as the interstitial fluid velocity (Mills 1997, Lawrence and Henry 1996, Harvey 1991 and references therein).

Of all of the parameters that influence bacterial transport, only those that are relatively easy and cost effectively manipulated are worth investigating for the development of competitive bioremediation technologies. The parameters that appear to be the easiest to manipulate are those related to the bacterial inoculum. Table 1 lists the various cell inoculum parameters that can affect microbial transport along with their respective references.

TABLE 1. Cell inoculum parameters that may affect bacterial transport.

Parameters	Description	Reference
Surface molecules	Proteins and Extra-cellular polysaccharide production	Williams and Fletcher 1996
Motility	Pili, filaments	McCaulou and Bales 1995, Camper et al. 1993
Cell morphology	Cell shape, size, buoyancy	Harvey et al. 1997, Weiss et al. 1995
Cell surface	Charge, hydrophobicity	Van Loosdrecht et al. 1987
Nutrition	Growth state, nutrient status, starvation	Sharp et al. 1998, Lappin-Scott et al. 1988, Brown et al. 1977
Surfactants	Natural and man-made	Jackson et al. 1994

Attempts to enhance transport by altering a number of these parameters have mostly resulted in a decrease in cell activity and viability, with little or no increase in transport efficiency (Gross and Logan 1995). However, starvation of bacteria has been shown to be an effective method for facilitating bacterial transport through porous media (Lappin-Scott et al. 1988, MacLeod et al. 1988).

Starved bacteria have been described in many extreme environments including the deep-sea and the deep terrestrial subsurface. Many of these bacteria have survived in a dormant, starved state for 100's of years (Amy and Haldeman 1997). Starvation of bacteria results in a decrease in cell size, a decrease in cell activity, and eventually cell dormancy (Kjelleberg 1993).

Studies presented and discussed here used starved bacterial cultures under numerous different environmental conditions and varying injection scenarios to demonstrate the efficiency in which starved bacterial cultures can be transported through porous media to distances greater than 50 feet. Although the specific reasons for the enhanced transport properties of starved bacteria are not known, some appear to be obvious; such as decreased size and reduced activity. Other possible characteristics of starved bacteria that may enhance their transportability will also be discussed.

Objective. The objective of this paper is to demonstrate the general effectiveness and potential applicability of using starvation as a mechanism for delivering a desired bacterial population into a subsurface formation for the purpose of bioaugmentation. In addition, this paper will discuss the possible characteristics of starved cells that make them an effective agent for enhancing bacterial transport and the importance of understanding these characteristics in the context of engineering design, modeling, and field application of bioaugmentation technologies.

MATERIALS AND METHODS

The transport of starved cells, as opposed to vegetative cells, in porous media has been studied using column studies by numerous researchers. The primary purpose of these studies was to compare the overall transport efficiencies of starved cells with respect to vegetative cells. The column studies were all conducted and analyzed in the same general manner described below.

Cell Preparation. Cells were grown aerobically to the late exponential growth phase at room temperature using the appropriate general growth media. Cells were harvested by centrifugation (5000 to 8000 rpm, 4 °C, 15 to 20 min.), washed 3 times , and then re-suspended in a phosphate buffered saline solution. The washed cells were either used directly (vegetative cells) or starved by aseptically stirring them at room temperature over a period of 7 to 12 weeks.

Column Set-Up and Operation. The continuous flow column systems used varied in diameter (0.5-in. to 6-in.) and length (1ft to 50 ft). Typically, the columns were made of PVC and were packed with F-70 silica sand or similar porous media. All columns were positioned vertically and were run in up-flow mode. Prior to inoculation many of the columns were disinfected using diluted chlorine and/or an appropriate antibiotic. Some of the columns were run with constant head, while others were run with constant flow.

Columns were inoculated with a known volume and concentration of either vegetative or starved cells. The vegetative and starved inoculum for each study had approximately the same concentration and total number of cells. The effluent from the columns was collected and analyzed for total cells using direct counts, total plate counts, and/or selective plate counts to determine total cell breakthrough. The columns were destructively sampled by removing a known mass of porous media from the columns at specific distances along their length. The cells were removed from the sand samples using one of the methods described by Camper et al. (1985). Cell distributions were calculated by determining number of cells per length of column for each section of column (i.e. cells/ft or cells/2-feet of column).

Resuscitation of the starved cells within the porous media columns was performed by chasing the inoculum with multiple pore volumes of nutrient solution (citrate, molasses etc.). Once the cells were revived *in situ* and allowed to grow, the columns were destructively sampled to determine cell distributions along the column length.

RESULTS AND DISCUSSION

Results from the various porous media column studies consistently show the same general trend in the degree of cell breakthrough and the distribution of starved and vegetative cells within the columns (Gerlach et al. 1998, Sharp et al., 1998, Lappin Scott et al. 1988, MacLeod et al. 1988). A typical cell distribution curve for A 50-ft porous media column is shown in Figure 1. These 50-ft column studies used a muccoid strain of *Klebsiella oxytoca.* The majority of the vegetative cells attach and accumulate in the first 4 to 6 feet of the columns, and the vegetative cell numbers decrease by many orders of magnitude further down the column length. The starved bacteria, however, are distributed relatively uniformly along the entire length of the columns, whether they were 1-ft. or 50-ft. in length.

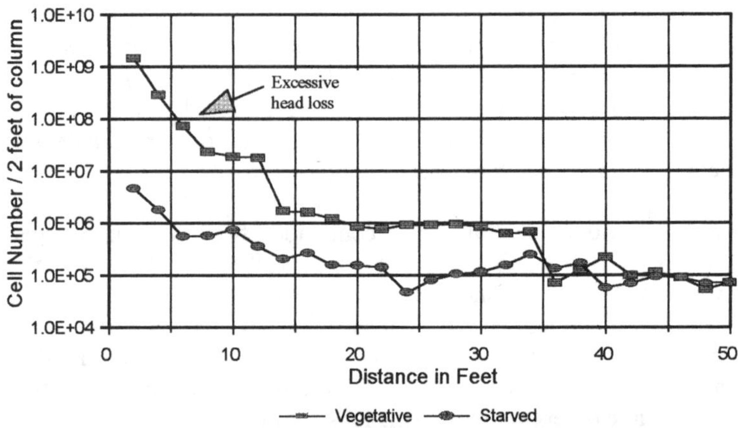

FIGURE 1. Distribution of adsorbed starved and vegetative *Klebsiella oxytoca* cells in 50-ft porous media columns with no nutrient addition.

The percentage of starved and vegetative cells that break through the columns varies with length and type of porous media. However, the starved cells breakthrough in numbers that are at least an order of magnitude greater than that of the vegetative cells. Typical results from the various column studies show that the starved cells break through the columns at relatively high percentages (4% to 37% breakthrough). In contrast, the overwhelming majority of the vegetative cells fail to breakthrough the column and concentrate near the injection point (4% to 0.0006% breakthrough).

In situ resuscitation of the starved bacterial cultures within the columns, to form a plug in the porous media has also been performed by numerous researchers (Sharp et al., 1998, Lappin Scott et al. 1988, MacLeod et al. 1988). Typical results from resuscitation experiments are shown for the same 50-ft column studies described above (Figure 2). The columns that were inoculated with vegetative and starved *Klebsiella oxytoca* cells, were fed citrate media and allowed to grow until the systems failed due to severe plugging and head-loss, or until a steady-state reduction in flow was achieved. As can be seen in Figure 2, the vegetative column prematurely plugged, allowing only the cells in the first 4-6 feet of the column to grow. However, the starved column had signs of abundant and relatively consistent growth throughout the whole length of the column. The uniform resuscitation and growth in the starved columns is attributed to the even distribution of the starved cells after inoculation and the relatively long lag time previous to the resuscitation of starved cells. The lag time allows the nutrients to reach cells toward the end of the columns before the columns plug near the injection point.

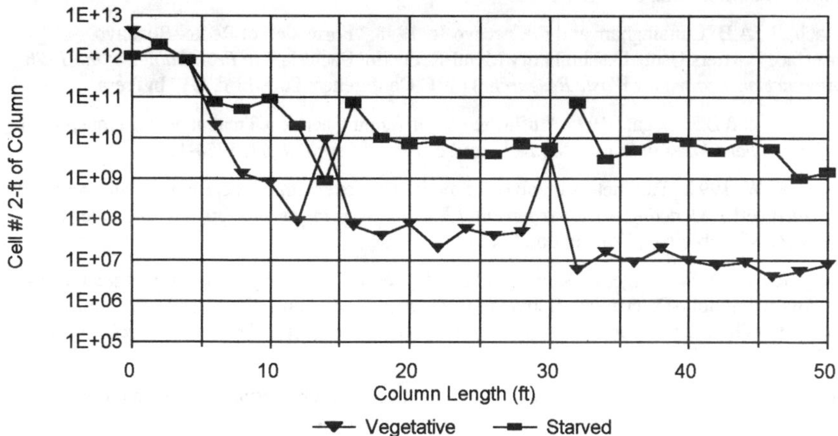

FIGURE 2. Distribution of resuscitated and growing cells in columns initially inoculated with starved and vegetative populations of *Klebsiella oxytoca.*

Essentially all of the other column studies using starved bacterial cultures demonstrate the same results. The overall conclusion is that starving bacteria enhances their transport through porous media. In addition, the longer column studies suggest that starved bacterial cultures can be transported within a subsurface formation over distances that are relevant at field scale. Starvation affects the properties of a cell in many ways, which could enhance their transportability. These effects include: Reduction in size, change in cell shape and morphology, decreased production of extra-cellular poly-saccharide or surface proteins, and possible changes in cell surface charge and hydrophobicity.

The ability to functionally transport and deliver starved bacterial populations (reactive or otherwise) to a specific area of an aquifer could lead to significant advances in efficacy and applicability of bioaugmentation technologies. Currently, starved bacteria are being used to develop subsurface biobarriers in the field. Subsurface biobarriers are used to reduce or control the hydraulic conductivity of a subsurface formation in order to attenuate plume development or redirect groundwater flow.

REFERENCES

Amy, P. S. and D. L. Haldeman (eds.). 1997. *The Microbiology of the Terrestrial Deep Subsurface.* CRC Press, Boca Raton, FL.

Brown, C. M., D. C. Ellwood and J. R. Hunter. 1977. "Growth of Bacteria at Surfaces: Influence of Nutrient Limitation." *FEMS Microbiol.Lett. 1*: 163-166.

Camper, A. K., M. W. LeChevallier, S. C. Broadaway and G.A. McFeters. 1985. "Evaluation of procedures to desorb bacteria from granular activated carbon." *J.Microbiol.Meth. 3*: 187-198.

Camper, A. K., J. T. Hayes, P. Sturman, W.L. Jones and A.B. Cunningham. 1993. "Effects of motility and adsorption rate coefficient on transport of bacteria through saturated porous media." *Appl.Environ.Microbiol. 59*: 3455-3462.

Gerlach, R., A.B. Cunningham and F. Caccavo Jr. 1998. "Formation of Redox-Reactive Subsurface Barriers Using Dissimilatory Metal-Reducing Bacteria." In *Proceedings of the 1998 Conference on Hazardous Waste Research.* HSRC Conference, Snowbird, UT. In Press.

Gross, M. J and B.E. Logan. 1995. "Influence of Different Chemical Treatments on Transport of *Alcaligenes paradoxus* in Porous Media." *Appl.Environ.Microbiol. 61*: 1750-1756.

Harvey, R. W. 1991. "Parameters involved in modeling movement of bacteria in groundwater", In C. J. Hurst (Ed.), *Modeling the Environmental Fate of Microorganisms*, pp. 89-114. American Society for Microbiology, Washington D.C.

Harvey, R. W., D. W. Metge, N. Kinner and N. Mayberry. 1997. "Physiological Considerations in Applying Laboratory-Determined Buoyant Densities to Predictions of Bacterial and Protozoan Transport in Groundwater: Results of In-Situ and Laboratory Tests." *Environ.Sci.Technol. 31*: 289-295.

Jackson, A., D. Roy and G. Breitenbeck. 1994. "Transport of a bacterial suspension through a soil matrix using water and an anionic surfactant." *Wat.Res. 28*: 943-949.

Kjelleberg, S. (ed.). 1993. *Starvation in Bacteria.* Plenum Press, New York.

Lappin-Scott, H. M., F. Cusack, F.A. MacLeod and J.W. Costerton. 1988. "Nutrient Resuscitation and Growth of Starved Cells in Sandstone Cores: A Novel Approach to Enhanced Oil Recovery." *Appl.Environ.Microbiol. 54*: 1373-1382.

Lawrence, J. R. and M. J. Hendry. 1996. "Transport of bacteria through geologic media." *Can.J.Microbiol. 42*: 410-422.

MacLeod, F. A., H. M. Lappin-Scott and J.W. Costerton. 1988. "Plugging of a Model Rock System by Using Starved Bacteria." *Appl.Environ.Microbiol. 54*: 1365-1372.

McCaulou, D. R. and C.B. Bales. 1995. "Effect of Temperature-Controlled Motility on Transport of Bacteria and Microspheres Through Saturated Sediment." *Wat.Resourc.Res. 31*: 271-280.

Mills, A. L. 1997. "Movement of Bacteria in the Subsurface", In P. S. Amy and D.L. Haldeman (Eds.), *The Microbiology of the Terrestrial Deep Subsurface*, pp. 225-243. CRC Press, Boca Raton, Florida.

Sharp, R.S. and G. Lewandowski. 1998. "Factors Affecting Ultra-Micro Bacteria Bio Barrier Design and Scale-up". HSMRC-NJIT Final Report.

VanLoosdrecht, M. C. M., J. Lyklema, W. Norde, G. Schraa, and A.J.B. Zehnder. 1987. "Electrophoretic mobility and hydrophobicity as a measure to predict the initial steps of bacterial adhesion." *Appl.Environ.Microbiol. 53*: 1898-1901.

Weiss, T. H., A.L. Mills, G.M. Hornberger, and J.S. Herman. 1995. "Effect of Bacterial Cell Shape on Transport of Bacteria in Porous Media." *Environ.Sci.Technol. 29*: 1737-1740.

Williams, V. and M. Fletcher. 1996. "Pseudomonas fluorescens Adhesion and Transport Through Porous Media are Affected by Lipopolysaccharide Composition." *Appl.Environ.Microbiol. 62*: 100-104.

COMETABOLIC DEGRADATION OF TRICHLOROETHYLENE USING A FULL-SCALE INTEGRATED BIOPROCESSING SYSTEM

Jian Xing, Ph.D. and Richard M. Raetz, PE
Global Remediation Technologies, Inc.
1235 Woodmere, Traverse City, MI 49686, USA.

ABSTRACT: In situ biosparging has been integrated with pressurized fluidized bed reactor (PFBR) and conventional pump and treat (Air Stripping) technology to enhance the removal efficiency of trichloroethylene (TCE) in Lansing, Michigan. In this system two PFBR's are alternatively fed with a phenol solution (the primary substrate) and TCE contaminated groundwater at 12 hour alternating intervals. Influent TCE concentrations varied from 700 µg/l to 50,0000 µg/l with an average value of 15,000 µg/l. A maximum TCE degradation capacity of 0.89 kg/m^3-d was achieved by the PFBR under an empty bed hydraulic retention time (HRT) of 30 minutes. Each PFBR generates microorganisms (which cometabolically degrade TCE) at observed effluent populations of 1.1x10^6 CFU/ml (as quantified by the substrate phenol). Microorganism generation occurs during the phenol feed cycle; while the alternate PFBR actively treats extracted groundwater containing the TCE (using microorganism grown during the previous cycle). Treated effluent exiting the PFBR system is reinjected back to the TCE impacted aquifer carrying microorganisms capable of degrading TCE in situ by process's of cometabolism. The effluent is injected into the saturated and unsaturated zone of the aquifer through a series of 15 injection laterals. Microorganisms enter the in situ cometabolic treatment window, which is equipped with biosparging and venting laterals. After seven months of continuous enrichment culture addition, the microorganism population (as quantified by phenol) increased from a low of 3.10x10^2 CFU/ml to 5.04x10^4 CFU/ml in the aquifer water. After fifteen months of operation, 131 kg of soluble phase TCE has been removed from the subsurface aquifer by this fully integrated system.

INTRODUCTION

TCE is a frequently found organic contaminant impacting ground water. TCE appears unusable as a primary energy source for bacteria metabolism, but can be destroyed through cometabolism under aerobic conditions. Enzymes (oxygenase) involved in the cometabolic degradation of TCE can be induced by various compounds including methane, ethylene, toluene and phenol. Oxygenase induced by microorganisms growing on the above substrates fortuitously oxidize TCE to TCE epoxide; which spontaneously degrades chemically to a variety of products that can be mineralized by mixed microbial communities. The oxygenase induced by aerobic microorganisms using phenol is particularly effective for TCE degradation (Hopkins et al., 1993). Therefore phenol was selected as the primary substrate to support bacterial growth in this field application.

Bioremediation describes the process whereby organic wastes are biologically degraded under controlled conditions; preferably to an innocuous state or to levels below concentration limits established by law. In situ bioremediation has been evaluated as a method of reducing the volume, toxicity, and mobility of on-site contaminants. Biological catalysts used to facilitate this process can include indigenous microbes and/or specially selected microbial inocula. The PFBR has been successfully used for the treatment of groundwater and process water contaminated by aromatic hydrocarbons (Raetz and Xing, 1996). Although cometabolic biodegradation of TCE has been intensively studied recently, and various lab-scale biological systems have reported effective treatment of TCE impacted groundwater (Hasegawa et al., 1997 and Segar et al., 1993), little is known about using PFBR technology for TCE contaminated groundwater treatment in full-scale field applications.

Characteristics of the Lansing, Michigan site including nature of contaminants, soil type and climate, make it amenable to bioremediation. Hence, in situ and ex situ bioremediation has been integrated with conventional pump and treat technologies to aggressively remove TCE from site soil and groundwater.

Objective. The objective of this study was to demonstrate the cometabolic removal efficiency of TCE by phenol-utilizing microorganisms in a full-scale pressurized fluidized bed reactor (PFBR) and in situ biosparging system while integrated with conventional pump and treat/venting technology.

Site Description. The remediation project was conducted in Lansing, Michigan at a site where TCE exist in both soil and groundwater. Investigation efforts revealed that the site TCE impact exists within a complex soil network composed of peat, clay, silty clay, sand with gravel, and silty sand lenses varying in thickness from 2.5 to 220 cm. Each lithologic unit is discontinuous and of extremely low permeability in both the saturated and unsaturated sections of the aquifer. A series of artificial drainage corridors, 12 feet plus in width, were created for effective aquifer water removal, treatment, and reinjection. The corridors were installed below the unsaturated zone, which is approximately 4 meters to 1/2 meter in thickness. The subject saturated zone terminates at 5 meters from the top of the unsaturated zone on a base clay layer.

Adsorbed-, pore trapped- and soluble-phase TCE was distributed over 4500 m^2 with TCE concentrations in soil ranging from 100 to 1,800 µg/kg. TCE concentrations in groundwater ranged from 12 to 440,000 µg/l.

REMEDIAL PROCESS DESCRIPTION

A simplified mechanical flow diagram of the integrated bioremediation system is shown in Figure 1. Groundwater is recovered from the source area plume, amended with balanced nutrients and oxygen, and passed through the PFBR or air stripper for treatment. Treated effluent from the PFBR containing micron size enrichment cultures is amended with nutrients a second time and injected back into the biosparging window through 15 injection laterals. Air stripper effluent is partially surface water discharged (50%) and partially mixed (50%) with PFBR effluent for reinjection. Oxygen is added to the biosparging system through a series of passive air sparge wells placed within the reinjection window. The operation of this integrated bioprocessing system is controlled remotely, and in part automatically, by a computer based monitoring and data collection system.

Pressurized Fluidized Bioreactors. The PFBR system consisted of two pressurized fluidized bed reactors. Each reactor column is a 2.4 meter tall by 0.6 meter diameter stainless steel vessel; having a working volume of 700 liters. Each reactor is filled with 200-liters of activated carbon as the biomass attachment media. The unexpanded height of the carbon bed is 0.7 meters. The two reactors are alternatively fed with a phenol solution (the primary substrate) and TCE impacted groundwater at 12-hour intervals to minimize competitive inhibition while the TCE is being cometabolized. The PFBR feeding rate for TCE contaminated groundwater was 23 L/min creating a HRT of 30 minutes. The PFBR feeding rate for the phenol solution was 11.3 l/min creating an HRT of 1 hour. Oxygen gas (purity 99%) was supplied to the bioreactor to maintain a dissolved oxygen (DO) effluent concentration of 4.8 mg/L. The PFBR effluent was amended with nitrogen and phosphorus and re-injected into the impacted aquifer after passing through a 35 to 50 micron filter.

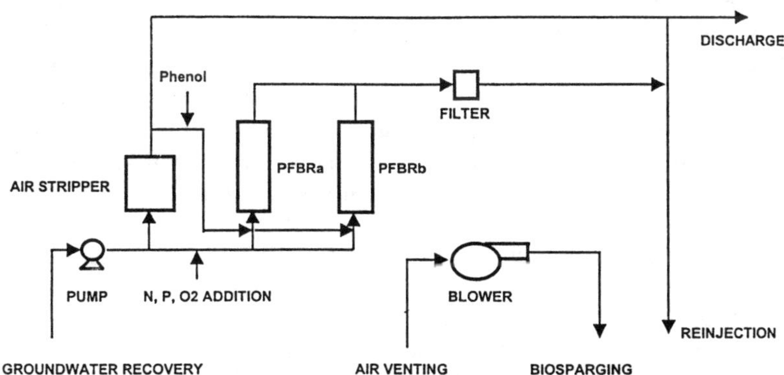

FIGURE 1. Integrated bioprocessing system

Biosparging-Venting System. Passive air sparging was used to provide oxygen to the aerobic bioprocess (biosparging). Oxygen was supplied to the TCE impacted area through sparge laterals 3.2 cm in diameter by 305 cm in length. Air sparge rates were generally 7 standard L/min for each lateral. Vent wells were constructed to create a negative pressure in the aquifer system while the biosparge system operated. Venting rates were approximately 17.5 standard L/min each.

Sampling and Analysis. Liquid samples were collected in 40 ml glass vials on a weekly to monthly basis. Quantitative analysis of liquid phase TCE and phenol was performed using gas chromatographic (GC) analysis. U.S. EPA Methods 8010 (TCE) and 604 (Phenol) were employed.

RESULTS AND DISCUSSION

PFBR Start up. The PFBR was initiated with mixed inocula (containing a known capacity to degrade TCE obtained from a batch reactor in Global Remediation Technologies Inc. laboratory) and activated sludge (from a local wastewater treatment plant). Nitrogen (N) and phosphorus (P) were added to the influent at a stoichiometric ratio of 100/5/1: COD/N/P. Microorganisms capable of cometabolizing TCE were cultured in the PFBR at a HRT of 1 hr and organic loading rate of 3 kg COD/m^3-day. After eight weeks of reactor operation the biomass concentration reached 20 g/l (as suspended solids) and the reactor achieved a 95% phenol removal efficiency.

PFBR Operation Results. When biomass was fully developed in each PFBR, TCE contaminated groundwater and stock phenol solution were alternatively fed into the each PFBR in 12 hour-intervals. The periodic shift of the PFBR feeding conditions was performed by using timer controlled solenoid valves. GRT's lab test results revealed that the phenol utilizing microorganisms, cultured in the PFBR reactors, could degrade TCE at a maximum degradation rate of 80 mg TCE/g-SS-day. During the field operational period the influent TCE concentration fluctuated from 700 μg/l to 50,000 μg/l with an average value of 15,000 μg/l. The corresponding TCE loading rate of the PFBRs varied from 0.1 to 1.9 kg/m^3-d. Based on operational data, the PFBR reactor continuously removed TCE at an influent TCE concentration up to 50 mg/l and a HRT of 30 minutes. The TCE removal efficiency varied with influent TCE concentrations and ranged from 31% to 68%. It appears that the TCE removal efficiency increased with a influent TCE concentration increase from 2 mg/l to about 25 mg/l. Beyond this range the TCE removal efficiency began to decrease as the influent TCE concentration increased. This observation is consistent with laboratory controlled experimental results (Figure 2). The observed TCE removal capacity of the PFBR is presented in Figure 3. As shown in this figure, at an influent TCE concentration of 33 mg/l a maximum TCE removal capacity of 0.89 kg/m^3-d was achieved. This field operations data also compared favorably with lab test results. During the field operational period the average effluent DO concentration was 4.8 mg/l, while the average pH was 7.6.

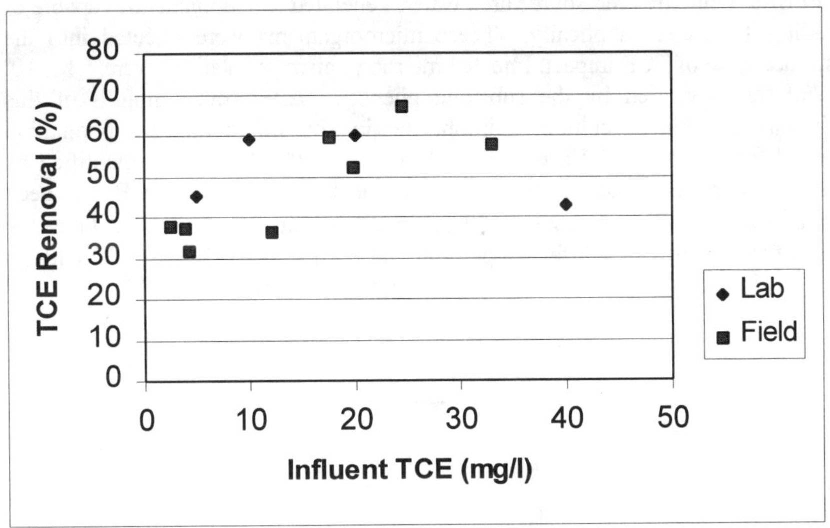

FIGURE 2. PCE removal vs. influent TCE concentration

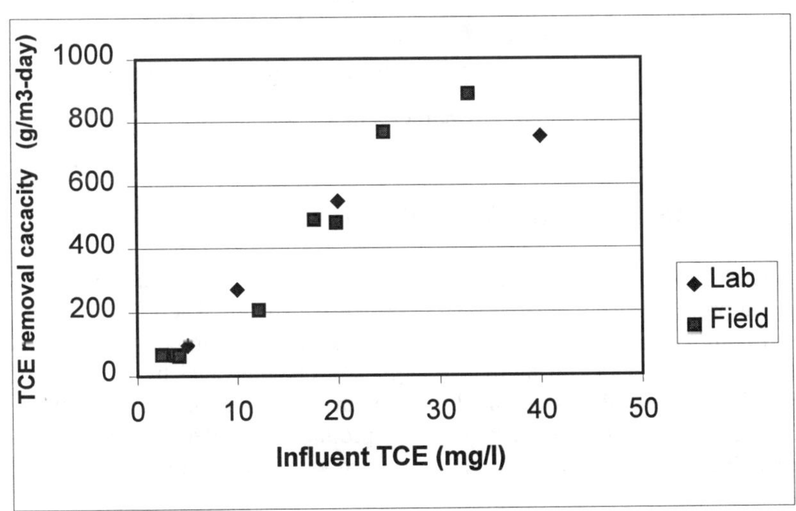

FIGURE 3. TCE removal capacity vs. influent TCE concentration

Overall Site Remediation Results. The fully integrated bio-processing system started operation in May of 1997. Due to the effective operation of the integrated system, the TCE concentration in the source area decreased continuously (Figure 4). The PFBRs, while treating source area water, generated microorganisms capable of degrading TCE cometabolically. These microorganisms were injected into the subsurface zone of TCE impact. Injected microorganism populations were 1.1 x 10^6 CFU/ml (as quantified by the substrate phenol). After seven months of this continuous enrichment culture addition, the in situ microorganism population increased from a low of 3.10x10^2 CFU/ml to 5.04 x 10^4 CFU/ml (as quantified by the substrate phenol). Based on field operational data, 131 kg TCE has been removed from the subsurface zone of TCE impact within fifteen months. Over the course of treatment, the aquifer temperature varied from 9.2 °C to 19.8 °C with an average value of 14.8 °C. The aquifer pH slightly fluctuated around 7.7.

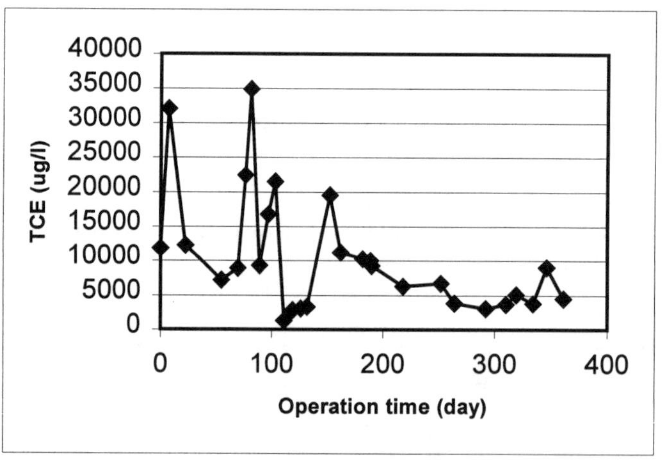

FIGURE 4. Recovery Well (RW-2) TCE concentration change

This study demonstrates the effectiveness and advantage of using integrated biological processes for industrial sites impacted with TCE. The PFBR system can be used both as a ex situ TCE treatment process and "most importantly" a inocula source for enhanced in situ bioremediation. As demonstrated by the operational results, integrating PFBR technologies with biosparging, air venting, and conventional pump and treat technology has produced the effective removal of 131 kg of soluble phase TCE from the aquifer system within fifteen months.

REFERENCES

Hasegawa, H., T. Shimomura, H. Uchiyama and O. Yagi. 1997. "Continuous Biodegradation of Trichloroethylene by *Methylocystis sp.* In a Membrane

Bioreactor." In Symposium Chairs and Andrea Leeson (Eds), *In Site and On-site Bioremediation: Volume 5*, pp. 31-36. Battelle Press, Columbus OH.

Hopkins, G. D., J. Munakata, L. Semprini and P. L. McCarty . 1993. "Trichloroethylene concentration effects on pilot field-scale in-situ groundwater bioremediation by phenol-oxidizing microorganisms." *Environ. Sci. & Technol.*, 27(12): 2542-2547.

Segar, R. L. Jr., Shelley L. De Wys, Gerald and E. Speitel, Jr. 1995. "Sustained Trichloroethylene Cometabolism by Phenol-Degrading Bacteria in Sequencing Biofilm Reactors." *Water Environment Research*, 67(5): 764-774.

Raetz R. and J. Xing,. 1997. "Accelerated BTEX and PNA removal rates using an integrated bioprocessing system." In Symposium Chairs and Andrea Leeson (Eds), *In Site and On-site Bioremediation: Volume 5*, pp. 63-69. Battelle Press, Columbus OH.

FULL-SCALE BIOREMEDIATION OF CHLORINATED ETHYLENES: COMBINING AN ANAEROBIC BIOREACTOR WITH IN SITU STIMULATION

M.J.C. Henssen, S. Keuning, A.W. van der Werf, J.J. van der Waarde (Bioclear Environmental Biotechnology, Groningen, The Netherlands)
C. Hubach, R. Blokzijl (DHV Environment & Infrastructure, Groningen, The Netherlands)
J. Mourik, C. Haasnoot, H. Bosgoed (Logisticon Water Treatment, Groot-Ammers, The Netherlands)
E. Meijerink (Province of Drenthe, Assen, The Netherlands)

ABSTRACT: Recently, a full-scale bioremediation effort at the Evenblij site in Hoogeveen, The Netherlands, was undertaken. The remediation process is based on a combination of several bioremediation techniques: In the highly contaminated core zone active bioremediation is performed, based on bioaugmentation combined with stimulation due to electron donor addition. Moderate stimulation with electron donor infiltration is used in the plume area of the contaminated zone, and natural attenuation is monitored in the contaminated peripheral plume area. From the laboratory experiments it was concluded that the core and plume area both contain the capacity to transform perchloroethylene (PCE) into *cis*-dichloroethylene (*c*-DCE), whereas the transformation of *c*-DCE to ethylene occurs only very slightly or not at all. Addition of lactate or acetate stimulated the dechlorination of PCE and trichloroethylene (TCE) to *c*-DCE. Bioaugmentation with bacteria from dechlorinating bioreactors accelerated the process of vinyl chloride (VC) and ethylene production from PCE. Natural biological attenuation hardly occurred due to lack of electron donor as dissolved organic content (DOC) in the water phase. However, from the groundwater characterization results, iron-reducing conditions seem to have dominated. These conditions are favorable for dechlorinating processes.

INTRODUCTION

Due to industrial processes at the site and spills of PCE, both soil and groundwater down to 40 m below ground level have been contaminated. Because conventional remediation techniques for this site would require huge investments, the possibilities for biological treatment are examined. At the same site a pilot-scale anaerobic bioreactor has been used successfully for the past 3 years to transform PCE partially to ethylene. Based on these results, it was concluded that the bioreactor would be usable for bioaugmentation at sites where natural attenuation of PCE into ethylene is not occurring due to lack of suitable biomass. Dechlorinating biomass within the effluent from the bioreactor can serve as inoculating material.

Recently, full-scale bioremediation has been started at Evenblij. Several bioremediation techniques are being combined (Total Concept Evenblij) (Figure 1). Active bioremediation takes place in the highly contaminated core zone, based on withdrawal of contaminated groundwater, treatment of the contaminated groundwater in an anaerobic bioreactor, and infiltration of the methanogenic effluent of the reactor in the subsoil. Due to this infiltration, the redox conditions are improved and the core zone is enriched with dechlorinating biomass. Stimulation of biodegradation is also enhanced by adding a suitable electron donor and nutrients to the infiltrated bioreactor effluent.

Moderate stimulation is induced in the contaminant plume by adding an electron donor and nutrients to the bioreactor effluent and infiltration of this effluent in the plume area. Intrinsic bioremediation is monitored in the tail of the (slightly) contaminated peripheral plume area.

Objective. Based on the results of this project, suitable remediation techniques can be combined efficiently from both a technical and an economical point of view.

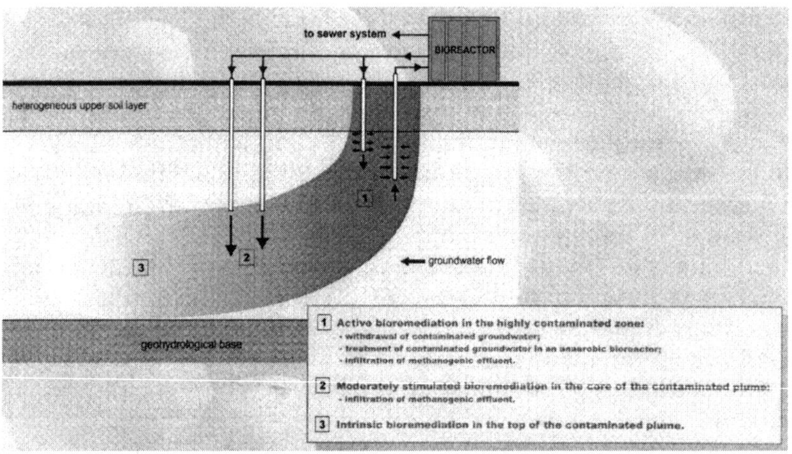

Figure 1. Schematic draft of Total Concept Evenblij

MATERIALS AND METHODS

Both laboratory-scale experiments and full-scale monitoring are used at the site to evaluate the (expected) effects and performance of the in situ bioremediation concept. Phase 1 of the project investigated the natural degradation capacity of the soil, possibilities for enhancement of biological dechlorination by adding carbon sources, the effect of bioaugmentation, and the dominating conditions in the groundwater.

Soil and Groundwater Sampling. Soil samples were taken from zone 1 and 2 (Figure 1) to perform laboratory experiments to determine the dechlorinating capacity of the soil under both natural and stimulated conditions. Anaerobic groundwater sampling combined with online analysis of water samples was carried out to gather information about the natural conditions in the water phase and to determine dominant redox conditions by analyses of all electron acceptors.

Laboratory Experiments. To obtain information concerning the biodegradation capacity and effects of stimulation with both bioaugmentation and carbon source dosing, anaerobic batch experiments were carried out. Soil from zone 2 was incubated under both natural conditions (no additions, batch 1) and stimulated conditions, by adding acetate (batch 2), lactate (batch 3), or methanol (batch 4). Soil samples from the core zone (zone 1) were incubated with lactate. Similar batches were fed with biomass from the dechlorinating bioreactors to simulate the effect of bioaugmentation.

RESULTS AND DISCUSSION

Dominating Redox Conditions of Groundwater. The groundwater contains almost no oxygen or nitrate. Although some water samples contain higher concentrations of sulfate and traces of sulfide, iron-reducing conditions seem to dominate. In most water samples, elevated iron(II) amounts of approximately 2 to 11 mg/L are detected, indicating iron reduction processes. This redox condition is not unfavorable for dechlorination processes. However, total organic carbon (TOC) concentrations are between <3 mg/L and 30 mg/L (average 15 mg/L) and are expected to be too low to maintain dechlorination processes. In conclusion, lack of a sufficient carbon source as an electron donor seems to limit the dechlorination process.

Natural Attenuation Capacity. Under natural conditions the soil seemed to have only limited capacity to transform PCE and TCE. Within 12 weeks PCE and TCE were degraded partly to *c*-DCE (Table 1). Only small traces of VC and/or ethylene were detected in the batches, including both zone 1 and zone 2. In most cases degradation started after a lag phase of approximately 4 weeks, perhaps indicating that the degradation capacity was not immediately present but had to be induced.

Table 1. Degradation of TCE under natural and stimulated conditions (TCE spiking)

	TCE concentration (mg/kg dw)			*c*-DCE concentration (μg/L gas)
Weeks	0	4	12	14
Acetate fed	5,8	7	6,5	1,240
Lactate fed	5,8	5,3	0,05	290
Methanol fed	5,8	6,6	1,1	3
No addition	5,8	5,7	3,9	14
Lactate fed 10°C	5,8	5,7	1,7	700
Abiotic batch	5,8	3	7,4	0

Enhanced Bioremediation. Adding lactate or acetate stimulated the degradation of PCE and TCE (Table 1). In comparison to the nonstimulated batches, degradation in the presence of lactate was much faster. Acetate also seemed to enhance the dechlorination. There was only a slight impact of temperature on the degradation.

In the TCE-spiked batches, acetate and lactate stimulated the formation of low-chlorinated and nonchlorinated components compared to the nonstimulated batches (Table 2). Although TCE was degraded in both nonstimulated and stimulated batches, higher amounts of VC and ethylene were present in the acetate and lactate batches.

Table 2. Degradation of TCE under natural and stimulated conditions (TCE spiking)

	TCE concentration (mg/kg dw)		VC concentration (µg/L gas phase)		ethylene concentration (µg/Lgas phase)	
Weeks	0	14	16	22	16	22
Acetate fed	2,6	0,17	0	14	48	31
Lactate fed	2,6	0,97	-	8	-	29
Methanol fed	2,6	0,24	-	0	-	13
No addition	2,6	0,2	-	0	-	16
Abiotic batch	2,6	1,5	0	0	0	0,3

When biomass was added, enhanced dechlorination to VC and ethylene occurred (Table 3). After adding dechlorinating biomass to the batches in week 6 of the incubation a significant difference in *c*-DCE, VC, and ethylene concentrations was visible when comparing augmentated and nonaugmentated batches.

Table 3. Degradation of PCE, TCE,and *c*-DCE with and without biomass addition (all batches lactate stimulated), concentrations of *c*-DCE, VC, and ethylene in headspace in week 15

Week 15 of incubation	*c*-DCE (µg/L gas)	VC (µg/L gas)	ethylene (µg/L gas)	methane (µg/L gas)
PCE-spiked batch				
- without biomass	2,080	0	2	45
- with biomass	1,440	654	11	445
- abiotic	0	0	0.1	18
TCE-spiked batch				
- without biomass	2,440	0	7.7	50
- with biomass	1,580	870	27	250
- abiotic	0	0	0	11
c-DCE-spiked batch				
- without biomass	1,170	0	1	33
- with biomass	960	410	7	320
- abiotic	1,230	0	0.5	18

Bioreactor. In November 1998, a 30-m^3 bioreactor (reactor 1) was inoculated with anaerobic sludge from an industrial upflow anaerobic sludge blanket (UASB) reactor. The sludge was tested prior to inoculation for its capacity to dechlorinate PCE. These tests showed that complete dechlorination to ethylene was possible with the anaerobic sludge. From November 1998 groundwater containing mostly PCE (approx. 30,000 µg/L) and TCE (approx. 10,000 µg/L) was used to adapt the biomass to the chlorinated ethylenes and develop dechlorinating capacity. PCE degraded to *c*-DCE, and small amounts of VC and ethylene were already detected.

PRELIMINARY CONCLUSIONS

From the results in phase 1 we concluded that under natural conditions dechlorination occurs only very slightly and ends primarily with *c*-DCE. Natural attenuation will not lead to significant biological dechlorination of PCE and TCE at the site, due mainly to a lack of electron donor concentration. This may also result in a lack of suitable biomass to transform *c*-DCE into ethylene. Dechlorination seems to be enhanced when adding carbon sources such as acetate and/or lactate. With addition of biomass and lactate, complete dechlorination is occurring.

ACKNOWLEDGMENT

This project is funded partly by NOBIS, the Dutch research program on biotechnological in situ soil remediation.

ENHANCING BIOCOLLOID TRANSPORT TO IMPROVE SUBSURFACE REMEDIATION

Bruce E. Logan and Terri A. Camesano (Penn State University, University Park, PA), Brock Rogers (URS Greiner Woodward Clyde, Phoenix, AZ)
Yan Fang (University of Arizona, Tucson, Arizona)

ABSTRACT: Bioaugmentation can be an effective method of subsurface remediation but well clogging must be minimized by reducing bacterial adhesion to soil particles in the vicinity of the well. In this paper, we review recent work performed in our laboratory to determine the most effective methods for increasing bacterial transport distances in groundwater aquifers. The presence of a non-aqueous phase liquid (NAPL) increased bacterial transport despite data that indicated bacteria favorably partitioned into hydrophobic NAPL phases. Gas sparging also increased bacterial transport distances compared to those obtained when bacteria were suspended in an artificial groundwater, but gas sparging was not as effective as using low ionic strength (IS) water or surfactants. Filtration models predict reduced bacterial transport at low versus high flow velocities, but the opposite effect was observed when bacteria were motile. For efficient bioaugmentation, we recommend that motile bacteria be suspended in low ionic strength (~10^{-5} M) water, and that they be injected at low pumping velocities (~1 m/d) to minimize bacterial attachment.

INTRODUCTION

Subsurface remediation can be enhanced by injecting bacteria into the aquifer via a procedure known as bioaugmentation, but bacteria-sized (~1 μm) colloidal particles readily adhere to soil particles and may not be transported more than one meter in soils leading to well clogging. In order to efficiently treat large volumes of soils by injecting bacteria from wells, bacterial transport distances of 10's of meters must be obtained (Gross and Logan, 1995). Measuring factors that could sufficiently reduce bacterial attachment to achieve these large transport distances would require laboratory columns 1 to 10 m long using a conventional approach of measuring cell concentrations in column breakthrough tests (Jewett *et al.*, 1993), so smaller-scale methods were developed to screen bacterial attachment properties.

Using the MARK method (Gross *et al.,* 1995) we have previously summarized the effects of solution IS, dissolved and sorbed organic matter, and various chemical additives on bacterial transport (Johnson et al., 1996). Attachment was quantified using filtration theory in terms of the collision efficiency, α, defined as the ratio of the rate that bacteria stick to a soil grain to the rate that they strike it. Of the chemicals examined, surfactants such as Tween-20 have been found to be the most effective, reducing α by 2.5 orders of magnitude on glass surfaces, while dissolved natural organic matter has little effect. The most consistent method of increasing bacterial transport in a variety of soils and porous media is decreased IS. Decreasing the IS from that of growth media (10^{-1} M) to that of ultrapure water (10^{-5}

M) decreased α of *Pseudomonas fluorescens P17* from 0.18 to 0.026 (Johnson et al., 1996).

We summarize here the results of more recent research in our laboratory to determine the effects of other factors, such as fluid velocity, cell motility and gas sparging. Because non-aqueous phase liquids (NAPLs) may be present at many heavily contaminated sites, we also examined the effect of NAPLs on bacterial transport.

METHODS

Column experiments were performed by adapting the MARK procedure, which consists of measuring the retention of radiolabeled cells in short columns (Gross *et al.,* 1995), to transport experiments using longer columns (7 to 15 cm). The bacterium used for all experiments was *Pseudomonas fluorescens* strain P17, a Gram negative motile rod. Cells were radiolabeled by incubating a cell suspension with ^3H leucine. In MARK tests cells are pulled by vacuum through an open-ended short column. For the longer column experiments reported here we pumped bacterial suspensions through capped glass or stainless steel columns packed with a sandy soil collected from the North Fallow Field at the University of Arizona farm, or cleaned quartz (average grain diameters of 127 μm and 200 μm, respectively).

After the cells were passed through a column, the column was rinsed and sliced into sections. The fraction of bacteria retained in each slice, R_i, was

$$R_i = N_i / \left(N_o - \sum N_{i-1}\right) \qquad \textbf{(1)}$$

where N_0 is the concentration of bacteria added to the sample (dpm), N_i is the concentration of bacteria in the slice, and N_{i-1} is the total number of bacteria retained in previous slices. To account for incomplete radiolabel recovery, samples were spiked with additional radiolabeled cells as previously described (Camesano and Logan, 1998). The extent of bacterial attachment was evaluated using a steady state clean bed filtration equation

$$C/C_o = \exp\,(-\alpha\,\lambda\,L) \qquad \textbf{(2)}$$

where C_0 and C are the concentrations of bacteria entering and leaving the column of length L, $\lambda = 3(1-\theta)\eta/2d_c$ is the filter coefficient, θ=0.40 the bed porosity, η the collector efficiency calculated using the RT model (Logan *et al.* 1996), d_c the soil grain diameter, and α the collision efficiency.

The fraction of bacteria removed in the column, R, is related to the concentration of bacteria in the column effluent, C/C_0, by $R=1-C/C_0$. Therefore, the collision efficiency for bacteria in each slice, α_i, can be calculated as a function of the thickness of each slice, L_i using $\alpha_i = [\ln(1-R_i)]/(\lambda L_i)$. The overall collision efficiency for the whole column was calculated from the total removal of cells in the column calculated by adding up the removal from all slices.

TABLE 1. Comparison of sticking coefficients for motile bacteria in AGW, motile bacteria in low IS water, and non-motile bacteria (adapted from Camesano and Logan 1998).

Pore velocity	Sticking coefficient (α)		
(m/d)	Motile; AGW	Motile; low IS	Non-motile; AGW
0.56	0.003	0.001	0.018
2.7	0.007	0.005	---
8.9, 11	0.031	0.013	0.020
20	0.068	0.023	---
54	0.24	0.11	---
110, 120	0.37	0.21	0.013
560, 590	0.43	0.23	0.020

RESULTS AND DISCUSSION

Average Flow Velocity. According to filtration theory, the fraction of bacteria retained should decrease as velocity increases (producing a constant α). Our experiments found the opposite to be true (Table 1). The collision efficiencies increased with velocity for velocities ranging from <1 to several hundred m/d. At a velocity of 0.56 m/d, the overall α was 0.003 for a 7-cm column. Under typical groundwater conditions, this would result in a one-log reduction in cell concentration after the bacteria had traveled 7 m. At >110 m/d, α was high enough to cause ~70% of the bacteria to be retained in just a 7-cm column, conditions that would render bioaugmentation difficult as continued cell injection at this high cell retention would eventually clog the well.

Cell Motility. Cell motility was likely the reason for the observed changes in cell retention with fluid velocity (Camesano and Logan 1998). Bacteria swim at typical velocities of 3.5 m/d (Mercer *et al.*, 1993). Thus, at low bulk fluid velocities, bacterial motility could allow bacteria to avoid collisions with soil grains. However, at velocities much larger than swimming speeds of the cells, we hypothesized that the bacteria would be trapped in the flow. We found that at high velocities α was essentially constant, results that supported our hypothesis. To test our theory that cell motility was responsible for extremely low collision efficiencies at low fluid velocities, we conducted experiments with bacteria made non-motile by staining with acridine orange (which also killed the cells). Collision efficiencies for non-motile bacteria had essentially constant α's over the range of fluid velocities (Table 1) supporting our conclusions regarding cell motility at low flow velocities.

Ionic Strength (IS). P17 has a zeta potential (z) of -40.45 mV and is also relatively hydrophobic. Due to its very negative surface charge, the attachment of P17 is likely governed by electrostatics. Because the soil and cells are both negatively charged they repel each other at the neutral pHs used in our experiments; this repulsive force is enhanced using low IS water. Low IS water (~10^{-5} M) reduced bacterial attachment

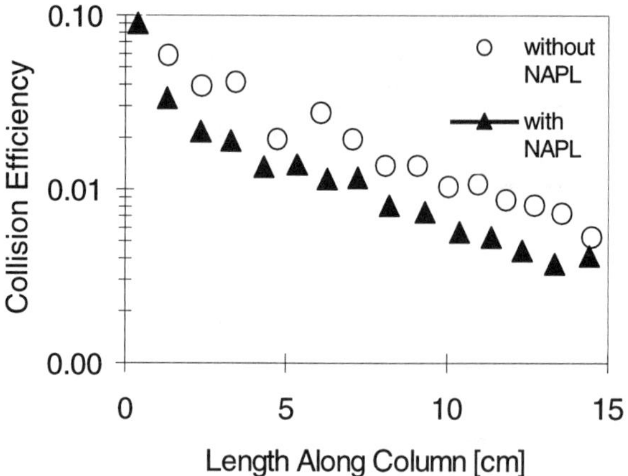

Figure 1. Effect of NAPL (tetrachloroethylene) on the fraction of bacteria retained in a soil column versus that predicted by filtration theory. Adapted from Rogers (1997).

by approximately a factor of 20 compared to that in a higher IS artificial groundwater (0.004 M) at higher fluid velocities (Table 1). The reduction in attachment under low IS conditions is consistent with the results obtained in previous studies using natural soils at high flow velocities in MARK tests (Johnson et al., 1996).

Effect of NAPLs on Transport. The presence of a NAPL residual (21.3 % of void spaces filled) increased the overall transport of bacteria by 22% (Figure 1). The cumulative fraction retained along the length of the column decreased from R=0.78 (α=0.025) to R=0.61 (α=0.016).

The reduction in bacterial attachment in the presence of a NAPL phase is a result of the occupation of stagnant flow regions, or the micropore spaces, by the NAPL. This had the effect of blocking a portion of the flow paths in the column, thus increasing the pore velocity of the bacterial suspension. Filtration theory predicts that an increase in velocity will decrease the collector efficiency (η), which would have the effect of reducing the fraction retained. An alternative explanation for the enhanced transport could be that the NAPL is blocking favorable sites on the collector surface. If these sites are inaccessible to the bacteria, the probability of their attachment will decrease thereby enhancing their transportability. However, sorbed PCE, in the absence of a residual phase, did not affect PCE transport (Rogers, 1997). A decreased attachment to the soil surfaces, due to hydrophobic repulsion between P17 and the NAPL was also considered as a possible explanation. However, hydrophobicity tests indicated that P17 favorably partitions into PCE. Therefore, the presence of a hydrophobic phase should have increased, not decreased bacterial retention in the column as observed in our experiments.

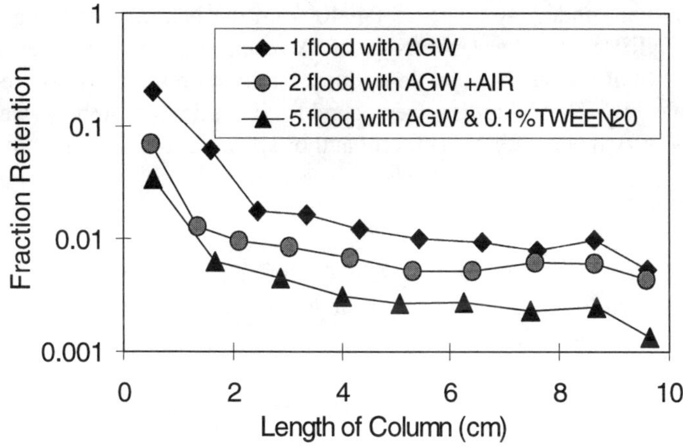

Figure 2. Gas sparging decreased the fraction of bacteria retained in a quartz-media column (data from Fang and Logan, 1999).

Gas Sparging. Bacterial transport can be enhanced for some bacterial species by gas (N_2) sparging, but the enhancement due to the gas flow for P17 is less than that obtained using surfactants (Figure 2). The fraction of bacteria retained in a 10-cm long quartz-media column was reduced from R=0.38 to 0.15 by gas sparging while water was simultaneously being pumped through the column. Addition of a non-ionic surfactant (0.1% v/v Tween 20) solution reduced the fraction of bacteria retained to R=0.06 (Fang and Logan, 1999).

The enhancement of bacteria transport by gas sparging is likely due to the mobile gas-water interface. Because P17 is a relatively hydrophobic bacterium we expect that cells will preferentially sorb to the gas-water interface. The sorption of bacteria on the gas-water interface is essentially irreversible since capillary energy provides a large attractive force to hold the bacteria on the gas-water interface. Thus, the movement of the gas-water interface during gas sparging results in the increased forward transport of the cells in comparison to saturated water conditions. However, the increase in bacterial transport due to sparging is not as large as that possible using surfactants or low IS water.

CONCLUSIONS

Bacterial transport can be increased in natural soils by using low IS water (~0.01 mM). For the motile species examined here cell attachment is minimized by pumping bacteria at low water velocities (~1 m/d) even though filtration models predict that attachment will be decreased with increased water velocity. Gas sparging will not be as effective as low IS water in increasing bacterial penetration in groundwater aquifers.

ACKNOWLEDGEMENTS
This work was funded by grant ES-04940 from the National Institute of Health Sciences, NIEHS, to the University of Arizona. Its contents are solely the responsibility of the authors and do not necessarily represent official views of the funding agency. These studies were conducted while the authors were at the University of Arizona, although current author affiliations are given above.

REFERENCES

Camesano, T.A. and B.E. Logan. 1998. "Influence of fluid velocity and cell concentration on the transport of motile and non-motile bacteria in porous media." *Environ. Sci. Technol.* 32(11):1699-1708.

Fang, Y. and B.E. Logan. 1999. "Bacterial transport in gas sparged porous media." *J. Environ. Engng.* In press.

Gross, M.J., O. Albinger, D.G. Jewett, B.E. Logan, R.C. Bales, and R.G. Arnold. 1995. "Measurement of bacterial collision efficiencies in porous media." *Wat. Res.* 29:1151-1158.

Gross, M.J. and B.E. Logan. 1995. "Influence of different chemical treatments on transport of *Alcaligenes paradoxus* in porous media." *Appl. Environ. Microbiol.* 61(5):1750-1756.

Jewett, D.G., R.C. Bales, B.E. Logan, and R.G. Arnold. 1993. Comment on "Application of clean-bed filtration theory to bacterial deposition in porous media". *Environ. Sci. Technol.* 27(5):984-985.

Johnson, W.P., M.J. Martin, M.J. Gross, and B.E. Logan. 1996. "Facilitation of bacterial transport through porous media by changes in solution and surface properties." *Colloids Surf. A.* 107:263-271.

Li, Q. and B.E. Logan. 1999. Enhancing bacterial transport for bioaugmentation of aquifers using low ionic strength solutions and surfactants. *Wat. Res.* 33(4):1090-1100.

Logan, B.E., D.G. Jewett, R.G. Arnold, E. Bouwer and C.R. O'Melia. 1995. "Clarification of clean-bed filtration models." *J. Environ. Eng.* 121(12):869-873.

Mercer, J.R., R.M. Ford, J.L. Stitz, and C. Bradbeer. 1993. "Growth rate and effects on fundamental transport properties of bacterial populations." *Biotechnol. Bioengin.* 42:1277-1286.

Rogers, B. 1997. "Bacterial transport in NAPL-contaminated porous media." M.S. Thesis, University of Arizona, Tucson.

PERFORMANCE OF SAND/SAWDUST MIXTURES IN PERMEABLE WALLS

J. G. Gundrum (PSU, University Park, PA)
Y. U. Kim (PSU, University Park, PA)
M.C. Wang (PSU, University Park, PA)

ABSTRACT: Use of permeable treatment walls to remove soluble contaminants from ground water is a new on-site remediation technology. The wall material is required to be more permeable than or at least as permeable as the surrounding soil throughout the entire service life of the wall. This study investigated the consolidation and permeability behaviors of sand/sawdust mixtures in the permeable treatment wall. The mixtures had 2, 4, 6, and 8% by weight of sawdust and were tested in the laboratory using 1-D consolidation and variable head permeability tests. The test results were analyzed using the conventional consolidation theory to evaluate permeability variation with time. The test results provided geotechnical properties of the mixtures necessary for the design of sand/sawdust permeable walls. Although possible bio-chemical degradation of the mixtures was not considered, the research findings are believed to be useful for the design of permeable walls constructed of a sand/sawdust mixture.

INTRODUCTION

Restoration of contaminated aquifer is vital to water supply for domestic and industrial use. For restoration, the contaminated ground water can be remediated by either containment of the contaminated plume or removal/destruction of the pollutant. Containment of a contaminated plume can be accomplished by changing the direction of ground water flow or by installing physical barriers such as slurry walls, grout curtains, or sheetpile walls.

As of today, the contaminated ground water is commonly remediated by the pump and treat technique. Because of its high cost and unnecessary waste of clean water extracted along the contaminated plume. The pump and treat method is not an effective long-term solution.

An alternative remediation method is the passive in-situ treatment using porous reactive and permeable materials that are installed in the aquifer to intercept the contaminated plume. The permeable walls are used to degrade/reduce the contaminant from the percolated ground water.

A commonly found contaminant in the ground water is nitrate which is released from domestic and some industrial waste water. Also, sulfide minerals generated from mine waste often contaminate both surface and ground water systems. The sulfide minerals can produce acid water which contains high concentration of dissolved metals.

Solid organic carbon materials such as sawdust, cellulose, and straw are available to remove the nitrates and sulfates from ground water (Vogan, 1993). They have been used as electron donors in the permeable walls for heterotrophic denitrification (Robertson et al., 1995; Blowes et al., 1994 and 1995). In applications, the reactive media are mixed with sand which provides structural matrix and also enhances the permeability of the wall. The permeability of the mixture must remain greater than or equal to the permeability of the replaced aquifer material. Both sawdust and sand are low-cost materials and can be constructed using conventional construction technique. These make the sawdust/sand permeable treatment walls cost effective and technically sound.

Factors critical to the performance of permeable walls include reactivity and longitivity. The reactive media must remain permeable enough to facilitate ground water flow through the wall and also allow sufficient residence time within the wall for chemical reaction. Meanwhile, the reactive media should not consolidate or produce biological accumulations that significantly reduce the permeability of the reactive media. These considerations indicated that the important geotechnical properties of reactive media for design and construction of permeable treatment walls are permeability and consolidation behavior. This paper presents the results of a study on permeability and consolidation properties of sand/sawdust mixtures.

TEST MATERAILS

The test materials were a sawdust and a sand. The sawdust was obtained from a hickory tree. Its particle size ranged between 0.250 mm and 4.750 mm with $D_{50} \approx 1.5$ mm and $D_{10} \approx 0.75$ mm. It had an air-dry water content of about 12%, and a specific gravity of approximately 0.7 (Wood Handbook, 1987). The test sand was well graded having a maximum particle size of 4.750 mm with $D_{50} \approx 0.8$ mm and $D_{10} \approx 0.45$ mm. Its specific gravity was 2.67.

The test mixtures had sawdust/sand ratio of 2, 4, 6, and 8% by weight. The gradation of the mixture with 8% sawdust had $D_{50} \approx 0.9$ mm and $D_{10} \approx 0.45$ mm. The mixtures were soaked overnight in a nitrate solution; the nitrate solution had 124 mg/L nitrate concentration and was prepared from sodium nitrate. The mixtures were kept in the solution during the testing.

LABOTATORY TESTING

The laboratory testing involved permeability and consolidation tests. The test specimen was contained in a 10-cm diameter steel mold, and had a specimen height of 10.1 cm. Inside the steel mold, the test specimen was sandwiched between two stainless steel screens to retain fine particles in the specimen. While remaining in the steel mold, each test specimen was subjected to both 1-D consolidation test and permeability test. The consolidation test followed the standard test procedures of ASTM 2435. The permeability test, which adopted the falling head principle, was performed at every level of consolidation pressure. For each test condition, a minimum of three replicate tests were performed to obtain a statistically representative results.

GEOTECHNICAL PROPERTIES

The specific gravity (G_s), saturated weight (γ_{sat}), initial void ratio (e_0), compression index (C_c), initial coefficient of permeability (k_0), and coefficient of consolidation (c_v) of the test mixtures are summarized in Table 1. The permeability (k) data are plotted against void ratio (e) in semi-logarithmic scale (log k vs. e) in Figure 1. The Consolidation curves which relate the logarithmic consolidation pressures (log p) and void ratio (e), i.e. e vs. log p curves are presented in Figure 2.

TABLE 1 Geotechnical Properties of Sand/Sawdust Mixtures

Sawdust Content (%)	G_s	γ_{sat} (lb/ft^3)	e_0	C_c	k_0 (cm/sec)	c_v (ft^2/day)
2	2.53	118.0	0.72	0.06	0.022	0.140
4	2.41	113.6	0.72	0.07	0.019	0.124
6	2.30	109.8	0.71	0.09	0.010	0.093
8	2.21	106.8	0.70	0.12	0.007	0.070

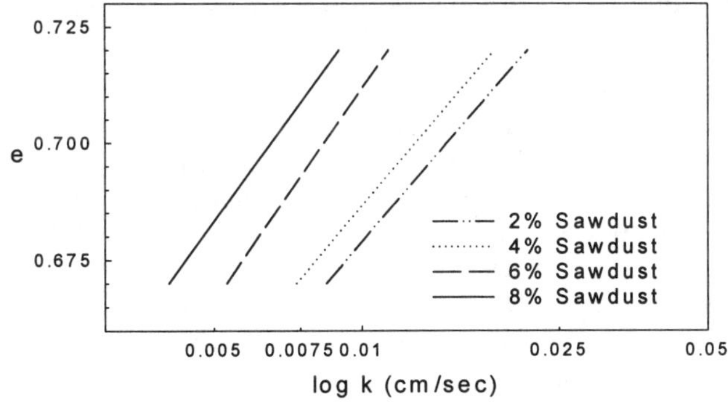

FIGURE 1 Relation Between Void Ratio and Coefficient of Permeability

The permeability of the test specimens increases with increasing void ratio as would be expected. The linear relation between log k and e shown in Figure 1 is similar to that for a great majority of natural soils (Lambe and Whitman, 1969). For the different test mixtures, permeability becomes smaller as the percentage of sawdust content increases. Also, the linear relations between log k and e for the various mixtures are not necessarily parallel to each other indicating that the rate of permeability increase with void ratio differs for different levels of sawdust content.

The slope of e vs. log p curves in Figure 2 becomes steeper as the sawdust content increases. Increasing sawdust content increases the compression index (C_c) but decreases the coefficient of consolidation (c_v) as shown in Table 1. Thus, for mixtures with higher sawdust contents, the magnitude of consolidation settlement will be greater, but the time rate of consolidation will be slower.

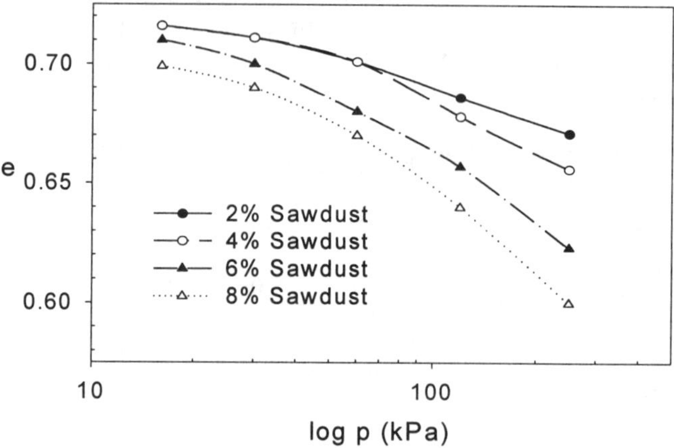

FIGURE 2 Relation Between Void Ratio and Consolidation Pressure

PERFORMANCE EVALUATION

The performance of sand/sawdust mixtures in a permeable treatment wall is evaluated in terms of its void ratio (or density) and permeability variation with time for different levels of sawdust contents. The treatment wall is assumed to be 5 ft (1.5 m) wide and 30 ft (9.1 m) deep of which the bottom 20 ft (6.1 m) is in the aquifer. The analysis was made for consolidation of the mixtures under the geostatic pressure using the 1-D consolidation theory. The saturated unit weight of the mixtures, though vary with the type of the sand and sawdust as well as the sawdust content, are not much different from twice the unit weight of water. Thus, the unit weight of the sand/sawdust mixtures and the overlying soil equal to twice the unit weight of water, i.e. $\gamma_{sat} \approx 124.8$ pcf (19.6 kN/m^3) are used in the analysis. Due to consolidation, the sand/sawdust mixture in the wall will undergo an ultimate (or final) change in void ratio (Δe_f), which can be computed from the following equation:

$$\Delta e_f = C_c \log_{10}\left[(p_0' + \Delta p)/p_0'\right] \tag{1}$$

in which C_c = compression index
p_0' = effective overburden pressure = $(\gamma_{sat} - \gamma_w)h \approx \gamma_w h$
Δp = consolidation pressure = $\gamma_{sat} h \approx 2\gamma_w h$
h = depth below the ground surface to the point of interest

Substituting the vertical pressure into Equation (1), the final (or ultimate) void ratio change becomes

$$\Delta e_f = C_c \log_{10}(3) \cong 0.477 C_c \cong 0.5 C_c \tag{2}$$

According to equation (2), the final void ratio change due to consolidation is independent of depth, and is approximately equal to 0.5 C_c throughout the entire depth of the permeable wall.

The void ratio change at any time can be estimated using the consolidation theory. Assuming that the sand/sawdust mixture is more permeable than the aquifer soil, the dimensionless time factor (T) equals

$$T = \frac{c_v t}{{H_{DP}}^2} = \frac{c_v t}{(20.0)^2} \cong 0.0025 c_v t \qquad (3)$$

in which c_v = coefficient of consolidation
t = time
H_{DP} = maximum length of drainage path

The degree of consolidation (U_t) corresponding to the computed dimensionless time factor (T) can be determined from the available tables such as that given in Lambe and Whitman (1969). From this degree of consolidation, the void ratio change at time t, i.e. Δe_t, equals

$$\Delta e_t = U_t \cdot \Delta e_f \qquad (4)$$

The results of analysis show that the final void ratio and the void ratio at the end of 5 years after wall construction are, respectively, 0.69 and 0.69, 0.69 and 0.69, 0.67 and 0.68, and 0.64 and 0.66 for 2, 4, 6, and 8% sawdust contents. Meanwhile, the final coefficient of permeability and the coefficient of permeability at the end of 5 years are, respectively, 0.012 and 0.012, 0.011 and 0.011, 0.0052 and 0.0062, and 0.0024 and 0.0034 cm/s for 2, 4, 6, 8% sawdust contents. It is seen that the sand/sawdust mixture in the permeable wall consolidates under their own weight resulting in a decrease in void ratio and permeability. The reduction in void ratio and permeability is greater for higher sawdust content, primarily because the sawdust particles are more compressible than the sand particles. Also, as the sawdust content increases, the duration of consolidation becomes longer.

It should be noted that the analysis is made with assumptions that the consolidation is due to vertical flow only, and that the soil surrounding the permeable wall is much less permeable than the sand/sawdust mixtures. With these assumptions, the longest flow path length (H_{DP}) used in Equation (3) equals the thickness of aquifer layer. If the surrounding soil is nearly as permeable as the sand/sawdust mixture, the longest flow path length will be smaller and the duration of consolidation will be shorter. Furthermore, one important factor that a possible change in permeability of the mixture due to bio-chemical degradation of sawdust is not considered in the study. Sawdust degradation quite possible will decrease the permeability of the mixtures. Thus, the combined effect of the assumption made and the factor not considered may result in the results of analysis not far from the actual performance of the sand/sawdust mixtures in the permeable wall.

SUMMARY AND COCLUSIONS

The consolidation and permeability behaviors of sand/sawdust mixtures were investigated in the laboratory. The mixtures consisted of a sand and a sawdust obtained from a hickory tree; four sawdust contents (2, 4, 6, and 8% by weight) were investigated. The change in void ratio of the mixtures due to the consolidation under its own weight was analyzed using the consolidation theory. The permeability of the mixtures was determined from both the consolidation test and the variable head permeability test. The results from both tests were averaged. The test results showed that the compression index increased but the coefficient of consolidation decreased with increasing sawdust content. The coefficient of permeability (k) varied with void ratio (e) linearly in e vs. log k plot. Also, increasing sawdust content decreased the coefficient of permeability of the mixtures.

The results of analysis demonstrated that due to self-weight consolidation, the sand/sawdust mixtures in the permeable wall will undergo an increase in density and a decrease in permeability. Although the study did not consider possible decrease in permeability due to bio-chemical degradation of the sawdust in the permeable wall, the results of the study probably will reflect the actual field performance because of the assumption made in the analysis. Therefore, the research findings should be useful for design/evaluation of permeable walls constructed of sand/sawdust mixtures.

REFERENCES

Blowes, D. W., W. D. Robertson, C. J. Ptacek, and C. Merkley. 1994. "Removal of Agricultural Nitrate from Tile-Drainage Effluent Water Using In-Line Bioreactors." *Journal of Contaminant Hydrology*. 15: 207-221.

Lambe, T., W., and R. V. Whitman. 1969. *Soil Mechanics*. John Wiley & Sons, Inc., New York.

Robertson, W. D., and J. A. Cherry. 1995. "In Situ Denitrification of Septic-System Nitrate Using Reactive Porous Media Barriers: Field Trials." Ground Water. 33(1): 99-111.

Vogan, J.. 1993. *The Use of Emplaced Denitrifying Layers to Promote Nitrate Removal From Septic Effluent*. M. Sc. Thesis. University of Waterloo. Ontario.

Wood Handbook: Wood as an Engineering Material, Rev. 1987. Prepared by the Forest Service Laboratory. U.S. Dapartment of Agriculture. Madison. WI.

ECOTOXICOLOGICAL ASSESSMENT AND BIODIVERSITY DETERMINATION OF BIOFILMS DEVELOPED IN GROUNDWATER

Nathalie Ross, Jacques Bureau, Louise Deschênes and Réjean Samson (École Polytechnique de Montréal, Montreal, Quebec, Canada)
Richard Villemur (INRS - Institut Armand-Frappier, Laval, Quebec, Canada)

ABSTRACT. Biofilms are increasingly studied for their use as *in situ* biological barriers. Their development in groundwater conditions was subjected to an ecotoxicological assessment. A biotest (Microtox®) and a biodiversity study were conducted during a biofilm formation from a groundwater indigenous microbial population, inoculated in semicontinuous reactors and maintained in darkness at 10°C. Biofilms developed to a maximum thickness of 800 μm after 10 days and the Microtox® test results showed no significant toxicity ($IC_{50} > 49.5$). Results from the 16S rRNA sequencing revealed a decrease in the biodiversity of biofilms when compared to the groundwater microbial population. These findings support the assumption of an innocuousness resulting from a biofilm development in groundwater conditions.

INTRODUCTION

Microbial barriers formed with micro-organisms and exopolymeric matter produced in an aquifer offer an excellent potential to prevent the spreading of a contaminant plume (Bellamy *et al.*, 1993). To develop such technology, the potential of groundwater indigenous micro-organisms to produce sufficient amount of exopolysaccharides (EPS) has been previously demonstrated (Ross et al. 1998). The effectiveness of ultramicrobacteria (UMBs) to recover their full vegetative size and to clog a porous media has been studied using isolates from oilwell water (Cusack *et al.*, 1992).

Before their application on the field, the innocuousness of bioremediation technologies has now to be proven (Environnement Canada, 1997). Ecotoxicological assessments are increasingly used in conjunction with chemical characterization to ensure that soil and groundwater matrixes are not only decontaminated, but also detoxified. Functional and genetic diversity techniques are used to detect population shifts during bioremediation and to predict a potential hazard (Power *et al.*, 1998). The objectives of this study was to determine a toxic potential and to observe the change in the biodiversity during a biofilm development in groundwater conditions.

MATERIAL AND METHODS

Biofilm Development Apparatus. Experiments were conducted in semi-continuous reactors (1 L) filled with a synthetic groundwater (500 ml) and maintained at 10 °C in darkness. The synthetic groundwater had a chemical composition similar to the local groundwater (mg/L): Na_2CO_3 (679), NaCl (617), $MgSO_4$ (19), $CaCO_3$ (15), K_2HPO_4 (11), KNO_3 (6) and $FeCl_2$ (1). Biofilms developed on a ceramic coupon (2.5 x 15 cm) immersed in the synthetic groundwater.

Ecotoxicological assessment and biodiversity determination were conducted according two different experimental designs. In the ecotoxicological assessment design, indigenous groundwater micro-organisms were cultured and 1 ml was inoculated (10^7 heterotrophic micro-organisms/ml) in synthetic groundwater. In the biodiversity determination design, indigenous microbial population has been previously filtrated on a 0.4 μm filter to separate UMBs from normal-size bacteria (NSBs) before their inoculation in different reactors. For both design, physicochemical conditions were the following: molasses fed at 20 mg m^{-2} min^{-1}, sparging aeration (4.2 cm^3/s), addition of calcium ions (100 mg/L) and a C:N:P ratio of 50:10:1.

Ecotoxicological Assessment. The ecotoxic potential was assessed on both groundwater and biofilm samples. The inhibition of *Vibrio fischeri* bioluminescence (Microtox®) was also measured on reactors to which 10 mg/L Na-PCP was added to simulate a contamination. The Microtox® test was conducted according to the method suggested by Le Bureau de normalisation du Québec, 1987 (Bureau de normalisation du Québec, 1987). Results are given as sample concentrations needed to inhibit 50% of the bioluminescence of a *Vibrio fischeri* population (IC_{50}).

Biodiversity Determination. Functional and genetic diversity was observed on groundwater UMBs, NSBs and biofilms. The most probable number method was used for the enumeration of total heterotrophs. Cells viability was determined using the Live/Dead BacLight™ epifluorescence kit (Taghi-Kilani *et al.*, 1996). Cell-surface hydrophobicity was measured according to the adherence of cells to hexadecane (Rosenberg, Gutnick, and Rosenberg, 1980). Groundwater and biofilm microbial DNA were extracted following a method described by Li *et al.* (1996). The PCR products were cloned in a T-vector according to the manufacturer specifications (pCRII, Invitrogene, San Diego, CA, USA). For several clones, small plasmid DNA extraction was performed (Sambrook, Fritsch, and Maniatis, 1989) and digested by *Eco*R1 and *Rsa*I restriction endonucleases as recommended by Pharmacia. The BLAST program (National Center for Biotechnology Information) was used to find, in several gene banks, the most similar 16S sequences of the isolated DNA sequences.

RESULTS AND DISCUSSION

Biofilm Development and Ecotoxcicological Assessment. The growth of indigenous groundwater micro-organisms showed a typical batch growth curve (FIGURE 1). The suspended EPS concentration curve resembled a logistic curve such as those observed in batch cultures and the biofilm thickness increased following a sigmoidal curve (Characklis and Marshall, 1990). After 12 days, the biofilm thickness reached a plateau of 800 μm.

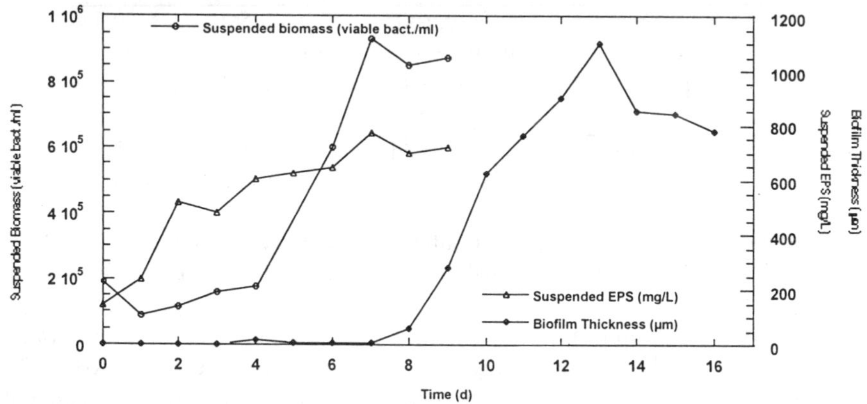

FIGURE 1. Microbial growth and biofilm development of a groundwater indigenous microbial population in a semicontinuously fed reactor.

Results from the Microtox® (*Vibrio fischeri*) test showed that controls I and II, which consisted of synthetic groundwater with and without molasses (3.2 mg/L) respectively, and the addition of the microbial indigenous culture to synthetic groundwater (no PCP, t=0) did not significantly affect bioluminescence of *Vibrio fischeri* (IC_{50} < 49.5) (TABLE 1).

After 8 days of culture, moderate inhibition of *V. fischeri* bioluminescence was measured ($IC_{50,\ 5min}$= 42.5). Results also indicated that addition of Na-PCP to the culture (t = 0 h) allowed IC_{50} to decrease by 4.8-fold. Even if the PCP concentration considerably decreased between 0 and 8 days, (from 10.0 to 1.6 mg/L), bioluminescence inhibition was not significantly different. Phenolic compounds are known to disrupt cell membranes, resulting in the release of nucleic material. proteins and other organic material in the medium (Dutka and Bitton 1986). Also, the intermediary metabolites of PCP degradation, the release of endotoxins, and the lysis of bacteria, those indigenous from groundwater as well as *V. fischeri*, may have also contributed to the inhibition of the bioluminescence. Biofilms detached from the ceramic were non toxic to *Vibrio fischeri*, possibly because the organic compounds were too diluted ($IC_{50.5\ min\ and\ 15\ min}$ >49.5).

TABLE 1. Bioluminescence inhibition of *Vibrio fischeri* exposed to water and biofilms.

Sample		IC_{50} ± C.C. (%)	
		5 min	15 min
Controls	I	> 49.5	> 49.5
	II	> 49.5	> 49.5
Groundwater & Microbial Population	no PCP, t=0	> 49.5	> 49.5
	with PCP, t=0	13.3 ± 1.2	8.4 ± 0.9
	no PCP, t=8d	42.5 ± 3.7	30.8 ± 1.4
	with PCP, t=8d	15.6 ± 1.2	11.0 ± 1.2
Biofilms	no PCP, t=8d	> 49.5	> 49.5
	with PCP, t=8d	> 49.5	> 49.5

IC: Inhibition Concentration, C.C.: Confidence Coefficient, Control I: synthetic groundwater, Control II: synthetic groundwater and molasses (3.2 mg/L)

Biodiversity Determination. Initially (t = 0d), the indigenous groundwater UMBs and NSBs showed to have a different functional diversity (TABLE 2). Indigenous heterotrophic UMBs represented 0.81 % of the groundwater microbial population and the cells were 16% more viable compared to the NSBs. The cell surface of UMBs was only 5% less hydrophobic than NSBs. It is recognized that UMBs have a modified membrane structure conferring them a reduced surface hydrophobicity (Cusack *et al.*, 1992). Moreover, groundwater is an oligotrophic milieu and either UMBs or NSBs were in some degree of starvation which can led to a low cell surface hydrophobicity when compared to cells hydrophobicity in a copiotrophe milieu (Kemp *et al.*, 1993).

TABLE 2. Functional diversity of groundwater indigenous UMBs and NSBs.

	NSBs	UMBs
Total Heterotrophs (MPN bact./100ml)	$2.01e^4$	$1.65e^2$
Viability (%)	62.9	78.7
Surface Hydrophobicity (% bact. adherent to hexadecane)	19.4	14.3

High diversity was observed in groundwater and significant difference was found between UMBs and NSBs (TABLE 3). Examination of cloned 16S rRNA sequence analysis confirmed this observation as non specific RsaI pattern was observed. Nucleotide sequences of some clones isolated were related to bacteria found in aquatic and terrestrial environments (Seviour *et al.*, 1997). Compared to the groundwater samples, the biofilms samples showed an important decrease of DNA fragments in SSCP patterns suggesting that a major change occurred in the microbial diversity (data not shown). Analysis of more than hundred clones revealed three major RsaI patterns. Among the nineteen representative clones that were sequenced, it appeared that they were related only to four bacterial groups: *Bacillus/Paenibacillus*, Enterobacteriaceae, Pseudomonads and Zoogloae.

TABLE 3. Affiliation of 16S rRNA cloned sequences obtained from groundwater and biofilms.

Groundwater samples/ Possible affiliation*	Biofilm samples/ Possible affiliation*
UMB	Strain OCS7 (*B* subdivision, proteobacteria)
Rhizobium spp.	*Bacillus insolitus*
Strain BD7-1 (unclassified bacteria)	*Paenibacillus amylolyticus*
Desulfocapsa thiozymogenes	*Paenibacillus macquariensis*
Strain 72chol (denitrifying bacteria)	*Rhanella* genosp. 3
NSB	*Pseudomonas* spp.
Cytophaga spp.	
Verrucomicrobium spp.	
Arcobacter spp.	

*: Clones homology were between 89 and 94 % for groundwater samples and 93 to 99% for biofilm samples

Although the watered biofilms did not affected significantly the *Vibrio fischeri* bioluminescence, the accumulation of by-products in groundwater could possibly be adverse to the ecosystem if accumulated on a local basis. The biodiversity study outlined a population shift during the development of a biofilm in groundwater conditions.

ACKNOWLEDGMENTS

The authors acknowledge the financial support from the industrial Chair partners: Alcan, Bodycote/Analex, Bell Canada, Browning-Ferris Industries, Cambior, Centre québécois de valorisation de la biomasse (CQVB), Hydro-Québec, Natural Science and Engineering Research Council (NSERC), Petro-Canada and SNC-Lavalin.

REFERENCES

Bellamy, K. L., N. de Lint, D. R. Cullimore, and A. Abiola. 1993. *In-situ Intercedent Biological Barriers for the Containment and Remediation of Contaminated Grounwater.* Regina: Droycon Bioconcepts Inc.

Bureau de normalisation du Québec. 1987. "Eaux - détermination de la toxicité méthode avec la bactérie bioluminescente *Vibrio fisheri*": BNQ.

Characklis, W. G., and K. C. Marshall. 1990. *Biofilms.* New York: John Wiley & Sons, Inc.

Cunningham, A. B., W. G. Characklis, F. Abedeen, and D. Crawford. 1991. "Influence of Biofilm Accumulation on Porous Media Hydrodynamics". *Environ. Sci. Technol. 25* (7):1305-1311.

Cusack, F. M., S. Singh, J. Novosad, M. Chmilar, S. A. Blenkinsopp, and J. W. Costerton. 1992. "The Use of Ultramicrobacteria for Selective Plugging in Oil Recovery by Waterflooding". Paper read at *SPE Internaltional Meeting on Petroleum Engineering,* 24-27 March, at Beijhing, China.

Dutka, B. J., and G. Bitton. 1986. *Toxicity Testing Using Microorganisms.* Vol. II. Boca Raton: CRC Press, Inc.

Environnement Canada. 1997. "Reglement modifiant le Reglement sur les renseignements concernant les substances nouvelles".

Janssen, P. H., A., Schuhmann, E. Morschel and F. A. Rainey. 1997. "Novel Aerobic Ultramicrobateria Belonging to the Verrucomicrobiales Lineage of Bacterial Descent Isolated by Dilution Culture from Anoxic Rice Paddy Soil". *Applied and Environmental Microbiology 63* (4):1382-1388.

Kemp, P. F., B. F. Sherr, E. B. Sherr, and J. J. Cole. 1993. *Handbook of Methods in Aquatic Microbial Ecology.* Boca Raton: Lewis Publishers.

Li, T., J. G. Bisaillon, R. Villemur, L. Létourneau, F. Lépine, and R. Beaudet. 1996. "Isolation and identification of a new bacterium carboxylating phenol to benzoic acid under anaeraobic conditions". *Journal of Bacteriology 178*:2551-2558.

Rao, K. R. 1978. *Pentachlorophenol - Chemistry, Pharmacology, and Environmental Toxicology.* Vol. 12, Environmental Science Research. New York: Plenum Press.

Ross, N., L. Deschenes, J. Bureau, B. Clément, Y. Comeau, and R. Samson. 1998. "Ecotoxicological Assessment and Effects of Physicochemical Factors on Biofilm Development in Groundwater Conditions". *Environmental Science and Technology 32* (8):1105-1111.

Sambrook, J., E. F. Fritsch, and T. Maniatis. 1989. *Molecular Biology, a Laboratory Manual.* 2nd ed. Cold Spring Harbor: Cold Spring Harbor Press.

Sanger, F., S. Nicklen, and A. R. Coulson. 1977. "DNA sequencing with chain terminating inhibotors". *Proc. Natl. Acad. Sci. USA 69*:1408-1412.

Seviour, E. M., L. L. Blackall, C. Christensson, P. Hugenholtz, M. A. Cunningham, D. Bradford, H. M. Stratton, and R. J. Seviour. 1997. "The filamentous morphotaype Eikelboom type 1863 is not a single genetic entity". *Journal of Applied Microbiology 82*:411-421.

Snair, J., R. Amann, I. Huber, Ludwig W., and K. H. Schleifer. 1997. "Phylogenic analysis and *in situ* identification of bacteria in activated sludge". *Applied and Environmental Microbiology 63*:2884-2896.

SHEARING FOR ENHANCED *IN SITU* TREATMENT OF CONTAMINATED SOIL

D. J. Walter (C-CORE, St. John's, NF, CA)
K. M. Kosar (EBA Engineering Consultants, Vancouver, BC, CA)
R. Phillips and J. I. Clark (C-CORE, St. John's, NF, CA)

ABSTRACT: Many technologies used for *in situ* remediation of contaminants in soil and groundwater promote the movement of fluids through the subsurface to either treat the contaminant in place or to facilitate removal of the contaminant for surface treatment. Where the soil requiring treatment is low in permeability, *in situ* technologies are ineffective and not often used. This paper describes a novel approach that is currently under development at Memorial University which may help to enhance the effectiveness of existing *in situ* techniques in some low permeability soils. The approach, called "soil shearing", uses geotechnical engineering methods to increase the porosity and, therefore, the permeability of these soils. A major advantage of soil shearing is that permeability changes can be realized using standard geotechnical drilling equipment and at virtually no additional cost to the field program. Typical results from soil shearing experiments carried out in a geotechnical centrifuge are presented and key aspects of the process are discussed.

INTRODUCTION

"Soil shearing" for enhanced *in situ* remediation of contaminated soil is a relatively new concept that is in an early stage of development. The shearing process is initiated by injecting a fluid into a soil deposit through a standard wellbore. As the fluid pressure within the soil increases, the effective stress changes and the soil yields. Many over-consolidated soils, including glacial till, will increase in volume (or dilate) when sheared. If the soil is dilatant, yielding will result in an increase in porosity and the development of shear bands in the soil. These physical changes to the soil structure increase the bulk permeability and can significantly increase connectivity between naturally occurring fractures within the soil.

The use of external methods to improve the fluid flow characteristics of the soil and enhance *in situ* treatment of contaminants is not new. Work in the late 1980's examined the use of hydraulic fracturing to achieve permeability enhancements in soil. Hydraulic fracturing involves a tensile parting mechanism whereby a highly viscous sand laden fluid is injected into a wellbore to initiate and propagate a fracture in the adjacent soil. Due to the manpower, equipment, and materials required for hydraulic fracturing, the approach is expensive and is used on relatively few projects. Soil shearing is an inexpensive process that can be carried out from standard well installations using conventional pumps and water as an injection fluid.

DESCRIPTION OF SOIL SHEARING PROCESS

Previous work has been carried out which indicates shearing to be a promising mechanism for enhancing the flow of fluids in soil. Walter et al (1998) presented the results of a field experiment in which a shear zone was created within a test bed constructed from recompacted glacial till. Mori and Tamura (1986) reported increases in permeability of almost three orders of magnitude during triaxial shear tests on grouted sand at low confining stresses. Kosar and Been (1991) presented the results of small and large scale tests carried out to investigate the geomechanical response of oil sands subjected to thermal bitumen recovery techniques. In this work, a permanent porosity increase due to shear dilatancy of oil sand was identified as a mechanism likely to have a significant influence on fluid flow within the reservoir.

Shear failure was a dominant mechanism in experiments carried out by Golder Associates (1994) as part of a joint industry study investigating hydraulic fracture propagation in oil sands. The size of the shear zone was found to be related to the volume of fluid injected and the distance the injected fluid traveled from the injection well during a test was 1 to 2 orders of magnitude greater than would occur if flow was through the pores of the sand. Chalaturnyk and Scott (1997) reported that shearing is a major geomechanical factor in the steam assisted gravity drainage process used for the enhanced recovery of heavy oil from oil sand reservoirs. Changes in reservoir fluid pressure during tests of the process at the Underground Test Facility near Fort McMurray resulted in shear induced volume strains, increases in porosity, and 30% to 50% increases in the absolute permeability of the formation.

CENTRIFUGE MODELING

Geotechnical centrifuge modeling is an effective and reliable method for examining full scale field processes, such as those associated with in situ treatment of contaminated soils, in a laboratory environment. The geotechnical centrifuge allows reduced scale physical model tests to be carried out where gravity affects are important. The mechanical response of soil is highly dependent on the stress level in the soil and the stress history of the soil. For reduced scale models tested at 1g, the stresses in the soil do not replicate those in the prototype, and the mechanical response of the soil due to loading is often quite different between the model and the prototype. This may cause difficulties with interpretation of data and extrapolation of scale model test results to full scale. The use of a centrifuge for physical modeling of geotechnical problems allows gravity dependent parameters in a reduced scale model to be controlled such that the response of the model is similar to that of a full scale prototype.

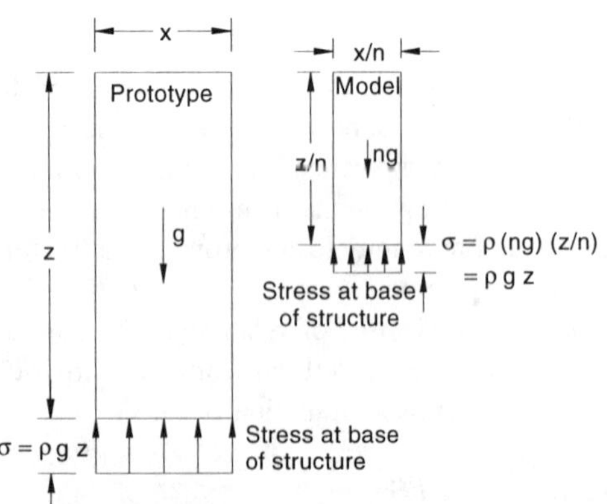

Figure 1. Stress at similar points in prototype and model

The rationale behind

centrifuge modeling is that the centripetal acceleration at the end of a rotating beam may be used to model the earth's gravitational acceleration. An often cited example from geotechnical engineering is that of an embankment. If a centrifuge model embankment is built from the same soil as the prototype embankment with every linear dimension in the prototype n times larger than in the model (scale of 1/n), and the centrifuge is operated such that the centripetal acceleration acting in the model is n gravities, then the stresses due to gravity at corresponding points in the model and the prototype, both with soil mass density, ρ, will be identical (Figure 1). If care is taken to ensure that any other boundary or applied stresses in the model are also made to correspond with the prototype, then the strain fields and hence deformations and behavior will be similar.

TEST SETUP AND METHODOLOGY

Eight soil shearing injection tests were carried out on the large 5.5 m radius geotechnical centrifuge located at the C-CORE Centrifuge Center in St. John's, NF. The centrifuge models were constructed and tested at a scale of 1:50 using Speswhite kaolin clay for the test soil. Speswhite kaolin is a quality clay with uniform geotechnical properties that have been well documented by numerous researchers including Al-Tabbaa (1987), Rossato et al (1992), and Ling (1995). The clay was mixed as a slurry at 120% water content, placed in a 0.9 m diameter strongbox, and pre-consolidated to 750 kPa vertical effective stress. When tested in the centrifuge at 50g, the OCR varied from about 8 at 100 mm to 4.2 at 200 mm depth (5 and 10 m prototype depths).

Four injection wells were installed in each test bed with the top of the screens located at depths of 100 mm and 200 mm, which corresponds to 5 and 10 m at prototype scale. The screened length of the wells was 25 mm (1.25 m prototype). Pore pressure transducers (PPTs) were buried at various depths and distances from the wells and were used to monitor pore pressure response in the soil. Two PPTs were used to monitor the injection pressure at the wellhead. Six displacement transducers (LVDTs) were used to monitor ground surface movements at various distances from the wellhead. The pump used was a piston and cylinder mounted in a vertical actuator (Figure 2) powered by a stepper motor. Water mixed with methylene blue dye was used as the injection fluid.

The test package was placed in the centrifuge and spun at an acceleration of 50g for several hours prior to the injection test to allow time for

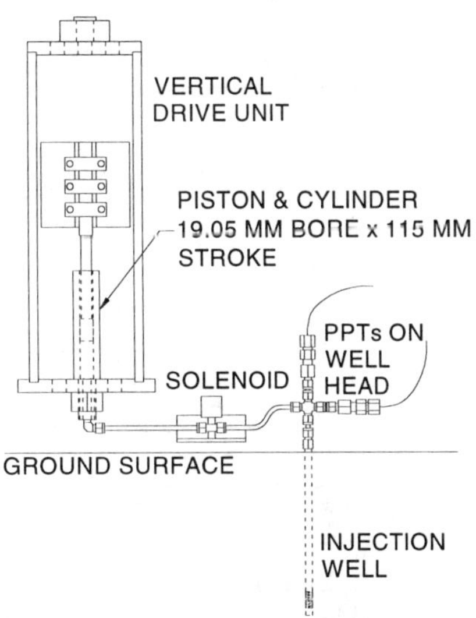

Figure 2. Injection pump and well

instrumentation to stabilize. Following each injection test, the centrifuge was stopped and the injection pump and displacement transducers were positioned at the next test location. The centrifuge was then spun back up to the test acceleration and the next test was carried out. After completing the tests, the package was removed from the centrifuge to the laboratory where the clay was excavated and the locations of dye traces in the soil were mapped.

INJECTION TEST RESULTS

The shape of the injection pressure-volume (P-V) curves was similar for all tests. The P-V curve for test CCFS01B is shown in Figure 3. Results from all tests are summarized in Table 1. The peak injection pressure appears to be influenced much more by the well depth (overburden stress) than by the injection rate. Following an initial steep pressure rise and peak, the injection pressure tended to drop off with increasing injection volume and approached a steady-state injection pressure, P_{ss}.

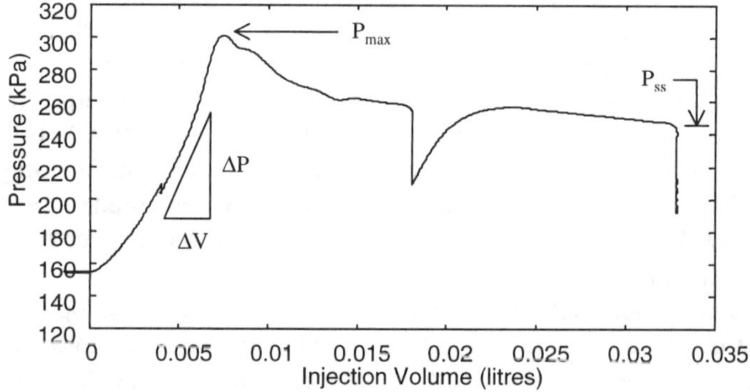

Figure 3. Injection pressure versus volume for test CCFS01B

Table 1. Test parameters and injection test results

Test ID	Well Depth (mm)	Inj. Rate (mm^3/s)	ΔP/ΔV (mean) (MPa/m^3)	P_{max} (kPa)	P_{ss} (kPa)	D_i (mm)	σ_{vo}' at D_i (kPa)	Extent of Dye (mm^2)
CCFS1A	200	28.50	30,000	286.3	250	240/200	211/176	10,600
B	200	285.0	30,200	301.5	248	200	176	33,800
C	100	28.50	19,100	175.3	136	80	70	13,050
D	100	285.0	23,800	237.9	138	135/105	119/92	35,100
CCFS2E	200	2850	12,300	311.6	187	100	88	27,800
F	200	14.25	13,800	293.0	263	230	202	4,200
G	100	14.25	*1.33×10^6	199.6	151	135	119	10,500
H	100	14.25	3,100	206.1	142	70	62	15,900

Note: * During test G, the piston was moved before the solenoid was opened, resulting in a pressure buildup in the pump and tubing. When the solenoid was opened, the pressure was rapidly applied to the fluid in the injection well, resulting in a very large value of ΔP/ΔV.

The depth, D_i, at which the dye traces began to extend laterally away from the well varied between tests. The areal extent of the dye traces is tabulated in Table 1 and shown in Figure 4. The observed dye pattern was not generally symmetrical about the injection well, however, the size of the dyed zone appears to be dependent on the injection rate and depth of well. The dye extent decreases in shallow wells and increases as the injection rate is increased. In the deep wells, the dye traces were found as far as about 8 m (prototype scale) from the injection well. The dye traces were mapped and compiled to create a 3D graphical image for each well. The dye trace image for CCFS01D is shown in Figure 5, note the two distinct levels of dye.

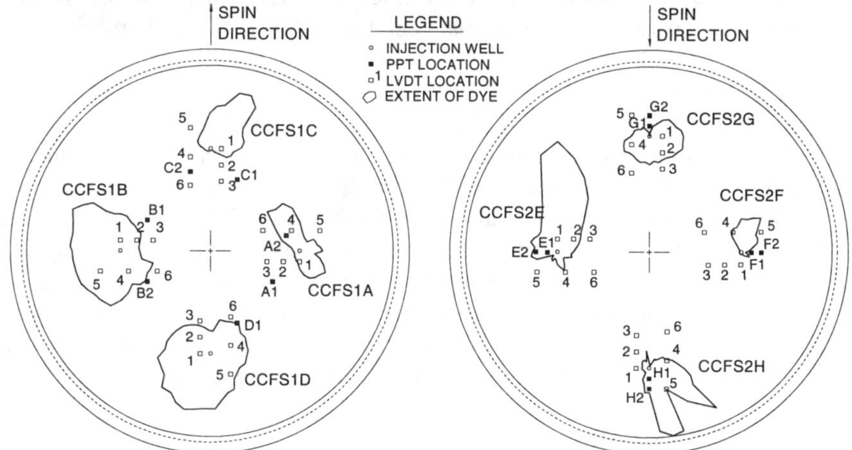

Figure 4. Areal extent of dye traces and location of instrumentation

DISCUSSION

The results from this test program and other work carried out, show that small volumes of fluid injected into low permeability soil result in the formation of higher permeability discontinuities in the soil surrounding a well. The discontinuities are known to have a higher permeability than the surrounding soil because the injection fluid was found to have traveled much greater distances during the test than can be explained by flow through the pore space alone. The discontinuities may extend for significant distances from the point of injection, sometimes branching out in multiple levels. The development of discontinuities is believed to be primarily by deformation due to injection induced shear stresses.

The extent to which shearing occurs during the injection process is dependent on a number of factors

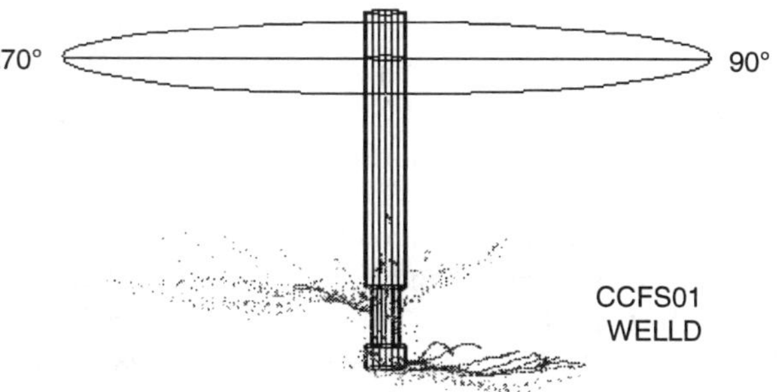

Figure 5. Isometric view of dye traces

including: injection fluid properties, absolute permeability of soil, well details, state of stress, soil constitutive behavior, and soil fabric. By developing a thorough understanding of how each of these factors may influence the shearing process, the technique can then be used to allow existing *in situ* remediation techniques to be applied more effectively in low permeability soil. Because the discontinuities formed by shearing are not held open with a proppant, *in situ* techniques that may benefit most through the use of shearing include those techniques that rely on injection of fluids into the soil rather than extraction. Promising applications include bioremediation, sparging, chemical treatment, thermal methods, and others.

CONCLUSIONS

Reduced scale centrifuge injection tests were carried out during this experimental program to examine the potential viability of using soil shearing to enhance the use of *in situ* remediation techniques in low permeable soils. The results from these initial tests are encouraging and will provide the basis for future laboratory and analytical studies aimed at developing better understanding of the mechanisms governing the development of shear zones during fluid injection. The ultimate goal of the work is to provide a low cost method of enhancing in situ techniques such as bioremediation, sparging, chemical treatment, thermal methods and others.

ACKNOWLEDGMENTS

The financial support of NSERC Canada is greatly appreciated.

REFERENCES

Al-Tabbaa, A. 1987. "Permeability and Stress-Strain Response of Speswhite Kaolin". PhD Thesis. University of Cambridge. November.

Chalaturnyk, R. and J. D. Scott. 1997. "Geomechanical Response of Heavy Oil Reservoirs to the Steam Assisted Gravity Drainage Process". SPE 37569.

Golder Associates Ltd. 1994. *Laboratory Study of Hydraulic Propagation in Oil Sands,* Phase III: Final Report to PERD and CANMET. July.

Kosar, K.M and K. Been. 1991. "The Effects of Geomechanical Behaviour on *In Situ* Recovery Processes in Oil Sands Reservoirs". 5th Int. Conf. on Heavy Crude and Tar Sands, UNITAR, Caracas, Venezuela, August 4 - 9.

Ling, L. 1995. "Strength Characteristics of a Modelling Silty Clay". Master's Thesis, Memorial University of Newfoundland, St. John's, NF, CA.

Mori, A. and M. Tamura. 1986. "Effect of Dilatancy on Permeability in Sands Stabilized by Chemical Grout". Soils and Foundations. Japanese Society of Soil Mechanics and Foundation Engineering. Vol. 26, No.1, pp. 96-104.

Rossato, G., N. L. Ninis, and R. J. Jardine. 1992. "Properties of Some Kaolin Based Model Clay Soils". ASTM GTJODJ, Vol. 15, No. 2, pp 166-179.

Walter, D. J., K. M. Kosar, and J. I. Clark. 1998. "Shearing of Contaminated Soils to Enhance *In Situ* Treatment". 51st Cdn. Geotech. Conf. Edmonton, October.

AIR FLOW IN FRACTURED BEDROCK FOR IN-SITU GROUNDWATER BIOREMEDIATION

Gregory L. Carter (Earth Tech, Roanoke, Virginia)
Jennifer C. Vincent and Barbara B. Lemos (Earth Tech, Concord, Massachusetts)
Rosann Kryczkowski (ITT Industries Night Vision, Roanoke, Virginia)

ABSTRACT: A pilot scale in-situ gaseous-phase injection test was implemented to evaluate the installation and operation of a methanotrophic bioremediation system for a fractured bedrock aquifer beneath an active manufacturing complex in Virginia. While the injection of necessary supplements, including oxygen, nutrients, and carbon sources, is rather routine in unconsolidated materials, it is quite complex in fractured bedrock. The site is underlain by fractured bedrock and is overlain by thin, clay-rich overburden, with groundwater generally encountered at the overburden-bedrock interface (5 to 10 ft (1.5 to 3m) below ground surface (BGS)). Contaminants of concern included chlorinated solvents such as trichloroethylene (TCE); 1,1,1-trichloroethane (TCA); and their breakdown products, as well as acetone and isopropanol.

To optimize and accelerate contaminant breakdown, the natural subsurface conditions were converted to aerobic conditions through the injection of air. Gaseous-phase nutrients and methane were also injected to further stimulate the growth of native microbial populations. As expected, heterogeneities in subsurface airflow have been observed during the pilot test. Variations in precipitation during the injection system operation resulted in varying backpressures and subsequent injection flow rates in the injection well. However, the successful conversion of a 75 to 150 ft (23 to 46 m) lateral zone of influence to aerobic conditions, stimulation of native microbial populations, and accelerated volatile organic compound (VOC) degradation in the zone of influence have been observed.

INTRODUCTION

An in-situ groundwater bioremediation project is being pilot tested at the ITT Industries Night Vision Division plant located in Roanoke, Virginia. The facility is an active manufacturing plant with production commitments that do not permit interruption of operations, and subsurface utilities preclude significant excavation work. When evaluating the technology options, particular emphasis was placed on innovative treatment technologies that could be applied in-situ given these site restrictions. After review of several technologies, in-situ enhanced bioremediation was selected as the technology best suited to the contaminants, geology, and logistical factors present at this site. The basis of the chosen technology, developed at the Westinghouse Savannah River Plant site (Hazen, 1995) and licensed by the U. S. Department of Energy, is an injection system used to deliver a gaseous phase mixture of air, nutrients (nitrous oxide and triethyl phosphate), and methane to the targeted subsurface area to stimulate the growth of methanotrophs. These bacteria produce enzymes which degrade the chlorinated solvents and their daughter products to non-hazardous constituents.

PROJECT OBJECTIVE

The purpose of this pilot test, which was implemented as a RCRA Interim Measure (IM), was to observe the reduction in mass of VOCs in the pilot test area. The degree of effectiveness of the pilot study would determine whether this technology could be expanded into the source area and applied to other source areas at the facility.

SITE DESCRIPTION

This facility is an active manufacturing complex consisting of three major buildings and associated infrastructure located on two parcels totaling 26.7 acres of land. The facility is situated on a small knoll, with the surrounding topography characterized by small hills and valleys and intermittent and continuous streams. Surrounding land uses include agricultural, residential, commercial, and industrial.

The contaminants of concern in groundwater at the Night Vision facility consist of chlorinated solvents including TCE, TCA and associated daughter products, as well as acetone and isopropanol. Remediation is complicated by the target VOCs occurring in the fractured bedrock. The VOC releases entered the very low permeability and hydraulic conductivity (10^{-4} to 10^{-7} cm/sec) clay overburden then migrated to the underlying bedrock. Groundwater flow and VOC migration primarily occurs via fractures in the bedrock, including high-transmissivity (10^{-2} to 10^{-4} cm/sec) fractures and "mud seams". Groundwater is typically encountered at 5 to 15 ft (1.5 to 4.6 m) BGS in this area. The logistical complexity of the study area was a key factor in the selection of this remedial technology. This complexity presents a concern because of the uncertainty of the injected gases flow paths in the fractured bedrock. However, gaseous phase nutrient injection was expected to affect a larger area than liquid nutrient injection.

METHODS

The primary components of the pilot test system consist of: an injection well; five monitoring wells; four soil gas monitoring points; and the air, nutrient and methane injection equipment. The monitoring wells were completed as 2-inch diameter PVC monitoring wells tapping the deeper water-bearing fracture zone encountered between 40 and 50 ft (12.2 and 15.2 m) BGS within a 6-inch diameter PVC casing which isolates the shallower (15 to 30 ft (4.5 to 9.1 m) BGS) water-bearing fracture zones. The injection well was constructed of 1-inch diameter steel. The soil gas points were installed with 2-inch diameter PVC intercepting the overburden and transition zones. This construction would allow for the soil gas points to be converted to soil vapor extraction wells, if needed.

The injection system, comprised of the air, nutrient and methane injection equipment, was housed in a temporary building located in the pilot test area utilizing an existing concrete pad. The air source was an existing facility air compressor which includes a condensate tank with a drain, an air line, coalescing filters and pressure regulator and valves. The methane and nitrous oxide are provided in standard air cylinders and are piped into the main air line using regulators and flow meters. The methane is stored in an attached shed and is piped into the main line through appropriate meters and regulators. The triethyl phosphate (TEP) is in liquid state stored in a steel tank. Air from the main line is diverted through the tank to volatilize the TEP for subsurface delivery. The air, nitrous oxide, and TEP are injected continuously while the methane is injected on a pulsed schedule of 8 hours per day, 5 days per week. The methane is closely monitored to ensure that the injection concentration does not exceed 4% by volume, thus avoiding the methane lower explosive limit of 5%.

The injection campaign was implemented in a phased approach to determine the optimum system operation conditions. Initially, only air injection was performed for 6 weeks until the groundwater data indicated limited nutrient availability. An air and nutrient injection phase was performed for 10 weeks, during which a steady decline of methane was observed. When methane levels became limited, methane was added to the injection stream to maintain an adequate amount of cometabolic carbon source.

During startup of the air injection, a mixture of approximately 5% helium and 95% air by volume was injected into the subsurface to evaluate the injection well zone of influence. Helium measurements were made in the surrounding monitoring wells and soil gas points, in addition to methane, carbon dioxide and oxygen measurements. Helium tests were subsequently performed to evaluate the flow path changes over time and at the various injection rates. Periodic headspace field screening was performed on the soil gas points and IM monitoring wells for the presence of methane, carbon dioxide, and oxygen. In addition, pressure readings at the monitoring points were made using magnehelic gauges. The soil gas was periodically sampled for laboratory analyses of VOCs, methane, and carbon dioxide.

The treatment system's effectiveness was monitored throughout the pilot test through low-flow groundwater sampling of the IM wells. Groundwater sampling events were performed pre-injection, post air-only injection, post air/nutrient injection, and post air/nutrient/methane injection to assess the performance of each of the injection campaigns. Groundwater samples were analyzed for VOCs, methanotroph (MPN) counts, phospholipid fatty acids (PLFA), biological oxygen demand, carbon dioxide, chloride, chemical oxygen demand, total and dissolved iron, nitrogen as nitrate, nitrite and ammonia, total phosphorous, sulfate, sulfide, and total organic carbon. Additionally, field equipment was used to monitor dissolved oxygen (DO), reduction/oxidation potential (redox), pH, conductivity, temperature, and depth to groundwater prior to sample collection. These field parameters were also monitored continuously in several key wells.

RESULTS AND DISCUSSION

Following the first two days of the system startup, the air injection delivery was established in the equipment building at 14 pounds per square inch (psi) and 30 standard cubic feet per hour (scfh). This resulted in an injection pressure at the well head of 12 psi. These operation parameters were based on back pressure occurrence at the injection well and to minimize air flow short circuiting and groundwater effervescence observed during startup. Minimal groundwater effervescence was desired to ensure that the observed groundwater treatment was a result of increased bioremediation and not the result of VOC volatilization from air sparging. Soil gas VOC concentrations were monitored, and no increase over background concentrations was observed following the low and high flow rate injection periods, indicating that VOC volatilization is negligible.

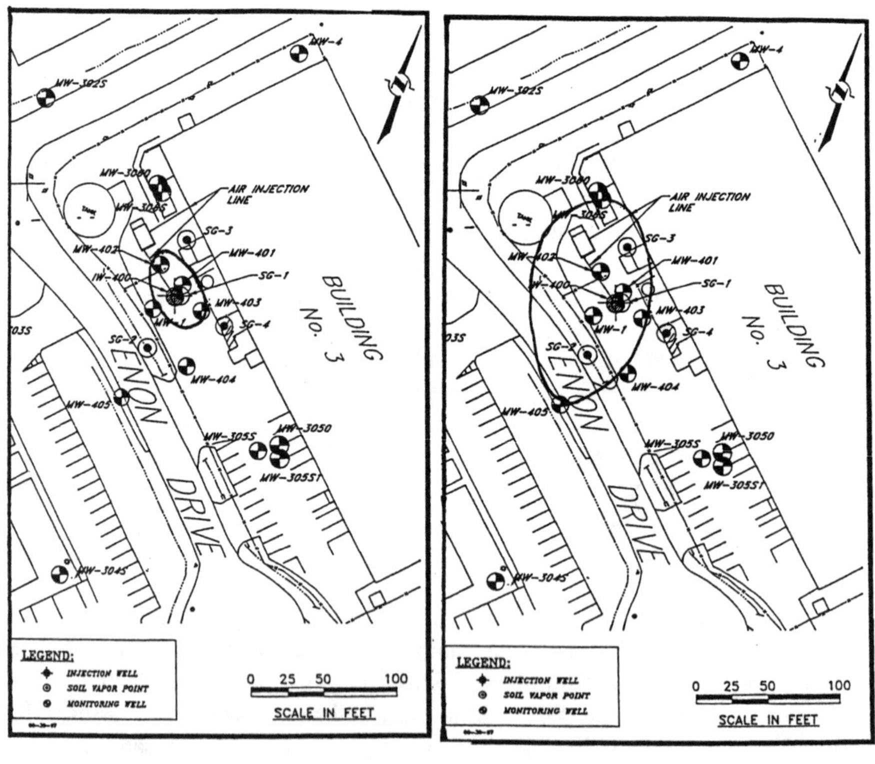

FIGURE 1 Site map showing pilot test zone of influence under (a) low air flow conditions (b) high air flow conditions.

Short-circuiting of air flow was observed in MW-401 and MW-403 following initial startup as a pocket of air was forced through the fracture sytem to the affected wellhead. This resulted in a preferential pathway which temporarily reduced airflow to other surrounding wells. This was remedied by sealing these monitoring wells.

Prior to the pilot test start-up, groundwater from shallow and deep zones in the vicinity of the pilot test appeared to be anaerobic as evidenced by DO and redox conditions. Monitoring well (MW)-401 and MW-404 began the test under anaerobic conditions and remained predominantly anaerobic during the air injection phase. However, the deep zone of MW-401 did exhibit aerobic conditions for approximately three days under low flow conditions following the startup short-circuiting. The shallow zones in the injection well (IW)-400, MW-401, MW-402, MW-403, and MW-1 appeared to be aerobic during air injection at the low flow rate (Figure 1a).

Visual observations and magnehelic readings confirmed the zone of influence and changes observed using the groundwater monitoring parameters. Air flow paths appeared to be affecting a significant lateral area from the injection well with a limited vertical extent. Air injected into the fracture at 45 ft (13.7 m) BGS flowed horizontally 25 to 40 ft (7.6 to 12.2 m) in fracture zones 10 to 15 ft (3 to 4.6 m) above the injection fracture at low air injection pressures and flow rates. This resulted in a somewhat conical zone of influence.

During the startup helium tracer test, helium was detected in multiple zones over a large area between MW-405 and MW-306S, approximately 75 ft (23 m) from the injection well. However, significant DO changes were not observed in many of these wells, particularly the deep zones in MW-401, MW-402, and MW-404, suggesting that oxygen was not being delivered to these wells. This has been attributed to a high degree of oxygen usage in proximity of the injection well which has resulted in a limited extent of oxygen diffusion in the deep zones.

At the start of the third phase of injection, the back pressure at the injection well appeared to have decreased since prior injection phases. This allowed for increased flow rates of air and gaseous supplement injection without the short circuiting previously observed. The air injection was increased to approximately 100 scfh and an injection pressure at the well head of 10 to 12 psi. The nitrous oxide and TEP injection continues at approximately the previous ratios. Nitrous oxide injection is between 0.7 to 1.0 scfh and the TEP injection is targeted at 0.007%. Limited additional groundwater mounding or effervescence was observed. This reduced back pressure coincided with a lower water table elevation which were approximately 6 to 12 ft (1.8 to 3.6 m) below those observed during the system startup and were attributed to precipitation decreases.

Based on the field parameters monitored, groundwater from the shallow and deep zones in the vicinity of the IM (MW-306S and MW-401) appears to have been converted from anaerobic to aerobic conditions in response to the increase in air flow delivery rate (Figure 1b). Headspace monitoring data from the methane injection phase confirms this zone of influence. MW-306S is 75 ft (23 m) from the injection well and taps shallower (20 ft (6.1 m) above injection point) fractures.

Other interconnectivity was observed in surrounding wells with a preferential pathway remaining to MW-403. Limited DO changes were observed in monitoring wells as close as 15 ft (4.6 m) from the injection well, indicating the extensiveness of the fracture heterogeneities. A follow-up helium tracer test confirmed the findings from previous tests that the injection well was affecting a 75 ft (23 m) zone of influence.

Baseline microbial results had shown that a diverse microbial population existed at the site; however, the microbial populations were environmentally stressed. Following the addition of nutrients and methane, microbial data indicated that the MPN and PLFA biomass counts have increased by 1 to 3 orders of magnitude. With the exception of MW-1, significant increases in biomass were not observed in the IM wells until the post-nutrient sampling round. The largest biomass population increases, greater than 1,000%, have been observed in MW-1, IW-400, MW-401S, MW-401, MW-403, MW-404S, and MW-306S. As with the VOC concentration decreases observed, these wells are generally shallow and closer to the injection well.

Throughout the pilot test, decreases in VOC concentrations, as compared with the lowest baseline data, have been observed in the majority of the area wells confirming the zone of influence of the treatment system. Following the air injection phase, VOC concentration decreases ranging from 10% in MW-402 to 98% in IW-400 and MW-404 were observed. Decreases ranging from 24% in MW-402 to 99% in MW-404S, as compared with the baseline data, were observed during the post-nutrient injection sampling round. The most recent sampling rounds, which were conducted following the initiation of methane injection, showed VOC concentration decreases of greater than 95% over baseline concentrations in IW-400, MW-401S, MW-404, MW-405, and MW-405S. Acetone and isopropanol concentrations have reduced the greatest amount throughout this IM. Vinyl chloride, chloroethane, 1,1-dichloroethane and 1,1-dichloroethene concentrations were also reduced. Based on the increased microbial populations coupled with the decreases in VOCs, the injection of air, nutrients and methane appears to have enhanced bioremediation at this site.

CONCLUSIONS

The data collected during this pilot scale enhanced bioremediation project indicated that a significant zone of influence can be obtained in fractured bedrock. Airflow at this site has resulted in an affected area that is somewhat conical as often seen in homogeneous soils. However, airflow in the fractured bedrock has shown definite preferential airflow pathways, and short circuiting has been observed in isolated areas.

Increased injection flow rates allowed more supplements to be delivered to a larger area which resulted in increases in biomass and reductions in VOC concentrations. Decreased VOC concentrations of up to 99% were observed in surrounding monitoring wells within 4 to 6 months of system operation. The treatment system affected monitoring wells 75 ft from the injection well.

Additional data is being collected to verify the thoroughness and completeness of the treatment process. This will be used to identify the extent that the technology will lower the VOC concentrations, given the observed distribution of air, nutrients and methane.

ACKNOWLEDGEMENTS

This pilot test was performed in conjunction with the USEPA Superfund Innovative Technology Evaluation (SITE) program. We appreciate their participation in the project and cooperative efforts regarding data and ideas.

REFERENCES

Hazen, T. C. 1995. *Preliminary Technology Report for the In Situ Bioremediation Demonstration (Methane Biostimulation) of the Savannah River Integrated Demonstration Project, DOE/OTD (u)*. U. S. Department of Energy Report, WSRC-TR-93-670, Westinghouse Savannah River Company, Aiken, South Carolina.

IN SITU BIOREMEDIATION UTILIZING HORIZONTAL LASAGNA™

Wendy J. Davis-Hoover, L. Taras Bryndzia [1], Michael H. Roulier, Lawrence C. Murdoch [2], Mark Kemper, Phillip Cluxton [3], Souhail Al-Abed, William W. Slack [4], and Stephen J. Vesper.
United States Environmental Protection Agency, Cincinnati, Ohio USA;
[1] Shell Oil, Houston, Texas, USA; [2] Clemson University, Clemson, South Carolina, USA; [3] Cluxton Instruments, Inc., Martinsville, Ohio, USA; [4] FR_X, Inc., Cincinnati, Ohio, USA

ABSTRACT: The application of *in situ* horizontal LASAGNA ™ combined with methanotrophic bioremediation of trichloroethene was demonstrated at Rickenbacker Air National Guard Base in Ohio. Electroosmosis was conducted between an anode of titanium mesh placed at the ground surface and a cathode composed of a horizontal graphite-filled fracture at a depth of 4 m. Biodegradation of the TCE by an inoculated consortium of methanotrophs was performed in a single layer consisting of granular activated carbon placed at a depth of 240 cm by hydraulic fracturing between the electrodes. Between January 15, 1997 and November 23, 1998, the concentration of TCE in the Biocell soil was reduced to near zero. During the same time period, in the untreated natural attenuation area, little change in TCE concentration was observed.

INTRODUCTION

Trichloroethene (TCE) is a chlorinated organic solvent that has been widely used in metal processing, paint, and electronics industries among others. After petroleum products, TCE is the most frequently reported contaminant at hazardous waste sites on the National Priority List of the U. S. Environmental Protection Agency (USEPA, 1991). Since TCE is a potential carcinogen, its persistence in the environment is a troubling problem.

Methanotrophic bacteria produce an enzyme, soluble methane monooxygenase, which is involved in methane utilization but which can, fortuitously, dechlorinate TCE. Methanotrophs have been used to destroy TCE in sandy aquifers (Semprini et al., 1992) but in impermeable clay soils, the TCE is bound such that it may be unavailable to methanotrophic bacteria. In addition, in deep clay soils, there is little methane available.

Electroosmosis, the application of a DC current, has been used since the 1930s to dewater soil for engineering purposes (Casagrande, 1949). When combined with bioremediation and hydraulic fracturing in remediation applications, LASAGNA™ (USEPA, 1997), has four advantages: methanotrophic bacteria can be placed in contaminated soils in proximity to the contaminant, the contaminant's movement can be directed into a remediation zone for degradation, degradative bacteria may be dispersed into the soil, and methane can be introduced into deep soil layers.

Objective. At a non-contaminated site, we had previously developed the horizontal LASAGNA™ technology (Murdoch and Chen, 1997). The objective of this work was to demonstrate the feasibility of utilizing the LASAGNA™ technology to biodegrade TCE with methanotrophic bacteria at an actual contaminated site.

MATERIALS AND METHODS

Site Description. The site selected for this demonstration is adjacent to a former chemical drum storage pad located at the Rickenbacker Air National Guard Base (formerly Rickenbacker Air Force Base) near Columbus, Ohio. Preliminary sampling showed that the TCE was fairly uniformly distributed between 0.5 to 2.5 m depth. The TCE had probably been there since the 1930s or 40s. The site is underlain by brown and gray silty clay glacial till with occasional sand seams. Soil permeability was about 10^{-6} cm/sec. The depth to ground water fluctuates between depths of 1 to 2 m.

Biocell Design. An area 4 by 4 m was selected to develop the Biocell and another area of comparable size was utilized as a control site or natural attenuation (NA) area. The NA was left undisturbed except for monitoring and sampling at the same times and depth as the Biocell.

The cell configuration was as previously described (Vesper et all. 1997). (See Figure 1.) Briefly, the soil surface and debris were removed to a depth of 20 cm. Titanium mesh (the anode) was laid on the surface and covered with sand bags. A cathode made of granular graphite was installed at a depth of 5 m. An automated water delivery and recovery system was installed in the Biocell to circulate the water from the cathode back to the anode, along with make-up water equivalent to 3 inches of rainfall per week .

The remediation zone was GAC (145 kg) mixed with a consortium of methnotrophs, 30 L of 1 x 10^7 cells per ml of NMS media (Bowman and Sayler, 1994), installed by HF on 11/14/96, at a depth of 240 cm (i.e. just under the contamination). Three percent methane was introduced into the remediation zone every other hour. A second inoculation of methanotrophic bacteria was done on July 10, 1997 by adding 30 L, 1 x 10^7 cells per ml NMS media, was pumped into the GAC fracture at a rate of about 3 cubic foot /hr.

Soil Sampling and TCE Analysis. Concentrations of TCE in soil were determined by taking continuous samples from the sufrace to a depth of 4 m at random locations within each quadrant of the cell. Pretreatment samples were collected with an Earthprobe ™ split tube sampler on a Simco drill rig; subsequent sampling was done with a Geoprobe™ macro-core sampler. The samplers were aseptically subsampled for about 2 grams for microbial analysis which were placed into sterile containers and refrigerated until analyzed the following day. The remainder of the sample was placed in 500 ml EPA approved amber bottles containing 400 ml of methanol. The samples were refrigerated and returned to the laboratory that day. All TCE analyses before October 1997 were by GC and, later samples, by GC/MS (US EPA Method

8260). All the results, expressed as μg TCE /g soil (ppm) for each sampling event were combined to calculate the median and standard deviation for the concentration of TCE in the soil profile (from the surface to a depth of 4 m.)

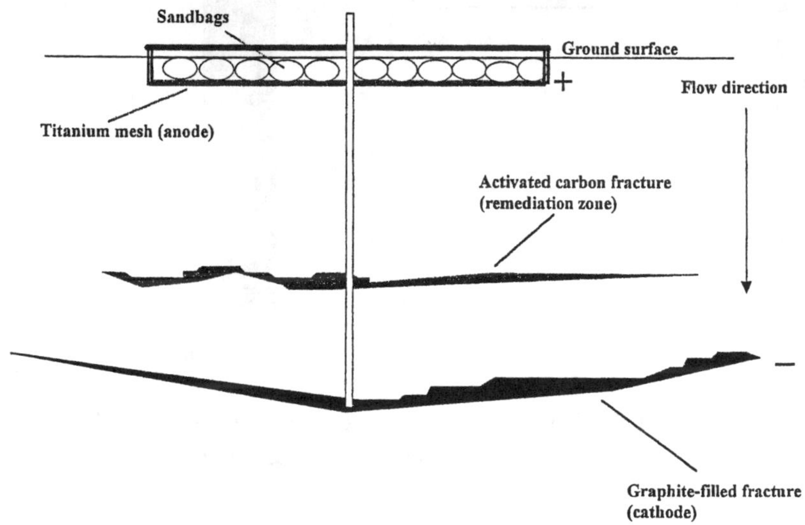

FIGURE 1. Schematic view of Biocell.

Quantifying Methanotrophic Bacteria. Duplicates or triplicates of 0.5 gm soil samples were diluted into sterile deionized water and plated in duplicates of NMS agar (Bowman and Sayler, 1994) medium using an Autoplate 4000(Spiral Biotech). The plates were incubated at 30°C in a glove box with 3 percent methane in air for 5 to 15 days, when the colony forming units (CFU) were enumerated and medians and standard deviations were determined.

RESULTS AND DISCUSSION

Soil Methanotrophic Bacterial Concentrations. Figures 2 and 3 demonstrate that we were able to successfully establish the introduced methanotropic consortium in the Biocell. The methanotrophic population is greatest near the GAC of the remediation zone at 240 cm. Thus, changing soil conditions during the electrokinetic treatment did not limit this population. In fact, the soil pH in the Biocell changed little during the process (initially 7.9 $\pm$ 0.3 to 8.4$\pm$ 0.4) and the temperature never rose above 31°C. There may be slightly more methanotrophic bacteria in the soil toward the anode side of the remediation zone. According to DeFaun and Condee

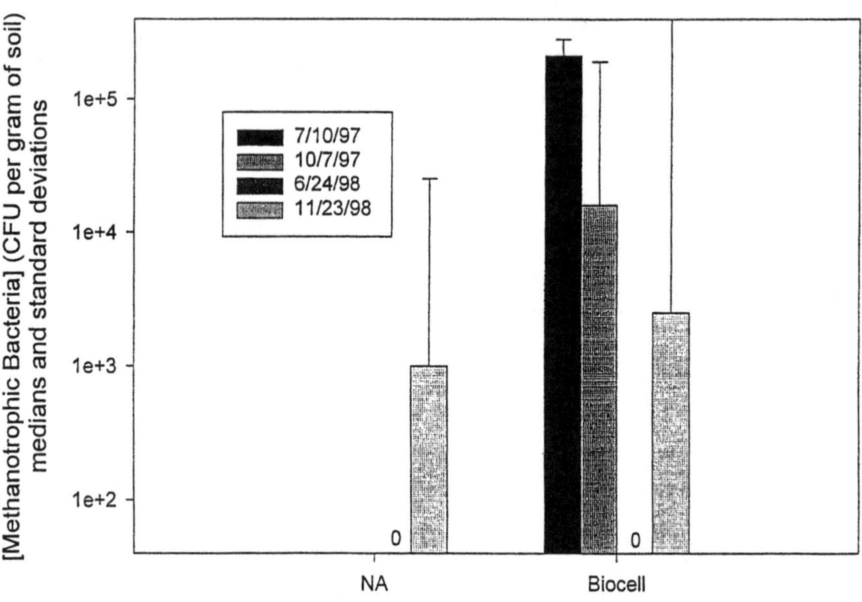

FIGURE 2. Concentrations of methanotrophic bacteria in soil in Natural Attenuation area and Biocell over time.

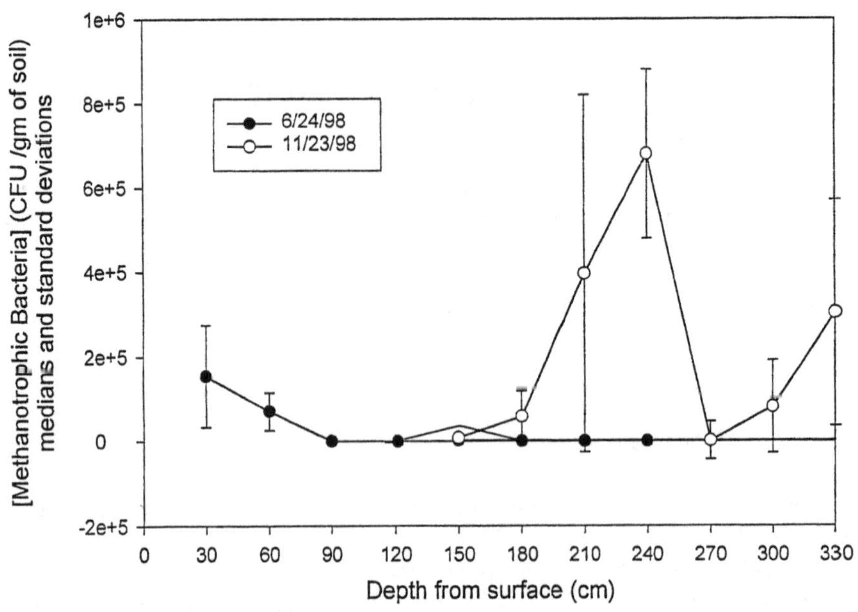

FIGURE 3. Concentrations of methanotrophic bacteria in the soil in the Biocell by depth over time.

(7), the bacteria would develop a more negative charge and move towards the anode. This may have happened here.

Soil TCE Analysis. Table 1 shows the median TCE concentrations and standard deviations in the Biocell compared to NA area between January 15, 1997 and November 23, 1998. There was a steady decline in TCE concentrations during the process of remediation, whereas the natural attenuation area showed no change in TCE concentration.

The fact that the methanotrophic population in the Biocell is still maintained as the TCE is disappearing suggests that bioremediation is occurring in the LASAGNA™ process.

TABLE 1 . TCE Concentrations in Biocell and Natural Attenuation

	SAMPLING DATES			
	1-15-97	**7-10-97**	**6-24-98**	**11-23-98**
Biocell*	2.83 ± 1.89	1.5 ± 2.61	0.46 ±1.00	0.00 ± 3.63
NA*	0.81 ± 2.07	1.33 ± 3.87	0.58 ± 9.06	0.93 ± 14.21

*Median ± 1 standard deviation (µg TCE /g soil)

REFERENCES

Bowman, J. P., and G. S. Sayler. 1994. "Optimization and maintenance of soluble methane monooxygenase activity in *Methylosinus trichosporium* OB3b."*Biodegradation*. 5:1-11.

Casagrande, L. 1949."Electroosmosis in soils". *Geotechnique* 1:159-166.

DeFlaun, M. F., and C. W. Condee. 1997. "Electrokinetic transport of bacteria."*J. Haz Mat*. 55:263-277.

Murdoch, L. C., and J-L Chen. 1997. "Effects of conductive fractures during in-situ electroosmosis." *J. Haz Mat* 55(1997) 239-262.

Semprini, L., Roberts, P.V., Hopkins, G. D., and McCarty, P.L. 1992. "Pilot scale studies of in situ bioremediation of chlorinated solvents." *J. Haz. Mat.* 32:145-162.

U.S. Environmental Protection Agency. 1997. "Lasagna™ Public-Private Partnership." Office of Research and Development and Office of Solid Waste and Emergency Response, Washington, D. C. 20460, EPA542-F-97-012A, November 1997.

U.S. Environmental Protection Agency.1991."Superfund NPL Characterization Project: National Results," Office of Emergency and Remedial Response, Washington, D. C. 20460, EPA/540/8-91/069, November 1991.

Vesper, S. J., S. R. Al-Abed, K. Catbas, and W. J. Davis-Hoover. 1997. "Use of methanotrophs to degrade TCE in electrokinetic field." *In Situ and On-Site Bioremediation:* 3:101-106.

AN ANALYSIS OF EXOGENOUS BACTERIA INJECTION FOR IMPROVED BIODEGRADATION

Sookyun Wang and M. Yavuz Corapcioglu
(Texas A&M University, College Station, Texas)

ABSTRACT: The degradation of organic hydrocarbons through microbial reactions in subsurface environments has been widely studied in the literature. Most studies, however, have focused on biostimulation, which aims to enhance biodegradation by indigenous bacteria. This study presents a mathematical model to simulate the fate and transport of a reactive contaminant degraded through cometabolism during *in situ* bioaugmentations involving the injection of nutrients and/or exogenous bacteria. We incorporate hydrogeologic factors affecting the transport of contaminant, as well as microbial metabolic reactions, into the model. Effects on the transport and biodegradation of an organic contaminant by both mobile bacteria in the aqueous phase and bacteria attached on soild surfaces are investigated by employing modified Monod kinetics and a microcolony concept. The permeability reduction, due to microbial accumulation in pore spaces, and its effect on the flow field are examined to simulate a field problem involving *in situ* biodegradations by exogenous bacteria. The effect of bacteria as biosorbents is also considered to investigate the biocolloid-facilitated transport of a contaminant. The two-dimensional governing equations are solved numerically using a fully implicit, finite-difference method with an alternating direction implicit scheme. The behavior of the model is demonstrated by applying it to a hypothetical case study.

A simulation results indicate that bioaugmentation by exogenous bacteria can enhance the effectiveness of the subsurface biodegradation of organic contaminants. The overall biodegradation rate is determined by the bioavailability of the contaminant in the aqueous phase, the bacterial population, the aqueous phase concentrations of a contaminant, a primary substrate, and an electron acceptor. The results also show that the placement of bacteria and/or nutrient injection wells, the ambient flow field, and the relative mobility of biodegrading bacteria, contaminants, and other components should be taken into consideration to achieve an effective *in situ* bioaugmentation scheme. Also, biomass accumulations near the injection wells need to be controlled in field operations.

INTRODUCTION

Recently, bioaugmentation techniques involving injections of exogenous bacteria and required nutrients have been developed to enhance intrinsic biodegradation in contaminated areas. A comprehensive understanding of the kinetics of biodegradation and the fate and transport of contaminants and biodegrading bacteria is required to establish an effective bioaugmentation technique. Although various mathematical models on biodegradation have contributed to the understanding of interactions between bacteria and organic contaminants, most of them have focused on bioreactions between contaminants and indigenous bacteria attached to soils and, therefore, do not address changes in the bacterial population due to the transport of bacteria and the biodegradation by suspended bacteria in the aqueous phase (Molz et al., 1986; Widdowson et al., 1988; Kindred and Celia, 1990; Semprini and McCarty, 1991; Zysset et al., 1994; Corapcioglu and Kim, 1995). However, Aamand et al. (1989) reported that, although most bacteria reside on the soil surface, noticeable numbers of bacteria are still active in the aqueous phase. The aqueous phase bacterial concentration can change the bacterial population in the solid phase. Therefore, the spatial and temporal distribution of bacteria in both the aqueous and solid phases should be considered in biodegradation modeling. It is especially critical to estimate the fate and transport of bacteria in the aqueous phase, the ensuing change of bacterial populations in both the aqueous and solid phases, and their effect on the fate and migration of contaminants in bioaugmentation projects involving an injection of exogenous bacteria into contaminated aquifers.

In this study, we propose a two-dimension mathematical model to investigate the effects of bacterial mobility on the fate and migration of an organic hydrocarbon contaminant in engineering biodegradation operations. The microcolony approach is applied to conceptualize the bacterial configuration in the solid phase and estimate the effects of bacterial accumulation on changes of hydrologic properties in a contaminated aquifer. We consider the effects of bacteria as biosorbents as well as biodegrading agents. In the model, a mass balance equation for a biodegradable organic contaminant in the subsurface is coupled with mass balance equations for a primary substrate, an electron acceptor, and bacteria in both the aqueous and solid phases. The fully implicit, finite-difference method and the altering direction iteration (ADI) scheme are used to solve the two-dimensional form of mass balance equations numerically. Model results for a hypothetical two-dimensional case study are presented to demonstrate the behavior and applicability of the model and to propose effective bioremediation strategies.

MODEL APPLICATION AND DISCUSSIONS

The success of *in situ* bioremediation projects depends mainly on how to control the contact of biodegrading microorganisms with target contaminants and how to supply the nutrients and electron acceptors required to stimulate microbial activities into contaminated regions. In engineering bioremediation involving the addition of specialized bacteria and nutrients into the subsurface, the relative mobility of bacteria to that of contaminants is an important factor in determining where and how to inject exogenous bacteria. If aqueous phase contaminants migrate fast with low retardation, contaminant spread in the form of a plume is a major concern in subsurface contaminant treatments. In this case, the injection of highly sorptive bacteria and supporting electron acceptors and nutrients into the aquifer should be conducted ahead of a plume front, so that a fixed-bed biobarrier composed of indigenous and injected bacteria will intercept contaminated groundwater passing through it. Duba et al. [1996] proposed a bioremediation strategy using an *in situ* microbial barrier constructed with exogenous bacteria, and demonstrated its applicability to the treatment of migrating TCE plumes. In order to demonstrate the applicability of the proposed model, we applied it to a hypothetical bioaugmentation operation for TCE plume using a fixed-bed biobarrier.

Physical configurations for the numerical simulation are illustrated in Figure 1. Before the bioaugmentation operation, indigenous bacteria were active on soil surfaces with a microcolony density of 500 cm^{-3}. To stimulate the construction of an *in situ* biobarrier, groundwater containing exogenous bacteria with strong sorption characteristics to soils is injected through W1 for five days. Groundwater containing high concentrations of methane and oxygen is continuously injected into the aquifer to enhance bacterial activities. The methane and oxygen injection well (W2) is installed at 4 m upstream from W1 to reduce the excessive biomass growth near W1 and to supply methane and oxygen to a wider area via dispersive transport.

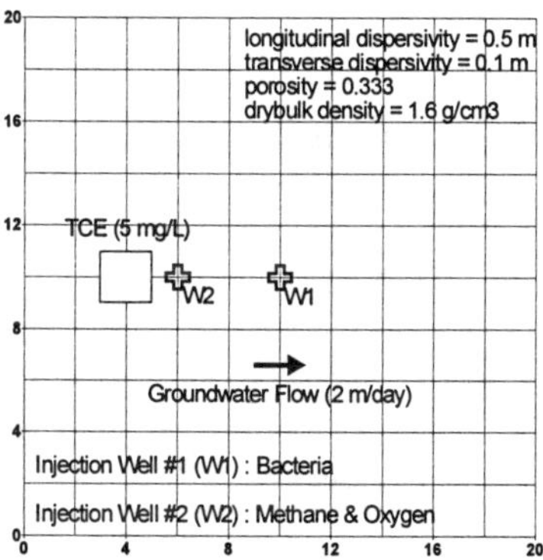

Figure 1. Areal view of model domain in the model simulation.

Figure 2 shows the areal distributions of methane and oxygen concentrations in the aqueous phase and the microcolony density at Day 5. Due to strong sorption characteristics, the injected bacteria are rapidly accumulated on the solid matrix near W1 and build up huge numbers of microcolonies. As bacterial injection continues, bacterial accumulation is spread toward surroundings, mainly downstream, and an elliptic biobarrier is constructed near W1. Any available methane and oxygen near the biobarrier are rapidly depleted by a large population of bacteria. Downstream of the biobarrier, only limited amounts of methane and oxygen are supplied by diffusive transport from surroundings because bacteria near the biobarrier utilize most of the methane and oxygen supplied from W2. Most of the biodegradation at a higher rate occurs within the biobarrier and in the region between W1 and W2 because of the high bacterial density and the replenishment of methane and oxygen from W2.

Figure 3 illustrates the fate and migration of the TCE plume and the growth of the biobarrier for 18 days after the introduction of exogenous bacteria and nutrients. At early time (Day 3), exogenous bacteria continue being injected into

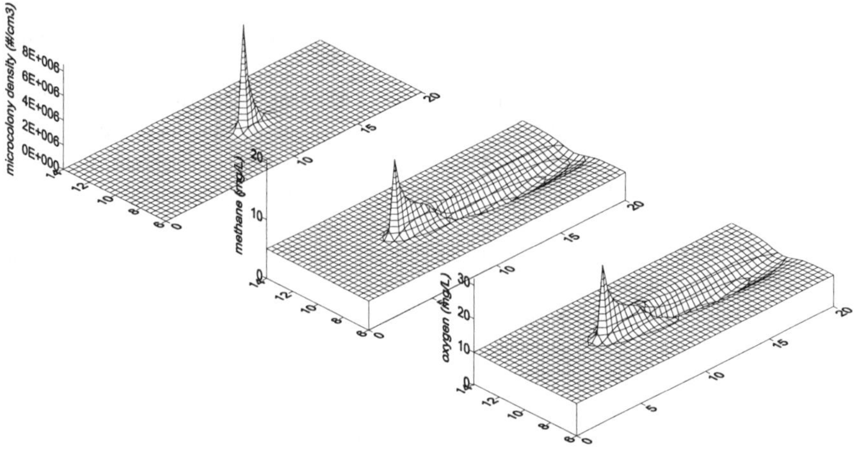

Figure 2. Areal distribution of aqueous phase methane and oxygen concentrations and microcolony density at Day 5.

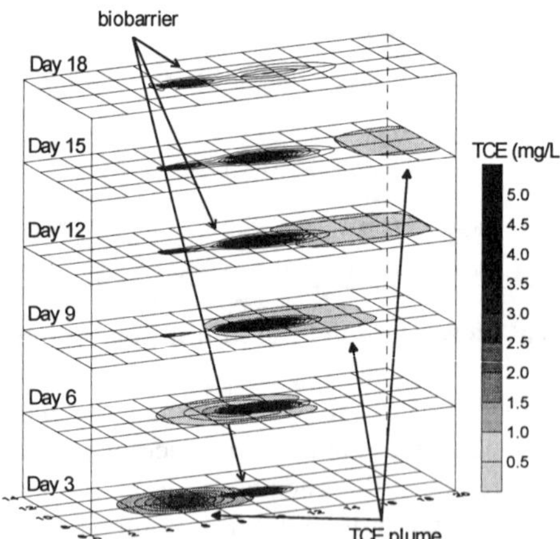

Figure 3. The fate and transport of TCE plume passing through the *in situ* biobarrier

the aquifer, and the biobarrier is under development. Since the TCE plume is far from the biobarrier, the overall rate of biodegradation in the entire domain is still very low. Due to the pulse injection of bacteria for the first five days, the biobarrier is fully developed as designed and the TCE plume is passing through the biobarrier (Days 6 and 9). Bacteria in the biobarrier uptake TCE at a high rate and the aqueous phase concentration of TCE significantly decreases when it passes through the biobarrier. The continuous injection of methane and oxygen through W2 creates a rich nutrient condition and stimulates the growth of indigenous bacteria near W2. As time goes on (Days 12 and 15), untreated TCE plume of quite low concentration passes by the biobarrier, and natural biodegradation by indigenous bacteria barely continues due to the depletion of methane and oxygen downstream of the biobarrier. The biomass growth near W2 becomes sufficient to limit the supply of methane and oxygen. Since the bacteria population in the biobarrier is too large to be supported without nutrient supply, it starts to decrease significantly. Finally (Day 18), most of the plume disappears out of the model domain and only the biomass accumulation near W2 continues.

Figure 4(a),(b) shows the relative mass of TCE transformed by biodegradation with and without exogenous bacteria injection and the change of the relative mass of TCE in each phase with exogenous bacteria injection. During the continuous injection of methane and oxygen over the entire simulation period, the biodegradation, with the addition of exogenous bacteria, transforms 84% more mass of TCE than with indigenous bacteria only. In the biobarrier operation involving the injection of bacteria, 64% of initial TCE in the aqueous and solid phases has been biodegraded by natural and exogenous bacteria for the entire simulation period. More than 70% of biodegradation occurs between Days 4-9 when the plume is passing through the biobarrier. After Day 10, biodegradation almost ends because of low methane and oxygen concentrations. Without bacteria injection, the biodegradation by natural bacteria continues up to Day 15 because, without the rapid depletion near W1, methane and oxygen can be supplied to downstream of W1. Because of the low rate of natural biodegradation, however, the overall biodegradation capacity is low.

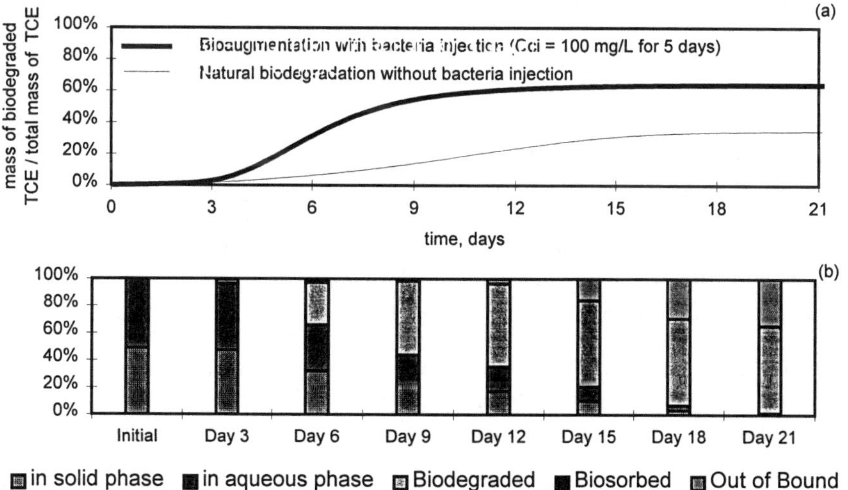

Figure 4. (a) Comparison of the efficiencies of biodegradation with and without exogenous bacteria injection, and (b) the change of the relative mass of TCE in each phase during bioaugmentation.

CONCLUSION

The modeling results indicate that the bioaugmentation involving the injection of exogenous bacteria with required nutrients and electron acceptors can enhance the effectiveness of the biodegradable contaminant treatment in the subsurface. The overall biodegradation rate is determined by the bioavailability of the contaminant, the bacterial population, and the concentrations of dissolved components required for biodegradation processes. The results also illustrate that the success of *in situ* bioaugmentation depends on controlling the contact of biodegrading microorganisms with target contaminants and supplying enough nutrients and electron acceptors to stimulate microbial activities into a contaminated zone. Therefore, to establish an effective scheme for an *in situ* bioaugmentation, bacteria and/or nutrient injection wells should be installed with consideration of the configurations of the entire flow field and the relative mobilities of biodegrading bacteria, biodegradable contaminants, and required components to enhance the microbial activities. In addition, a strategy to mitigate the biomass accumulation near bacteria and/or nutrient injection wells should be taken into account for a more effective *in situ* bioaugmentation scheme.

REFERENCES

Aamand, J., C. Jorgensen, E. Arvin, and B.K. Jensen. 1989. "Microbial adaptation of degradation of hydrocarbons in polluted and unpolluted groundwater." *J. Contam. Hydrol.* 4: 299-312.

Corapcioglu, M.Y. and S. Kim. 1995. "Modeling facilitated contaminant transport by mobile bacteria." *Water Resour. Res.* 31(11): 2639-2644.

Duba, A.G., K.J. Jackson, M.C. Jovanovich, R.B. Knapp, and R.T. Taylor. 1996. "TCE remediation using *in situ* resting-state bioremediation." *Environ. Sci. Technol.* 30(6): 1982-1989.

Kindred, J.S and M.A. Celia. 1989. "Contaminant transport and biodegradation 1. Conceptual model and test simulation." *Water Resour. Res.* 25(6): 1149-1159.

Molz, F.J., M.A. Widdowson, and L.D. Benefield. 1986. "Simulation of microbial growth dynamics coupled to nutrient and oxygen transport in porous media." *Water Resour. Res.* 22(8): 1207-1216.

Semprini, L and P.L. McCarty. 1991. "Comparison between model simulations and field results for in-situ biorestoration of chlorinated aliphatics, part 1. biostimulation of methanotrophic bacteria." *Ground Water* 29(3): 365-374.

Widdowson, M.A., F.J. Molz, and L.D. Benefield. 1988. "A numerical transport model for oxygen- and nitrate-based respiration linked to substrate and nutrient availability in porous media." *Water Resour. Res.* 24(9): 1553-1565.

Zysset, A., F. Stauffer, and T. Dracos. 1994. "Modeling of reactive groundwater transport governed by biodegradation." *Water Resour. Res.* 30(8): 2423-2434.

OPTIMIZATION OF *IN SITU* BIOREMEDIATION OF TRICHLOROETHYLENE USING GENETIC ALGORITHMS

Craig A. Garrett, Junqi Huang, and Mark N. Goltz
(Air Force Institute of Technology, Wright-Patterson AFB, Ohio)

ABSTRACT: A recent field evaluation of *in situ* aerobic cometabolic bioremediation demonstrated that this technology has promise to be an inexpensive and effective alternative to current technologies for the remediation of trichloroethylene (TCE)-contaminated groundwater. A fate and transport model has been developed to simulate the relevant processes that affect TCE fate and transport during technology application. Implementation of the technology requires that designers understand how to adjust engineered parameters in order to optimize system performance. While the fate and transport model can provide system performance information, the complexity of the system and the number of engineered parameters that can be varied make determining an optimal design problematic. Genetic algorithms may be used to determine those design parameter values that optimize the performance of this complex treatment system. By applying genetic algorithms, we hope to gain a better understanding of the technology, and ultimately, learn how to better design and implement *in situ* aerobic cometabolic biotreatment systems.

INTRODUCTION

A field-scale evaluation at Edwards Air Force Base (AFB), California demonstrated that aerobic cometabolic bioremediation is an effective technology for containment of trichloroethylene (TCE)-contaminated groundwater plumes (McCarty et al., 1998). The technology, as applied during the Edwards AFB study, relied on groundwater circulation wells to bring a mixture of TCE, primary substrate (such as phenol or toluene), and oxygen into the proximity of a population of indigenous microorganisms. The microorganisms, while aerobically metabolizing the primary substrate, fortuitously cometabolized the TCE to innocuous end products. This technology has many advantages over traditional methods of containing TCE plumes. Chief among these advantages is that the technology destroys the contaminants *in situ*, without the need to bring contaminated groundwater to the surface.

A fate and transport model that simulates the complex flow conditions of groundwater circulation wells, coupled with a kinetic description of cometabolic biodegradation, is a useful tool to help us understand the technology. Given a description of hydrogeology and contamination characteristics for a site, and a specified treatment system (number and location of treatment wells, well pumping rates, primary substrate feed schedule, etc.) the model can be used to simulate contaminant, dissolved oxygen, primary substrate, and microbial concentrations over space and time.

Implementation of the technology requires that designers understand how to adjust engineered parameters in order to optimize system performance. While the model can provide system performance information, the complexity of the system and the number of engineered parameters that can be varied make determining an optimal design problematic. In this paper, after briefly describing the technology and presenting a model that incorporates both physical and biochemical processes, we will demonstrate how genetic algorithms (GAs) can be applied to determine those design parameter values that optimize the performance of this complex treatment system. Our goal is to use modeling and GAs to gain a better understanding of the processes affecting technology performance, and ultimately, learn how to better design and implement *in situ* aerobic cometabolic biotreatment systems.

TECHNOLOGY DESCRIPTION

Figure 1 depicts how *in situ* aerobic cometabolic bioremediation was implemented during the Edwards AFB evaluation (McCarty et al., 1998).

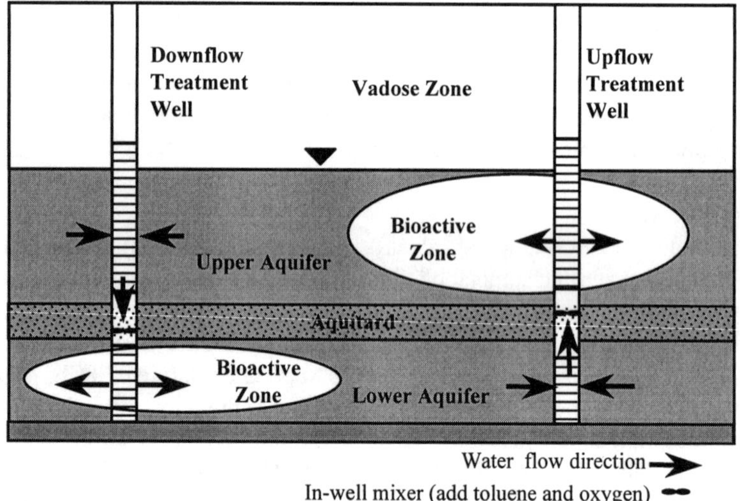

FIGURE 1. *In situ* aerobic cometabolism implementation at Edwards AFB (after McCarty et al., 1998)

The novel remediation system consisted of a pair of treatment wells that penetrated a confining zone between two aquifers. Each well had two screens, one in each aquifer. A submersible pump installed between the two screens of each well drew TCE contaminated water into the well at one of the screened intervals. Toluene, the primary substrate and enzyme inducer was pulsed into each well and an oxygen source was added continuously. Pulsing of the toluene was done to reduce biogrowth near the injection screen as well as to minimize competitive inhibition effects, whereby the primary substrate and the TCE compete for the same enzyme. The resultant mixture of dissolved TCE, toluene, and oxygen was discharged into the aquifer from the second screened intervals. This created *in situ* bioactive zones around the injection screens, where toluene was metabolized and TCE cometabolized. One of the wells operated in an upflow mode, the other in a downflow mode such that a circulation pattern was created between the two wells and the two aquifers. Due to this circulation pattern, contaminated water passed through the bioactive zones multiple times.

The demonstration at Edwards AFB lasted 410 days. TCE concentrations upgradient of the treatment system were approximately 1000 μg/L, while the water leaving the system had TCE concentrations averaging from 18-24 μg/L, for an overall removal of 97-98% (McCarty et al., 1998). Monitoring established that TCE removal for a single pass through one of the bioactive zones was about 83-87%. The overall removal rate of 97-98% was achieved because the recirculation between the wells caused contaminated water to pass through the bioactive zones several times, as noted above.

FATE AND TRANSPORT MODEL

The partial differential equations describing aerobic cometabolic biodegradation in one-dimension are presented in Semprini and McCarty (1991, 1992). In the current work, these one-dimensional equations are incorporated into a three-dimensional model and numerically solved using an efficient self-adaptive partial-implicit approach. The three-dimensional model permits simulation of bioremediation in the complex groundwater flow field induced by the two dual-screened treatment wells depicted in Figure 1. The model used in this work incorporates the following processes: (1) multi-dimensional flow, (2) advection and dispersion, (3) equilibrium or rate-limited sorption, and (4) biodegradation. Equilibrium sorption may be described using linear, Freundlich, or Langmuir isotherms. Rate-limited sorption may be described using either first-order kinetics or Fickian diffusion through various simple geometries. Biodegradation is described using either Monod kinetics for the primary substrate or cometabolic transformation kinetics for the TCE. Fate and transport of contaminant, primary substrate, oxygen, and microorganisms are described by a set of nonlinear partial differential equations. The equations are numerically solved using a customized self-adaptive partial-implicit approach. This numerical approach examines each term in the governing equations at every cell and time step. If the term is rapidly changing in space or time, an implicit solution method is used. An explicit method is used when the term is stable. The approach permits use of much larger time steps than would otherwise be allowed.

GENETIC ALGORITHMS (GAs)

Genetic Algorithms (GAs) are stochastic search algorithms inspired by the processes of natural selection and evolution. GAs maintain and "evolve" a population of individuals or *chromosomes*, each chromosome representing a possible solution to the problem, or in the case discussed here, a specific remediation strategy (defined as a particular pumping rate, injected toluene concentration, primary substrate and oxygen injection schedule, etc.). For instance, to minimize an objective function whose value is affected by three parameters (or *genes*), a single chromosome would represent a particular combination of values for those three parameters. The chromosome can be evaluated for *fitness* by decoding the representation (usually a binary string) into real values and applying the function to it, with a low value being desirable for a minimization problem. Good chromosomes are rewarded and are recombined or "mated" with other chromosomes, passing their "genetic material" on to subsequent generations. Recombination is also known as crossover. To help maintain diversity in the chromosome population, each gene in the population has the chance to "mutate" or randomly change value. The mutation rate is the probability that any one gene will randomly change value and is a parameter that may be specified by the modeller. While a full discussion of how GAs work is beyond the scope of this paper, a good introduction to GAs may be found in Goldberg (1989).

GAs are useful when trying to obtain the optimal solution for an objective function that is complex, multi-modal, and/or highly dimensional. In the case of optimizing *in situ* bioremediation, where we are dealing with a complex, non-linear system, it is likely that a GA can more effectively search the "landscape" of the objective function and find a better solution than a deterministic algorithm.

To solve our bioremediation problem, a real-valued GA was used, meaning that the representations of the variables in the chromosome are real values as opposed to binary strings. Since we used real-valued genes in the chromosomes, we applied the crossover and mutation operators designed for real-valued genes by Michalewicz et al. (1994) in their Genocop III GA software package.

The scenario that was modeled was generally based upon the Edwards AFB evaluation. Two biotreatment wells were used to remediate a TCE-plume. Initial TCE concentrations were set at 1000 μg/L in the entire aquifer, with a constant source of TCE along the upgradient boundary of 1000 μg/L. The model simulated 1000 days of treatment.

The chromosome was constructed as follows:

Gene 1: Pumping rate for each treatment well (m^3/d)
Gene 2: Horizontal distance between the two treatment wells (m)
Gene 3: Rate of oxygen injection (g/d)
Gene 4: Rate of toluene injection (g/d)
Gene 5: Time period oxygen is being injected (hr)
Gene 6: Time period oxygen is not being injected (hr)
Gene 7: Time period toluene is being injected (hr)
Gene 8: Time period toluene is not being injected (hr)

Upper and lower limits for each of the eight variables were established based on numerical tests and estimates of what was felt to be reasonable. The objective was to minimize cost per length of aquifer transverse to the direction of groundwater flow while adhering to regulatory constraints of no greater than 5 µg/L of TCE and 20 µg/L of toluene approximately 10 meters downgradient of the treatment wells. Constraints were incorporated into the objective function by imposition of a penalty (in dollars) for constraint violation. The objective function also included annualized capital costs, as well as costs for pumping, toluene, oxygen, and periodic well-redevelopment (to deal with well screen clogging).

RESULTS

The GA used a population of 50 chromosomes, ran for 42 generations, and had a mutation rate of 0.015. An elitist strategy was used (that is, new chromosomes were only kept if they were better than their parents) to force fast convergence and minimize run time.

The GA converged upon a solution that met the downgradient concentration constraints for both toluene and TCE. The optimal solution was:

Gene 1: Pumping rate for each treatment well = 90 m^3/d
Gene 2: Horizontal distance between the two treatment wells = 24 m
Gene 3: Rate of oxygen injection = 887 g/d
Gene 4: Rate of toluene injection = 407 g/d
Gene 5: Time period oxygen is being injected = 96 hr
Gene 6: Time period oxygen is not being injected = 0 hr
Gene 7: Time period toluene is being injected = 0.25 hr
Gene 8: Time period toluene is not being injected = 3.25 hr

The annual cost for operating the system was calculated at a little over $8,000. Interestingly, the optimal solution calls for continuous introduction of oxygen and pulsed injection of toluene, which is the strategy that was actually used during the Edwards AFB evaluation. Using the optimal strategy, TCE and toluene concentrations 10 m downgradient of the treatment system are 4.25 and 0.1 µg/L, respectively. This also, at least qualitatively, agrees with the results of the Edwards AFB evaluation, which, as had been expected, showed that toluene removals were much greater than TCE removals.

A few words are in order regarding what the "optimal" solution represents. Note that the solution tabulated above is for a particular objective function. While the objective function includes treatment system capital and operating costs that are reasonably well-quantified, the function also includes a penalty for violation of downgradient concentration constraints that is much more difficult to quantify. The optimal remediation strategy can change significantly, depending on how these constraints are quantified and incorporated into the objective function. Also, as with any optimization scheme being applied to a nonlinear system, there is no guarantee of convergence on a globally optimal solution. The definition of the objective function, and how the weights of the objective function components (operating cost versus penalty cost) may change as the GA searches for a solution are key in determining the optimal solution that is ultimately found. Knowledge of the system performance characteristics and the solution space are both important when applying a GA to help optimize a complex treatment system.

CONCLUSIONS

GAs are powerful tools for optimizing complex systems. The results from application of the GA to optimize aerobic cometabolic bioremediation are encouraging. The GA successfully found a solution that met downgradient concentration constraints at a cost that was at least locally optimal. We believe that the combination of the bioremediation fate and transport model with a GA, as discussed above, represents an important step towards helping us better understand how to design and implement *in situ* aerobic cometabolic bioremediation systems.

ACKNOWLEDGEMENTS

This work was sponsored in part by the Air Force Office of Scientific Research, USAF, under grant number FQ8671-9700520CS and the Air Force Research Laboratory, Airbase and Environmental Technology Division.

REFERENCES

Goldberg, David E. 1989. *Genetic Algorithms in Search, Optimization, and Learning*. Addison-Wesley, Reading, MA.

McCarty, P.L., M.N. Goltz, G.D. Hopkins, M.E. Dolan, J.P. Allan, B.T. Kawakami, and T.J. Carrothers. 1998. "Full-Scale Evaluation of *In Situ* Cometabolic Degradation of Trichloroethylene in Groundwater through Toluene Injection." *Environmental Science & Technology 32*(1):88-100.

Michalewicz, Z., T.D. Logan, , and S. Swaminathan. 1994. "Evolutionary Operators for Continuous Convex Parameter Spaces." In A.V. Sebald and L.J. Fogel (Eds.), *Proceedings of the 3rd Annual Conference on Evolutionary Programming*, pp. 84-97. World Scientific Publishing, River Edge, NJ.

Semprini, L. and P. L. McCarty. 1991. "Comparison between Model Simulations and Field Results for In-situ Biorestoration of Chlorinated Aliphatics: Part 1. Biostimulation of Methanotrophic Bacteria." *Ground Water 29*(3):365-374.

Semprini, L. and P. L. McCarty. 1992. "Comparison between Model Simulations and Field Results for In-situ Biorestoration of Chlorinated Aliphatics: Part 2. Cometabolic Transformations." *Ground Water 30*(1):37-44.

COMPUTER MODEL OF CO-METABOLIC BIOVENTING FOR CHLORINATED SOLVENT BIOREMEDIATION

T. Franz and A. Mason (Beatty Franz and Associates Ltd., Bolton, Ontario, Canada)
J. Zaidel and D. DeMarco (Waterloo Hydrogeologic, Inc., Waterloo, Ontario, Canada)
P. Morgan (ICI Technology, Runcorn, Cheshire, UK)
G. Sayles (US EPA National Risk Management Research Laboratory, Cincinnati, Ohio, USA)
L. Moser and D. Gannon (Zeneca Corp., Sheridan Park Environmental Laboratory, Mississauga, Ontario, Canada)

ABSTRACT: This paper summarises the background to, development and validation of new stirred tank and quasi-3-dimensional computer models of co-metabolic bioventing for the remediation of vadose zone soil contaminated with chlorinated solvent compounds.

INTRODUCTION

Most chlorinated solvents, including trichloroethene (TCE), 1,1,1-trichloroethane (TCA), and dichloromethane (DCM) can be degraded co-metabolically by bacteria growing aerobically on substrates including methane, butane, propane, and toluene (Semprini, 1997). This degradation has been applied on the pilot- and field-scale for the bioremediation of chlorinated solvent contamination, principally groundwater plumes (e.g., Hazen et al., 1994; Hopkins and McCarty, 1995; McCarty et al., 1998; Semprini et al., 1992; Suffin and Ramey, 1997) but there is also evidence that co-metabolic biodegradation can be effective for the remediation of vadose soil (Hazen et al., 1994; Suffin and Ramey, 1997).

Co-metabolic bioventing is a technology specifically designed for remediation of soil contaminated with chlorinated solvents. It involves enhancing co-metabolic degradation by the controlled supply of air (i.e., oxygen) and appropriate growth substrate(s).

As part of the program of the Remediation Technologies Development Forum (RTDF) Bioremediation of Chlorinated Solvents Consortium, co-metabolic bioventing has been tested on the pilot scale in the field at 2 locations: Dover Air Force Base (AFB), Delaware, and Hill AFB, Utah. In support of these projects, a quasi-3-dimensional (radial-symmetric vertical) computer model of co-metabolic bioventing has been developed and validated against laboratory and field data.

OVERVIEW OF FIELD PILOT-TEST PROGRAM

Test Sites. The test at Dover AFB, Delaware was situated adjacent to Building 719 (jet engine maintenance; see: Morgan et al., 1998; Sayles et al., 1997). Soil (sandy loam) properties and contaminant concentrations are summarised in Table 1. The facility at Hill AFB, Utah was situated within the former waste disposal area designated OU-2. This soil was classified as a fine sandy loam and its properties are summarised in Table 2.

TABLE 1. Soil properties and principle contaminant concentrations characteristic of co-metabolic bioventing test area, Dover AFB. TKN = total Kjeldahl nitrogen; TOC = total organic carbon.

Parameter	Value	Contaminant	Concentration
pH	8.0	TCE	5 mg/kg
TKN	220 mg/kg	*cis*-1,2-DCE	18 mg/kg
Total phosphate	36 mg/kg	1,1,1-TCA	150 mg/kg
TOC	<0.5 wt%	Xylene	14 mg/kg

TABLE 2. Soil properties and principle contaminant concentrations characteristic of co-metabolic bioventing test area, Hill AFB. TKN = total Kjeldahl nitrogen.

Parameter	Value	Contaminant	Concentration
pH	7.9	PCE	0.5 mg/kg
TKN	<100 mg/kg	TCE	7.5 mg/kg
Nitrate	5.3 mg/kg	TCA	2.0 mg/kg
Ammonium	1.4 mg/kg	Toluene	0.1 mg/kg
Total phosphate	36 mg/kg		
Organic matter	0.1 wt%		

Treatability Evaluation. The evaluation and design of co-metabolic bioventing was undertaken using laboratory microcosm and soil column tests (Morgan et al., 1998; Moser et al., 1997; Sayles et al., 1997). For both sites, the most effective and stable degradation of chlorinated solvent contaminants was achieved with propane (long-term rate for TCE up to 16 mg TCE/kg dry weight soil.year).

Pilot-Test Design. Based on the laboratory results, the field pilot test units were designed to operate with automated pulse injection of propane and air. Routine monitoring of soil gas composition (O_2, CO_2, propane, TCE, TCA) and soil temperature was backed up by an intensive sampling and analysis programme at the start and shut-down of operation to determine contaminant mass balance and changes in microbial population.

EVALUATION AND SELECTION OF BIOLOGICAL MODEL

There have been a number of attempts to develop models both for co-metabolism itself and as applied to chlorinated solvent biodegradation (e.g., Arcangeli and Arvin, 1997; Chang and Alvarez-Cohen, 1995; Criddle, 1993; Lang et al., 1997; Smith et al., 1997; Travis and Rosenburg, 1994, 1997). A detailed evaluation of these led us to select, inter-link and code appropriate equations from published models rather than develop *de novo* relationships. The biological parameters incorporated and modelled are summarised in Table 3.

Table 3. List of biological parameters employed in modelling.

Parameter	Description	Units
k_g	Max. degradation rate of growth substrate	mg growth substrate/mg cells.day
$K_{s,g}$	Half-saturation constant of growth substrate	mg/L
k_c	Max. degradation rate of co-metabolite	mg contaminant/mg cells.day
$K_{s,c}$	Half-saturation constant of growth substrate	mg/L
K_r	Half-saturation constant of NAD(P)H	mmol e^-/L
Y	Net cell yield	mg cells/mg growth substrate
T_c	Transformation capacity caused by toxicity	mg substrate/mg cells
B	Cell decay rate	1/day
R_0/X_0	Endogenous electron equivalents in cells	mmol e^-/mg cells
α_g	Stoichiometric coefficient of NAD(P)H generation/use from growth substrate	mmol e^-/mg growth substrate
α_c	Stoichiometric coefficient of NAD(P)H generation/use from co-metabolite	mmol e^-/mg growth substrate
R	Available reducing energy (NAD(P)H)	mmol e^-/ L
X	Available active biomass in solution	mg cells/L

DEVELOPMENT AND VALIDATION OF COMPUTER MODELS

Stirred Tank Model. Co-metabolic kinetics were programmed as a module for subsequent inclusion in a hydraulic model (see below). The model simulates substrate degradation kinetics, competitive inhibition between growth and non-growth substrates, cell growth and decay, product toxicity, reducing energy (NAD(P)H) limitation and regeneration, growth substrate pulsing, and partitioning between liquid & gas phases. It has been implemented in a Windows95^{TM} graphical user interface. Model output consists of mass of growth substrate and contaminant, energy, and biomass as functions of time. The model results were compared with data from various sources, including Chang and Alvarez-Cohen (1995). Figure 1 shows computer screen displays of the stirred tank model.

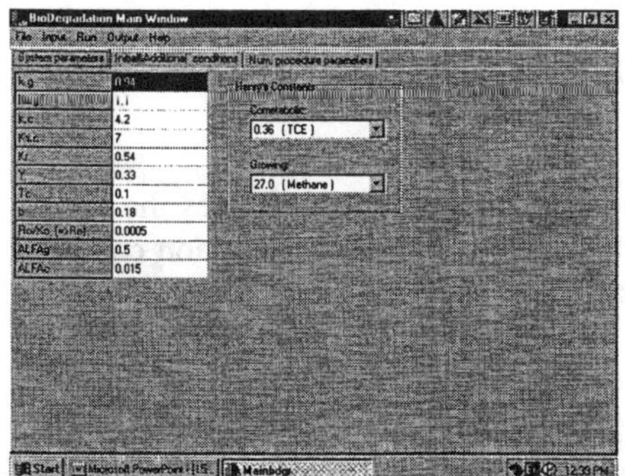

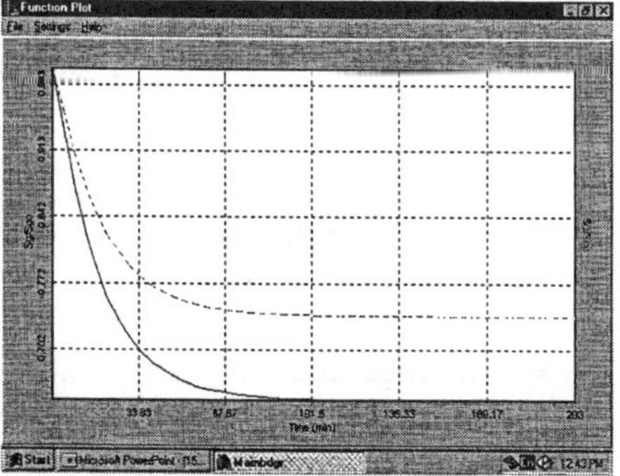

FIGURE 1. Screen shot showing input data menu (left) and growth and co-metabolic substrate vs. time (right).

3-Dimensional Co-Metabolic Bioventing Model. This was implemented by integrating the biological code from the stirred tank model into the established AIRFLOW/SVE™ vadose zone fate and transport model (Waterloo Hydrogeologic, Inc.). This provides a quasi-3-dimensional (radial-symmetric vertical) model which operates under DOS and Microsoft Windows™. The model is based on the finite difference method and calculates: 1) air pressure distributions; 2) spatial distributions of mass and vapour concentrations remaining in the system (at specified times); 3) mass and vapour as functions of time at specified monitoring points; 4) mass biodegraded; 5) mass stripped (due to volatilization during the venting process). The model distinguishes between total mass, and mass in the non-aqueous, adsorbed, dissolved, and vapour phases for both the contaminant and growth substrate. Figure 2 compares simulation results for homogeneous and heterogeneous soils. The results shown are mass of TCE remaining in the soil, mass of TCE stripped due to volatilization, and mass of TCE biodegraded as functions of time. The model is presently being validated against data from the RTDF pilot projects

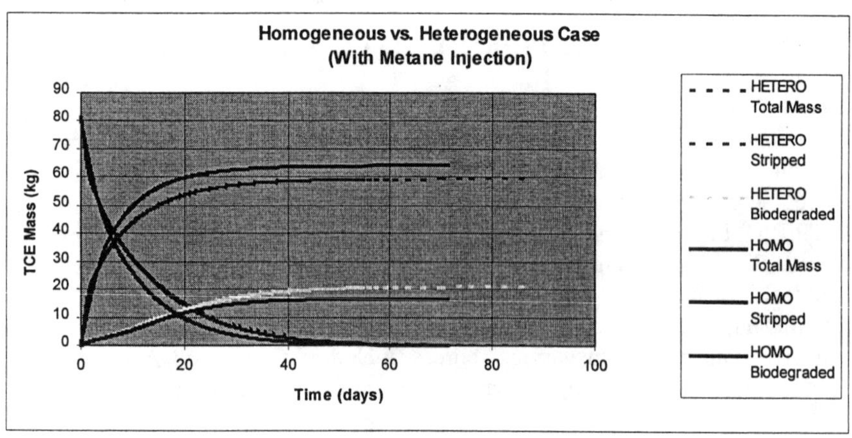

FIGURE 2. Comparison between co-metabolic bioventing results for example homogeneous and heterogeneous soils.

DISCUSSION

The final model incorporated into AIRFLOW/SVE™ couples the above-described biological code with a well-established soil vapour extraction code that enables modelling of co-metabolic bioventing, including:

- Heterogeneous and anisotropic soil conditions;
- Variable soil moisture content;
- Soil contamination (NAPL) and phase transfer into all relevant compartments;
- Non-equilibrium sorption;
- Contaminant partitioning to groundwater (modelled as mass sink);

- Continuous and pulsed (intermittent) injection of air and growth substrates;
- Estimation of flow rates, pressure distribution and velocity field;
- Degradation kinetics of primary substrate;
- Contaminant stripping and degradation;
- Competitive inhibition and toxicity;
- Cell growth and decay.

The stirred tank and 3-D models, FORTRAN source code, and user information (including operating instructions, details of the governing equations and their numerical implementation, and the validation examples) will be placed in the public domain.

ACKNOWLEDGEMENTS

The work described in this paper is part of the collaborative program undertaken by the RTDF (Remediation Technology Development Forum) Bioremediation of Chlorinated Solvents Consortium. The support and input of all members of the Consortium (Ciba; DuPont; Dow; General Electric; Geosyntec; ICI; Zeneca; US Air Force; US Dept. of Energy; US Environmental Protection Agency) is gratefully acknowledged. Particular thanks are due to Dover AFB, Delaware and Hill AFB, Utah for hosting the field tests.

REFERENCES

Arcangeli, J. P. and E. Arvin. 1997. “Modeling of the cometabolic biodegradation of trichloroethylene by toluene oxidizing bacteria in a biofilm system.” *Environ. Sci. Technol. 31*: 3044-3052.

Chang, H.-L. and L. Alvarez-Cohen. 1995. “Model for the cometabolic biodegradation of chlorinated organics.” *Environ. Sci. Technol. 29*: 2357-2367.

Criddle, C. S. 1993. “The kinetics of cometabolism.” *Biotechnol. Bioeng. 41*: 1048-1056.

Hazen, T. C., K. H. Lombard, B. B. Looney, M. V. Enzien, J. M. Dougherty, C. B. Fliermans, J. Wear and C. A. Eddy-Dilek. 1994. “Summary of in-situ bioremediation demonstration (methane biostimulation) via horizontal wells at the Savannah River site integrated demonstration project.” In G. W. Gee and N. R. Wing (Eds.), *In-situ Remediation: Scientific Basis for Current and Future Technologies*, pp. 137-150. Battelle Press, Columbus, OH.

Hopkins, G. D. and P. L. McCarty. 1995. “Field evaluation of in situ aerobic cometabolism of trichloroethylene and three dichloroethylene isomers using phenol and toluene as the primary substrates.” *Environ. Sci. Technol. 29*: 1628-1637.

Lang, M. M., P. V. Roberts and L. Semprini. 1997. "Model simulations in support of field scale design and operation of bioremediation based on cometabolic degradation." *Ground Water 35*: 565-573.

McCarty, P. L., M. N. Goltz, G. D. Hopkins, M. E. Dolan, J. P. Allan, B. T. Kawakami and T. J. Carrothers. 1998. "Full-scale evaluation of in situ cometabolic degradation of trichloroethylene in groundwater through toluene injection." *Environ. Sci. Technol. 32*: 88-100.

Morgan, P., G. D. Sayles, L. E. Moser, D. J. Gannon, D. H. Kampbell, T. Franz, C. M. Vogel, D. W. Major, M. J. Bell and M.W. Holmes. 1998. "Laboratory and field pilot evaluation of co-metabolic bioventing for the remediation of soil contaminated with trichloroethene (TCE) and trichloroethane (TCA)". In *Contaminated Soil '98*, pp. 1187-1188. Thomas Telford Publishing, London.

Moser, L. E., G. D. Sayles, D. J. Gannon, M. D. Lee, D. H. Kampbell and C. M. Vogel. 1997. "Comparison of methodologies for cometabolic bioventing treatability studies." In B. C. Alleman and A. Leeson, (Eds.) *In Situ and On-Site Bioremediation: Volume 3*, p. 299. Battelle Press, Columbus OH.

Sayles, G. D., L. E. Moser, D. J. Gannon, D. H. Kampbell, and C. M. Vogel. 1997. "Development of cometabolic bioventing for the in-situ bioremediation of chlorinated solvents." In B. C. Alleman and A. Leeson, (Eds.) *In Situ and On-Site Bioremediation: Volume 3*, p. 285. Battelle Press, Columbus OH.

Semprini, L. 1997. "Strategies for the aerobic co-metabolism of chlorinated solvents." *Curr. Opin. Microbiol. 8*: 296-308.

Semprini, L., G. D. Hopkins, P. V. Roberts and P. L. McCarty, P.L. 1992. "Pilot scale field studies of in situ bioremediation of chlorinated solvents." *J. Hazard. Mater. 32*, 145-162.

Smith, L. H., P. K. Kitanidis and P. L. McCarty. 1997. "Numerical modeling and uncertainties in rate coefficients for methane utilization and TCE cometabolism by a methane-oxidizing mixed culture." *Biotechnol. Bioeng. 53*: 320-331.

Suffin, J. A. and D. Ramey. 1997. "In situ biological treatment of TCE-impacted soil and groundwater: demonstration results." *Environ. Prog. 16*: 287-296.

Travis, B. J. and N. D. Rosenberg. 1994. *Numerical simulations in support of the in situ bioremediation demonstration at Savannah River*. Report LA-12789-MS, Los Alamos National Laboratory, Los Alamos, NM.

Travis, B. J. and N. D. Rosenberg. 1997. "Modeling in situ bioremediation of TCE at Savannah River: Effects of product toxicity and microbial interactions on TCE degradation." *Environ. Sci. Technol. 31*: 3093-3102.

A CHEMICAL DYNAMICS MODEL FOR CAH REMEDIATION WITH POLYLACTATE ESTERS

William A. Farone, Applied Power Concepts, Orange, CA, USA
Stephen S. Koenigsberg, Regenesis, San Juan Capistrano, CA, USA
Joseph Hughes, Rice University, Houston, TX, USA

Abstract: The polylactate esters of sorbitol, xylitol and glycerol are a class of new chemical compounds that slowly release lactic acid into groundwater. The lactic acid is biotransformed into pyruvic acid and subsequently into acetic acid releasing hydrogen in both steps. The hydrogen is used by naturally occurring dechlorinating bacteria to reduce PCE to ethene by way of TCE, DCE and VC. A chemical mechanistic model of all of the reactions involved including absorption and desorption of all the chemical species is postulated based on coupled elementary rate equations. The equations are solved simultaneously to match laboratory remediation data in closed systems using solutions containing initial concentrations of TCE up to 100 mg/L. A key feature of the model formulation that was required to simulate the laboratory data is the importance of microbial populations in controlling the rate of biotransformation of the chlorinated hydrocarbons as well as controlling the rate of hydrolysis of the esters and the rate of production of hydrogen from the lactic acid. The results support the hypothesis that the wide variation reported for the rate constants of CAH degradation is related to microbial populations and hydrogen availability.

INTRODUCTION

It is well known that the anaerobic biodegradation rates for a single chlorinated compound can vary over four orders of magnitude. When a new chemical compound was developed that could potentially increase the contaminant biodegradation rates it was decided that a study of mechanisms of the chemical reactions involved would be useful to help understand the key features of the process. A chemical model has been developed that can be applied to microcosm and larger scale simulations in the laboratory, as well as results from the field.

As with any sequence of chemical reactions, the model has to be specific for the study of the biodegradation of each chlorinated compound. The new chemical compound that will assist anaerobic biodegradation is a polylactate ester of sorbitol, xylitol or glycerol. These polylactate esters slowly release lactic acid into groundwater over a period of many months. The lactic acid is biotransformed into pyruvic acid and subsequently into acetic acid releasing hydrogen in both steps. The ultimate fate of the acetic acid is probably to methane and carbon dioxide by methanogenesis. The acids can also be used as as carbon source for bacterial cell growth.

It would appear that the new compounds could be used wherever enhanced anaerobic biodegradation is desired. To study at least one of these cases in detail we selected the biodegradation of 1,1,2-trichloroethylene (TCE). The work is being extended to tetrachloroethylene (PCE) and pentachlorophenol (PCP). The TCE study is somewhat simpler to analyze and document since there fewer chemical steps involved.

MATERIALS AND METHODS

The experimental data on TCE degradation is obtained by injection of known amounts of TCE in water into soil samples that are suspected to contain bacteria that could dechlorinate TCE. The lactic acid releasing polymer was added to the soil and water in various amounts. Generally TCE and its degradation products dichloroethylene (DCE) and vinyl chloride (VC) were measured by extracting an aqueous phase sample and mixing it with an internal standard of toluene dissolved in water. The CAHs and the toluene are then measured by headspace analysis using gas chromatography and both a photoionization detector (PID) and finally by a flame ionization detector (FID). Detection limits of the PID are 5 μg/L based on the concentration in the original liquid sample being analyzed.

Lactic acid, acetic acid and pyruvic acid are measured by high performance liquid chromatography using a UV detector. Citric acid is used as an internal standard and the limits of detection are 1 mg/L in the original sample. Bacterial counts were made through use of standard pour plate techniques. Total Plate Counts (Colony Forming Units per ml) were measured both aerobically and anaerobically.

A positive control was made by using liquid from a bioreactor that had been continuously degrading TCE at the rate of 5 mg/(L-day). A series of two ten-fold dilutions were made and the bacterial counts of the starting liquid was also made. The diluted solutions were placed in a system with sterile soil and known amounts of TCE and the lactic acid releasing ester.

The main test system is 200 ml test tubes containing 10 grams of soil and 160 ml of liquid at concentrations of 25 mg/L and 10 mg/L of TCE. Concentrations of TCE as high as 100 mg/L have been used. An alternative pilot systems uses 33 L of soil in a tube 1.83 m long and 15.25 cm in diameter. In this system the lactate releasing material in injected into the soil and the contaminated water flows through the tube at rates similar to groundwater flow rates. Sample ports in the tube occur every 15 cm.

Numerical analysis of the coupled differential equations was carried out by the Runga-Kutta-Gill algorithm. The model was programmed in Visual Basic for Applications as a series of macros in an Excel spreadsheet.

RESULTS AND DISCUSSION

The system of equations to be described here can be generalized to any number of species limited only by the programmer's patience and the availability of sufficient data to allow meaningful interpretation of the parameters. The large number of parameters involved leads to the accusation that it is always possible to make the model fit the data if the parameters are selected correctly. The value of the model, however, is not in fitting the data.

The model provides a rationale for selection of various mechanisms. A good model will work over a wide range of experimental parameters. Changes in the behavior of the system will be easily understood in terms of the model equations.

The chemical sequence is:

TCE => DCE => VC => ethylene

There are three groups of bacteria that we must consider:

1. bacteria that metabolize TCE (B1)
2. bacteria that produce hydrogen (H_2) from LA (B2)
3. bacteria that produce lipase for polylactate cleavage (B3)

These groups are not distinct. By this we mean that group B3 contains group B1 and B2 and in general will be larger than the sum of B1 and B2. Group B2 will be smaller than B3 but larger than B1. Since these bacteria catalyze the reactions we will have to keep track of their amounts during the course of the analysis. Bacteria can be modeled by first order kinetics:

$$\frac{d[B1]}{dt} = k_{B1}[B1] \quad (1)$$

$$\frac{d[B2]}{dt} = k_{B2}[B2] \quad (2)$$

$$\frac{d[B3]}{dt} = k_{B3}[B3] \quad (3)$$

The square brackets denote concentration and, in general, will be expressed in millimoles per L (mm/L). Bacteria are typically reported as numbers (CFU) per ml. In order to keep units consistent, bacterial concentration will be given as the plate count per liter and all rates will be per day. Thus, the k_i have units of day^{-1}.

If we now consider what happens in an aquifer or a test tube we will use a "control volume" that contains soil; water and some TCE both dissolved in the water and adsorbed on the soil. A "control volume" is a convenient way to keep track of changes that occur. In this defined volume we can keep track of everything that moves into the volume, the changes that occur in the volume and everything that moves out of the volume. This allows us to perform overall mass and species balance on the volume. The control volume used here does not allow any of the unused lactic acid (LA) or TCE to escape. In an aquifer this is essentially the volume downstream of the polylactate ester (HRC) injection site that encompasses the LA and metabolic products. The actual total amounts of material, e.g. in pounds, is calculated by multiplying the total volume by the concentration. In the aquifer case the volumes will be on the order of thousands of liters. In the test tube it is 170 ml. However, we keep the model very general by using a control volume that simply encompasses all of the aquifer of interest and is based on the concentration contained in that volume.

We place HRC into this volume and several things happen. HRC breaks down to LA. The bacterial group B3, the largest of the groups, catalyzes this. The LA can then be used to produce hydrogen. Unfortunately it can also be used to produce more bacteria that do not produce hydrogen and it can react chemically. Further LA can be adsorbed and previously adsorbed LA can be released. TCE can move into the control volume. TCE can be degraded and it can be adsorbed. TCE that was previously adsorbed can be released. In this model we will assume that TCE does not react chemically. The TCE equation is:

$$\frac{d[TCE]}{dt} = RTCEI - k_{aTCE}[S][TCE] + k_{dTCE}[TCEA] - k_1[B1]^{a1}[H_2][TCE] \quad (4)$$

where RTCEI is the rate at which TCE enters the volume in mm/(L day), [S] is the number of sites associated with the soil for the adsorbtion of TCE (mm/L) and $k\alpha_{TCE}$ is rate for TCE adsorbtion L/(mm day).

Note that [S] may be combined into $k\alpha_{TCE}$ for all practical purposes since the combination will be measured together. The units would then be day^{-1}.

[TCEA] is the amount of TCE adsorbed on the soil at any time (mm/L), k_{dTCE} is the rate of desorbtion (day^{-1}), k_1 is the rate of remediation with units of l^2/(mm day), a1 is the exponential catalyst (bacterial) dependence.

RTCEI is the input flow rate of TCE. The model assumes that there is some TCE initially adsorbed on the soil and some in the dissolved phase. RTCEI accounts for new TCE coming into the treatment volume. We need to account for the change in adsorbed TCE with time:

$$\frac{d[TCEA]}{dt} = -k_{dTCE}[TCEA] + k_{aTCE}[S][TCE] \quad (5)$$

Adding the rest of the equations in similar notation we have:

$$\frac{d[DCE]}{dt} = RDCEI - k_{aDCE}[S][DCE] + k_{dDCE}[DCEA] - k_5[B1]^{a5}[H_2][DCE] \quad (6)$$

$$\frac{d[DCEA]}{dt} = -k_{dDCE}[DCEA] + k_{aDCE}[S][DCE] \quad (7)$$

$$\frac{d[VC]}{dt} = RVCI - k_{aVC}[S][VC] + k_{dVC}[VCA] - k_6[B1]^{a6}[H_2][VC] \quad (8)$$

$$\frac{d[VCA]}{dt} = -k_{dVC}[VCA] + k_{aVC}[S][VC] \quad (9)$$

$$\frac{d[H_2]}{dt} = k_3[B_2]^{a3}[LA] - k_1[B1]^{a1}[H_2][TCE] - k_h[H_2] \quad (10)$$

$$\frac{d[LA]}{dt} = k_4[B3]^{a4}[HRC] - k_{aLA}[S][LA] + k_{dLA}[LAA] - k_g[B3]^{ag}[LA] - k_3[B2]^{a3}[LA] - k_2[LA] \quad (11)$$

$$\frac{d[HRC]}{dt} = -k_4[B3]^{a4}[HRC] \quad (12)$$

$$\frac{d[LAA]}{dt} = k_{aLA}[S][LA] - k_{dLA}[LAA] \quad (13)$$

The term A indicates the adsorbed species; for example, VCA is adsorbed VC.

Figure 1 shows typical results for the use of the model in explaining the reduction of TCE in a microcosm of active bacteria at around 10,000 CFUs per ml. Typical soils are less than 1,000 CFUs per ml. With bacterial levels the same, reduction can take 25 days. Figure 2 is a typical result that illustrates the utilization of lactic acid when the bacterial count is very high and the polylactate ester is limited. This example also shows the model can predict the maximum lactic acid concentration.

CONCLUSION

Standard chemical kinetic formulations are useful in predicting concentrations of chemical species in active dechlorinating systems using a polylactate ester that releases lactic acid. Modeling the effect of bacterial concentration as a catalyst with the order determined by the data appears to provide a reasonable understanding of the experimental data.

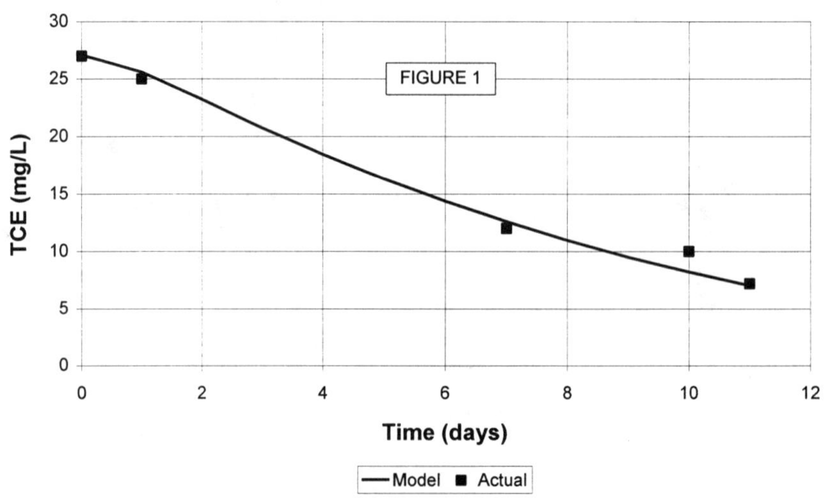

FIGURE 1 Results of TCE Reduction Model vs. Experiment

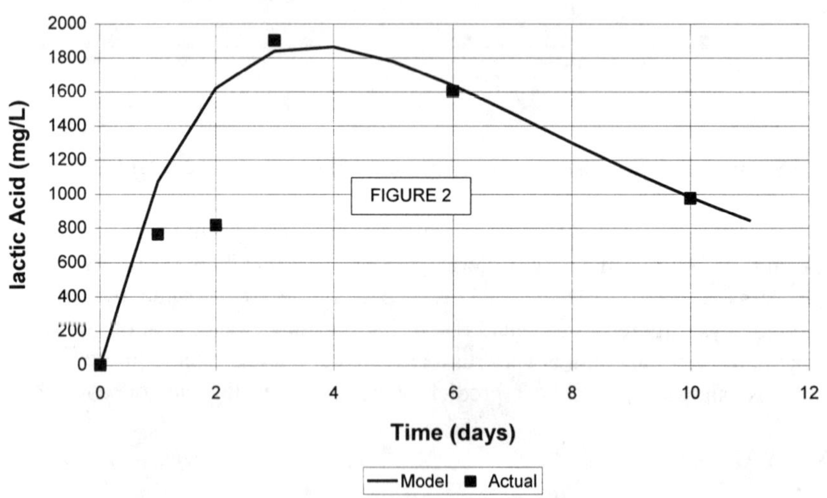

FIGURE 2 Results of Lactic Acid Production Model vs. Experiment

BIOAUGMENTATION FOR TRICHLOROETHYLENE REMEDIATION: WILD TYPE AND CONSTITUTIVE MUTANT STRAINS

Lori A. Swadley (Colorado School of Mines, Golden, Colorado)
Junko Munakata-Marr* (Colorado School of Mines, Golden, Colorado)
William R. Mahaffey (Pelorus Environmental and Biotechnology Corporation, Evergreen, Colorado)

Abstract: Bioaugmentation holds great potential for remediation of trichloroethylene (TCE)-contaminated sites. The degradation behaviors of the bacterium *Burkholderia cepacia* G4 (wild type) and two constitutive mutant strains were investigated in laboratory studies to optimize their application for *in situ* contaminant biodegradation, specifically in hydraulically emplaced fractures. Initially, the correlation between a colorimetric enzyme assay and TCE degradation was defined by measuring G4 activity toward both 3-trifluoromethyl phenol (TFMP) and TCE at different growth stages. This correlation determines whether the TFMP assay is a valid, timesaving measure of TCE degradation activity. Based on this relationship, the stage of growth for the highest TCE degradation activity in the different strains of *B. cepacia* was also determined, allowing the application of the most active organisms to TCE remediation. These studies greatly enhance the potential application of bioaugmentation for remediation of TCE-contaminated sites.

INTRODUCTION

Trichloroethylene (TCE), a widely used industrial solvent, is a known animal and suspected human carcinogen (Miller and Guengerich, 1982). TCE is a prevalent pollutant in drinking water supplies taken from groundwater sources (Westrick et al., 1984). Bioremediation holds great potential for the remediation of these sites by harnessing the activity of microorganisms to detoxify the TCE *in situ.*

While anaerobic degradation of TCE yields unwanted intermediates that can be even more dangerous than TCE (such as vinyl chloride), aerobic degradation leads to the production of harmless byproducts, making it an appealing choice for bioremediation. However, TCE degradation is a cometabolic process under aerobic conditions, requiring a primary substrate such as methane (e.g. Fogel et al., 1986; Little et al., 1988; Wilson and Wilson, 1985) or aromatics (e.g. Nelson et al., 1986; Wackett and Gibson, 1988). The enzyme that oxidizes the TCE has broad specificity and thus acts on the TCE as well as the primary substrate. The transformation of TCE proceeds without benefit, and can actually be harmful, to the organism performing the degradation.

Stimulation of indigenous populations with specific primary substrates has thus far shown modest remediation potential and certainly raises regulatory concerns with substrates such as phenol. The addition of specific substrates does

*corresponding author

not necessarily induce the growth of a population capable of cometabolizing TCE. The addition of cultures known to transform TCE can greatly enhance *in situ*bioremediation in areas where no indigenous, TCE-degrading microbes are present (Munakata-Marr et al., 1996).

The bacteria *Burkholderia cepacia* G4 (wild type), $PR1_{301}$ and PEL-1010PC (two constitutive mutant strains) produce the enzyme toluene *ortho*-monooxygenase (Tom) that transforms TCE. In the wild type strain G4, Tom is normally induced by phenol or toluene (Shields et al., 1991). Mutant *B. cepacia* strains that constitutively express Tom when grown on substrates such as lactate can, in theory, be maintained and stimulated to degrade TCE *in situ* with harmless substrates. However, previous studies using $PR1_{301}$ in aquifer microcosms showed limited TCE degradation (Munakata-Marr et al., 1996). Our current studies indicate that Tom activities differ significantly at different stages of growth in the different strains, which may play a critical role in applications for bioremediation.

Hydraulically formed fractures have been used to chemically oxide contaminants including TCE. Bioaugmentation of these hydraulically emplaced reactive fractures is an innovative technology that has been successfully used for *in situ* remediation of contaminants such as petroleum hydrocarbons (Stavnes et al., 1996). The use of this technique for *in situ* biological treatment of TCE will be investigated using proppant inoculated with *B. cepacia*, based on the results of the current study.

Objective. The primary objective of this study was to determine the growth stages at which Tom activity is optimal for TCE degradation for application in bioaugmentation of hydraulic fracture media.

MATERIALS AND METHODS

Initially, the correlation between microbial activities toward TCE and 3-triflouromethyl phenol (TFMP) was tested. The production of 7,7,7-trifluoro-2-hydroxy-6-oxo-2,4-heptadienoic acid (TFHA) from TFMP has been used regularly as a measure of Tom activity (Shields et al., 1991; Shields and Reagin, 1992). As hydroxylation of the aromatic ring is typically the rate-limiting step in aromatic catabolism (Dagley, 1986; Shingler et al., 1989), the rate of TFHA production is assumed to be limited by Tom activity. However, this relationship between TFHA production and TCE degradation has not been previously demonstrated.

To validate the use of the TFMP test as a measure of Tom activity, TCE degradation and TFHA formation by samples from batch-grown, phenol-fed G4 cultures were monitored simultaneously. Culture dry-mass concentrations were determined from the weight loss during combustion of sample solids retained on a Whatman GF/F glass fiber filter. Phenol grown cultures were centrifuged for 3 minutes at 14,000 rpm and then resuspended in a basal salts mineral medium (Stanier et al., 1966) with TCE. TCE degradation was measured at 20°C in amber glass bottles capped with Mininert valves. TCE concentrations were measured in hexane-extracted samples with a Hewlett-Packard HP6890+ gas chromatograph equipped with an electron capture detector and an autosampler. TFMP degradation was monitored in culture samples by pelleting the cells, resuspending in a TFMP solution and incubating at 35°C. TFHA production was determined by centrifuging the TFMP solution and measuring the absorbance of the supernatant at λ = 386 nm, where TFHA absorbs strongly, with a Perkin-Elmer UV/Vis Lambda4 spectrophotometer.

In order to better understand enzyme production under induced as compared with constitutive expression, enzyme activities at different growth stages and cell densities were investigated. The basal salts mineral medium was amended with low, medium and high concentrations of phenol or lactate and inoculated with G4, $PR1_{301}$, or PEL-1010PC. Cultures were grown at 20°C on a shaker table. Growth was monitored by measuring optical density at λ= 600 nm with the Perkin-Elmer spectrophotometer.

PRELIMINARY RESULTS

Preliminary laboratory studies investigated TFHA production at different points for G4 and $PR1_{301}$ cultures grown on phenol or lactate in groundwater. Autoclaved or filtered (0.2 µm) groundwater was amended with 130 mg/L phenol or 300 mg/L lactate. Aliquots (20 mL) of the amended groundwater were added to sidearm Erlenmeyer flasks (70 mL total volume) and inoculated with 1 mL of G4 or $PR1_{301}$ cultures that were grown on mineral medium (Shingler, et al., 1989) with 170 mg/L phenol or 190 mg/L lactate. Flasks were stored at 20°C, either stationary or shaken on a shaker table at 220 rpm. Growth was monitored by measuring optical density with a Manostat colorimeter with a red filter. G4 and $PR1_{301}$ demonstrated similar patterns in growth (data not shown), while phenol, lactate, and groundwater controls showed little increase in turbidity.

Figure 1 shows the amount of TFHA produced by the cultures at different times during their growth. The data indicate that the groundwater sterilization method used, the amount of aeration by shaking, and the inocula growth substrate did not affect subsequent culture growth or enzyme expression in groundwater, as the curves for each of these variables overlapped. Though normalization of TFHA production by cell mass may seem appropriate, large variations in normalized TFHA production caused by measurements at low turbidity make the data more difficult to interpret. As the differences in the growth curves are relatively small, direct comparison of TFHA production numbers was preferable. TFHA production by phenol-grown G4 reached a maximum of approximately 2.6 µM 2-3 days after inoculation, just before entering stationary phase. In contrast, lactate- and phenol-grown $PR1_{301}$ produced a maximum of around 1.7 µM TFHA 4-6 days after inoculation, after reaching stationary phase. Maximum activities were higher in G4, but activity persisted for a longer period of time in $PR1_{301}$.

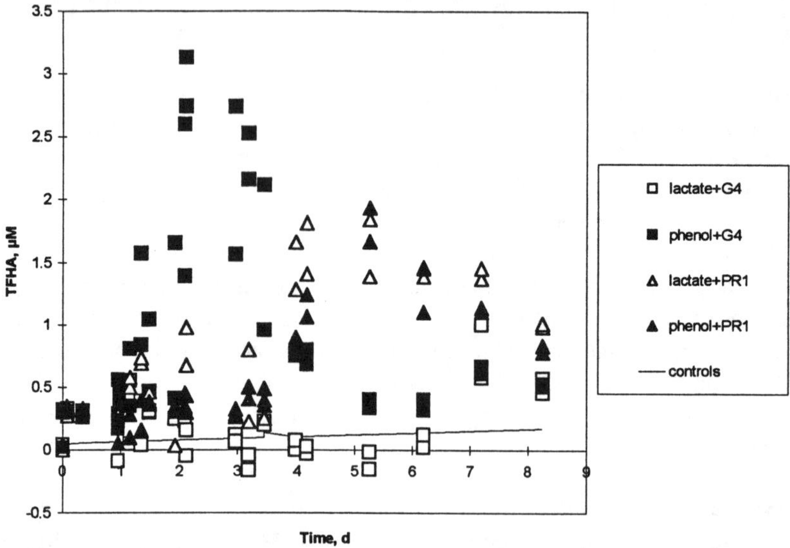

FIGURE 1. TFHA Production by G4 and $PR1_{301}$ in Groundwater.

DISCUSSION

The differences in enzyme activity seen between G4 and $PR1_{301}$ in different growth phases are important considerations in assessing cometabolic capabilities. Often, studies simply measure cometabolism at a specific point in the culture growth curve. The results presented here, however, demonstrate significant differences between the wild type and constitutive strains in cometabolic capabilities as a function of growth phase. Although results from previous batch culture studies indicated greater activity in constitutive strains grown on lactate than in either wild type or constitutive strains grown on phenol, the observed difference in activity may have been due to a difference in Tom expression as displayed in Figure 1. If, based on the results of the earlier batch studies, a 2-day old $PR1_{301}$ culture were later applied to a remediation system, the outcome could be disappointing. This study emphasizes the need to understand the factors affecting enzyme expression in the practical application of cometabolic processes.

REFERENCES

Dagley, S. 1986. "Biochemistry of Aromatic Hydrocarbon Degradation in Pseudomonads." In J. R. Sokatch (Ed.), *The Biology of Pseudomonas*, pp. 527-555. Academic Press, Orlando, FL.

Fogel, M. M., A. R. Taddeo and S. Fogel. 1986. "Biodegradation of Chlorinated Ethenes by a Methane-Utilizing Mixed Culture." *Appl. Environ. Microbiol. 51*(4):720-724.

Little, C. D., A.V. Palumbo, S. E. Herbes, M. E. Lidstrom, R. L. Tyndall and P. J. Gilmer. 1988. "Trichloroethylene Biodegradation by a Methane-Oxidizing Bacterium." *Appl. Environ. Microbiol. 54*(4):951-956.

Miller, R. E. and F. P. Guengerich. 1982. "Oxidation of Trichloroethylene by Liver Microsomal Cytochrome P-450: Evidence for Chlorine Migration in a Transition State Not Involoving Trichloroethylene Oxide." *Biochemistry.* 21: 1090-1097.

Munakata-Marr, J., P. L. McCarty, M. S. Sheilds, M. Reagin, and S. C. Francesconi. 1996. "Enhancement of Trichloroethylene Degradation in Aquifer Microcosms Bioaugmented with Wild Type and Genetically Altered *Burkholderia (Pseudomonas) cepacia* G4 and PR1." *Environ. Sci. and Tech. 30*(6):2045-2052.

Nelson, M. J. K., S. O. Montgomery, E. J. O'Neill and P. H. Pritchard. 1986. "Aerobic Metabolism of Trichloroethylene by a Bacterial Isolate." *Appl. Environ. Microbiol. 55*(2):383-384.

Shields, M. S., S. O. Montgomery, S. M. Cuskey, P. J. Chapman and P. H. Pritchard. 1991. "Mutants of *Pseudomonas cepacia* G4 Defective in Catabolism of Aromatic Compounds and Trichloroethylene." *Appl. Environ. Microbiol. 57*(7):1935-1941.

Shields, M. S. and M. J. Reagin. 1992. "Selection of a *Pseudomonas cepacia* Strain Constitutive for the Degradation of Trichloroethylene." *Appl. Environ. Microbiol. 58*(12):3977-3983.

Shingler, V., F. C. H. Franklin, M. Tsuda, D. Holroyd and M. Bagdasarian. 1989. "Molecular Analysis of a Plasmid-Encoded Phenol Hydroxylase from *Pseudomonas* CF600." *J. Gen. Microbiol. 135*:1083-1092.

Stanier, R. Y., N. J. Palleroni, and M. Doudoroff. 1966. "The Aerobic Pseudomonads: a Taxonomic Study." *J. Gen. Microbiol. 43*(1):159-271.

Stavnes, S., A. Yorke and L. Thompson. 1996. "In Situ Bioremediation of Petroleum in Tight Soils Using Hydraulic Fracturing." In *HazWaste World/Superfund XVII Conference*. Washington, D.C.

Wackett, L. P. and D. T. Gibson. 1988. "Degradation of Trichloroethylene by Toluene Dioxygenase in Whole-Cell Studies with *Pseudomonas putida* F1." *Appl. Environ. Microbiol. 54*(7):1703-1708.

AUTHOR INDEX

This index contains names, affiliations, and volume/page citations for all authors who contributed to the eight-volume proceedings of the Fifth International In Situ and On-Site Bioremediation Symposium (San Diego, California, April 19–22, 1999). Ordering information is provided on the back cover of this book.

The citations reference the eight volumes as follows:

5(1): Alleman, B.C., and A. Leeson (Eds.), *Natural Attenuation of Chlorinated Solvents, Petroleum Hydrocarbons, and Other Organic Compounds*. Battelle Press, Columbus, OH, 1999. 402 pp.

5(2): Leeson, A., and B.C. Alleman (Eds.), *Engineered Approaches for In Situ Bioremediation of Chlorinated Solvent Contamination*. Battelle Press, Columbus, OH, 1999. 336 pp.

5(3): Alleman, B.C., and A. Leeson (Eds.), *In Situ Bioremediation of Petroleum Hydrocarbon and Other Organic Compounds*. Battelle Press, Columbus, OH, 1999. 588 pp.

5(4): Leeson, A., and B.C. Alleman (Eds.), *Bioremediation of Metals and Inorganic Compounds*. Battelle Press, Columbus, OH, 1999. 190 pp.

5(5): Alleman, B.C., and A. Leeson (Eds.), *Bioreactor and Ex Situ Biological Treatment Technologies*. Battelle Press, Columbus, OH, 1999. 256 pp.

5(6): Leeson, A., and B.C. Alleman (Eds.), *Phytoremediation and Innovative Strategies for Specialized Remedial Applications*. Battelle Press, Columbus, OH, 1999. 340 pp.

5(7): Alleman, B.C., and A. Leeson (Eds.), *Bioremediation of Nitroaromatic and Haloaromatic Compounds*. Battelle Press, Columbus, OH, 1999. 302 pp.

5(8): Leeson, A., and B.C. Alleman (Eds.), *Bioremediation Technologies for Polycyclic Aromatic Hydrocarbon Compounds*. Battelle Press, Columbus, OH, 1999. 358 pp.

KEYWORD INDEX

This index contains keyword terms assigned to the articles in the eight-volume proceedings of the Fifth International In Situ and On-Site Bioremediation Symposium (San Diego, California, April 19-22, 1999). Ordering information is provided on the back cover of this book.

In assigning the terms that appear in this index, no attempt was made to reference all subjects addressed. Instead, terms were assigned to each article to reflect the primary topics covered by that article. Authors' suggestions were taken into consideration and expanded or revised as necessary. The citations reference the eight volumes as follows:

5(1): Alleman, B.C., and A. Leeson (Eds.), *Natural Attenuation of Chlorinated Solvents, Petroleum Hydrocarbons, and Other Organic Compounds*. Battelle Press, Columbus, OH, 1999. 402 pp.

5(2): Leeson, A., and B.C. Alleman (Eds.), *Engineered Approaches for In Situ Bioremediation of Chlorinated Solvent Contamination*. Battelle Press, Columbus, OH, 1999. 336 pp.

5(3): Alleman, B.C., and A. Leeson (Eds.), *In Situ Bioremediation of Petroleum Hydrocarbon and Other Organic Compounds*. Battelle Press, Columbus, OH, 1999. 588 pp.

5(4): Leeson, A., and B.C. Alleman (Eds.), *Bioremediation of Metals and Inorganic Compounds*. Battelle Press, Columbus, OH, 1999. 190 pp.

5(5): Alleman, B.C., and A. Leeson (Eds.), *Bioreactor and Ex Situ Biological Treatment Technologies*. Battelle Press, Columbus, OH, 1999. 256 pp.

5(6): Leeson, A., and B.C. Alleman (Eds.), *Phytoremediation and Innovative Strategies for Specialized Remedial Applications*. Battelle Press, Columbus, OH, 1999. 340 pp.

5(7): Alleman, B.C., and A. Leeson (Eds.), *Bioremediation of Nitroaromatic and Haloaromatic Compounds*. Battelle Press, Columbus, OH, 1999. 302 pp.

5(8): Leeson, A., and B.C. Alleman (Eds.), *Bioremediation Technologies for Polycyclic Aromatic Hydrocarbon Compounds*. Battelle Press, Columbus, OH, 1999. 358 pp.

A

E

F

G

H

I

J

K

L

M

Q

R

S

T

U

V

W

Z